Heavy Weather Sailing

Heavy Weather Sailing

K Adlard Coles   Revised by Peter Bruce

Fourth Edition

International Marine
Camden, Maine

International Marine/
Ragged Mountain Press

A Division of The **McGraw·Hill** Companies

First published in Great Britain in 1967.
First published in the U.S.A. in 1968.

Reprinted 1969, 1970, 1971, 1972; Revised and Enlarged Edition 1975;
Reprinted 1976; Third Revised Edition 1981 (three printings); Fourth
Edition 1992; First Paperback Edition 1996.

4   6   8   10   9   7   5   3

A CIP catalog record for this book is
available from the Library of Congress.

ISBN 0-07-011732-2

Questions regarding the content of this book should be addressed to:

International Marine
P.O. Box 220
Camden, ME 04843
207-236-4837

Questions regarding the ordering of this book should be addressed to:

The McGraw-Hill Companies
Customer Service Department
P.O. Box 547
Blacklick, OH 43004
Retail Customers: 1-800-262-4729; Bookstores: 1-800-722-4726

Contents

PART 1

Heavy Weather Experiences

PART 2
Expert Advice

FOREWORD

**By Peter Blake, Skipper of *Steinlager II* ,
winner of the Whitbread Round the World Race 1990.**

In 400,000 ocean miles to date, I have been in many gales – in small yachts, in large yachts and in a wing-masted multihull, this latter occasion being probably the most worrying time of them all.

My first major gale at sea was in the south-west Pacific in 1969 with my parents in our 34 ft ketch *Ladybird*. It was my first time in really heavy weather with screaming winds and seas that came crashing over the decks. The yacht was hove-to under storm jib and fully reefed mizzen. Each big one sounded like an express train approaching, followed by a couple of seconds of quiet; then a crash that threw the yacht over on her side, at the same time as the light coming through the ports went frothy green when the crest submerged the decks. We stayed below, playing cards and eating corned beef hash and soup. The sails shuddered violently even though they were very small and sheeted tight, as the sailmaker had not installed large enough leech cords. Although we were relatively safe in a well-found vessel we were distinctly relieved when conditions eased.

A more recent and equally memorable experience took place early in 1980, when we were caught in the Tasman Sea by the tropical cyclone 'David' as we headed towards Auckland in the maxi-yacht *Condor*. We had plenty of warning, were well away from land, and took precautions very early on.

Despite much criticism from one of the crew, who thought we should still be sailing, the yacht was hove-to under trysail and storm jib as the wind rose above 30 knots. Then, with the wind going off the clock for long periods, the jib was soon down. The surface of the sea was smoking as though it was on fire. All the small waves had disappeared – just blown away, and we felt that being hove-to was the best way of riding out this storm. As conditions worsened, we discussed but did not employ what we felt was our only other option which was to run before the seas, trailing warps. We felt that lying a-hull would have been very dangerous. Sitting harnessed on the cockpit floor we took note of the time taken to go from trough to crest of each sea – between five and six seconds – and they were the most enormous waves imaginable. When the wind eased eighteen hours later, which it did very quickly leaving us with a towering and confused cross-sea, the feeling of relief was immense. We were lucky and came through unscathed.

A gale, or a storm at sea, usually seems worse when cruising rather than racing. When racing in a well-found yacht there are generally tactics to occupy one's mind, and quality of life for the crew matters little. But when cruising, a degree of comfort and pleasure are the reasons for 'doing it'; thus severe weather takes on a different meaning.

Gales, and occasional storm, are a part of sailing across an ocean. It should be understood that one must give with the weather as one cannot beat it. When coastal cruising, a prudent eye on the weather forecast will generally see one in harbour in time to avoid the worst. If not, then conditions on board can quickly get worse than in mid-ocean. The blasé coastal sailor may end up on a lee shore, ill-prepared, and it is under these conditions that most tragedies occur. To be well clear of land removes one level of anxiety and danger. Whenever I am at sea and the weather is deteriorating I begin to get a rather hollow feeling in my stomach. Questions spin round in my head. What will the next twenty-four hours or so bring? Is the yacht properly prepared? Have we, the crew, done everything we should have done? Will the forecast (if there is one) be right? Have we enough sea room? Should we have a hot meal now before the seas get up? Is everything properly stowed on deck? And so on.

No two situations at sea are ever going to be the same. No two boats handle in the same way. For example some heave-to well, some do not. Sometimes the heavy weather has been forecast, sometimes not. Weather forecasts must not be relied upon 100%; it is important to 'look out of the window' and make one's own assessment of the situation. But what is most important, whatever vessel one is in, is to be prepared. It is much easier to have thought about the consequences of heavy weather before leaving the comfort of the marina, mooring, or quiet anchorage. One must add that being able to rely upon the structural integrity of one's yacht, her rig, sails, engine, hatches, general mechanical equipment and the ability of the crew is equally important.

This book should be read and re-read by every yachtsman and yachtswoman, whether they have aspirations to cross an ocean or simply potter about the coastline, whether they are skipper or crew. Not every idea examined will apply to every situation, but the knowledge gained may one day be invaluable.

To put it in a nutshell, there is no substitute for forward planning and knowledge of what to do in heavy weather.

Peter J Blake MBE OBE
January 1991

Preface to the 4th Edition

Adlard Coles OBE died in 1985. He left behind him a multitude of friends and ad-
mirers, many of whom knew him only through his books. Apart from his ability to
write vividly and clearly, Adlard was an extraordinary man. In spite of being a
diabetic, a quiet and very gentle person with something of a poet's eye, he was
incredibly tough, courageous and determined. Thus it was very often that Adlard's
*Cohoe*s appeared at the top of the lists after a really stormy race.

When reading Adlard's enchanting prose one also needs to take into account that he
was supremely modest, and his scant reference to some major achievement belies the
effort that must have been required. It is clear that Adlard minimizes everything which
does him credit. As an example, the wind strengths he gives in his own accounts in
Heavy Weather Sailing are manifestly objective and warrant no risk of being accused of
exaggeration for narrative effect.

Revision of this important book is a task which should have been carried out by
Adlard himself had this been possible. With the support of Adlard's widow, Mamie,
and son Ross, my father Erroll Bruce, a lifelong friend of Adlard, and the publishers, I
have done my best to do what I believe Adlard would have liked to have done himself.

Only those who, like Adlard Coles, have hove-to in a force 6 wind to obtain a
sounding by lead line in 15 fathoms of water, and those who have struggled past rocks
in fog using a radio direction-finding set will be able fully to appreciate the blessings of
the recent developments in electronic navigation aids. Clearly things have also moved
on in every aspect of seafaring since the earlier editions of this book, and yet there is
still much to be learnt from them. Consequently this edition retains much of the
original material, then leads on chronologically from where Adlard left off with some
more recent accounts of storm experiences. The appendices from previous editions
have been extensively revised and formed into a second part of the book to provide a
source of more specialised information. The authors of these articles have been chosen
because they represent responsible and experienced views which are likely to stand the
test of time, and I am specially grateful to them for their contribution.

The earlier editions of *Heavy Weather Sailing* were more about prevention than cure
and the same theme is maintained in this edition. Thus I have not gone into detail on
matters such as rescue and survival which are covered by other books. I should also
mention that, as we are in an age where some have learnt Imperial units and some
metric, there is a mixture of both. I have left these as found, being pursuaded that any
alternative might be worse.

Peter Bruce
Lymington

Acknowledgements

The three previous editions of *Heavy Weather Sailing* have become such cornerstones of yachting lore that it would have been a formidable task for any ordinary person to try to carry out a revision single handed. I have sought the most eminent and wise counsel that I could find to help me with the task, and have had the script read by as many yachtsmen of scholarship and long experience as possible.

Two such clear-minded people in particular, who have given tremendous time and patience to the revision, are Leonard Wesson MA (Oxon) and Commander Sandy Watson MA (Cantab) RN, both members of, amongst other august organisations, the Royal Cruising Club. My sincere thanks go to these two, but I must also name other distinguished readers from the Royal Yacht Squadron, The Royal Ocean Racing Club, the Royal Naval Sailing Association and the Royal Yachting Association. These are my father, Commander Erroll Bruce RN, Commander Bill Anderson RN, Dr Nicholas Davis, Alan Green, Tom and Vicki Jackson, Group Captain Howard Lewis, Peter Nicholson, John Power, Michael Richey, Captain Simon van der Byl RN, and Hugo Walford, all of whom have produced many useful suggestions.

The second part of the book has benefited magnificently from the wealth of knowledge and experience provided by the principal contributors. These have not only each produced their own section, but have, in some cases, been good enough to read other chapters, enabling them to provide useful comment. They are:

Olin J Stephens II, who must rate amongst the most admired yacht designers of the century.

John Powell MRINA, who has built up a great reputation for good seamanlike design and manufacture of masts, spars and rigging.

Peter Haward, who has kindly updated his experiences of heavy weather in power craft during over 700 delivery voyages, with frank remarks which ring as true as ever.

John Shuttleworth MSc, one of the world's current leading multihull designers, who brings much new information to his esoteric subject, probably not seen in print before.

Alan Watts BSc FRMetS, another well known and strong contributor to the earlier editions of *Heavy Weather Sailing*, who has meticulously updated his weather chapter to take account of recent storms as well as the latest meteorological thinking.

Sheldon Bacon BA (Oxon), from the Institute of Oceanographic Sciences Deacon Laboratory, who has converted a scientist's perception of waves into plain words likely to be understood and enjoyed by people who, like him, own small vessels.

Andrew Claughton BSc CEng MRINA, from the Wolfson Unit for Marine Technology at the University of Southampton, whose chapter enables comparisons to be made between design characteristics of yachts in a way that was previously achieved by subjective means.

Captain Colin McMullen DSC RN, a former commodore of the Royal Cruising Club and for many years a marine consultant following his naval career, who provides food for thought when writing on the subject of drogues – and a hint of what may lie in the future.

Dr Dick Allan MB BS MFOM, Director of the Army Personnel Research Establishment at Farnborough, whose chapters on clothing and seasickness are a masterly combination of profound scientific knowledge and practical common sense.

I must also thank those who have enlivened and enriched the text by a contribution from their own experiences. Amongst these are:

Martin Bowdler
Lieutenant Commander Peter Braley BA FRMetS RN
Warren Brown

Alby Burgin
Andrew Cassell
Captain Roy Clare MNI RN
Major Richard Clifford MBE RM
Chris Dunning MBE
Group Captain Geoffrey Francis DSO DFC
John Channon
John Irving
Robin Knox-Johnson CBE RD
Captain Graham Laslett CBE RN
Air Commodore Brian Macnamara CBE DSO
Allan McLaughlan
Professor Dudley Norman
Stuart Quarrie
Michael Richey MBE
Brian and Pam Saffery Cooper
Jeff Taylor
George Tinley
Charles Watson
Alan and Kathy Webb
Harry Whale
Commander Simon Wilkinson RN

I should also thank others who have sent in material for which space has not been found. Much of this material has, nevertheless, been valuable in shaping the conclusions.

One cannot conclude without paying respect to the original chapters written by Adlard Coles and his contributors. Bearing in mind that the first edition was published in 1967, and the most recent edition in 1980, the worth of their work can best be judged by the fact that so little has needed to be deleted or changed with the passage of time.

The yacht Carnival of Dee *leaving Port Patrick, N Ireland on 14 October 1990 in heavy weather.*

PART 1
Heavy Weather Experiences

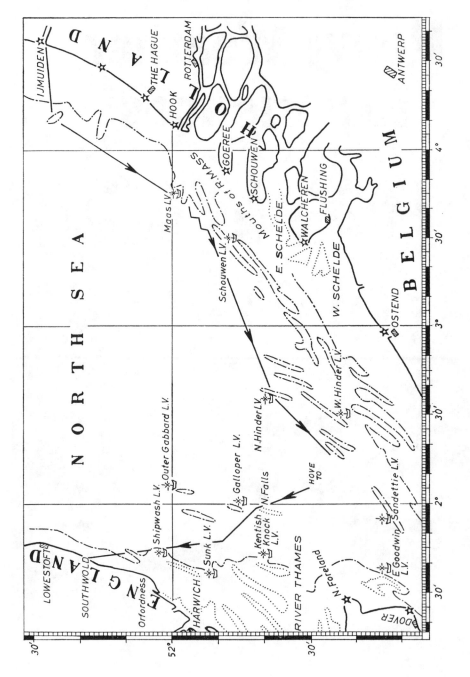

Fig 1.1. Course of Annette II in North Sea Gale, September 1925.

1 NORTH SEA GALE

Adlard Coles

The first occasions on which I got properly 'caught out' in gales occurred in 1925. That is not to say that I had no previous experience of gales, for in 1923 I had made a long cruise with two undergraduate friends in my 7-tonner *Annette* to the Baltic. It happened to be a year of particularly bad weather, and as result we experienced a number of gales in which we broke the boom, parted a shroud, pulled out the bobstay, sprang the mast and suffered much minor damage. But all this occurred near coastal waters and we were never caught in the open sea beyond reach of harbour.

In 1925 my wife and I bought a 12-ton gaff ketch at Riga, which we renamed *Annette II*. She was a heavy double-ender of Scandinavian design measuring 29 ft 7 in overall and no less than 11 ft 4 in in beam. The draft was 3 ft 9 in or 7 ft 9 in, with the centreboard down, and the sail area was 430 sq ft She was fitted with a hot-bulb semi-diesel engine, which I could rarely get to work. Everything about this yacht was heavy and solid. She seemed huge after the first *Annette*, which was of light displacement and only 19 ft on the waterline.

My wife and I had a wonderful cruise in *Anette II*, sailing from the strange historic port of Riga to Gotland and Öland, and then to Sweden and Denmark, before passing through the Kiel Canal to the North Sea, and westwards past the Frisian islands to Ymuiden in Holland.

We had a fair amount of bad weather during the cruise and were hove-to west of Öland Rev lightvessel in a gale, but it only lasted about six hours. The real gale was reserved for the last lap of the voyage, when at dawn on 18 September 1925 *Annette II* sailed out of Ymuiden on the sandy coast of Holland bound for Dover. Her staysail, mainsail and mizzen were set close-hauled to a light southerly breeze and she could just lay a little south of west. The yacht sailed steadily seawards over the grey waves, the Dutch coast gradually fading to a pencilled line, until it was finally lost as the distance increased and the hands of the cabin clock marked the passing hours.

During the morning, the wind backed and freshened. Occasionally, spray whipped across the deck and tumbled along the lee scuppers before running back into the sea. The backing wind, an ominous sign, had the merit of enabling the course to be altered to the south towards the Maas lightvessel, and progress was good.

At sunset *Annette II* passed the lightvessel, some fifteen miles WSW of the Hook of Holland, leaving her a few miles to the eastward. The evening meal was prepared and eaten, and by the time the crockery had been washed and stowed away, night was upon us. The wind moderated and headed and the tide was foul for six hours, so, although the yacht sailed well throughout the night, she failed to bring the Schouwen lightship abeam by morning.

Despite the light wind and sea it had been by no means an idle night, for we had to keep a vigilant watch as we sailed through a fleet of fishing boats and from time to time ships crossed our course, so for much of the night both of us were on deck. The yacht's navigation lights were inefficient, but I had bought two hurricane lamps at Ymuiden, one with red glass and the other green, which were kept in the cockpit and shown as required.

At dawn (19 September) my wife was at the helm and it was my turn on watch. I have never shared a poet's love of the dawn at sea. It is then that the long sleepless hours make themselves felt. The dawn is grey, the sea is grey. It is cold and it is damp and one gets hungry.

Annette II was just ploughing over a waste of empty sea, for the low Dutch islands were far below the horizon. The glass was falling steadily and the wind was freshening again. We had no wireless and, therefore, no weather forecasts, but the conditions were a warning in themselves. The only cheerful prospect was that the wind had backed so we could lay the course and were making fine progress with a fair tide.

Throughout the morning *Annette II* sailed in a welter of foam on her course for the North Hinder lightvessel. After lunch I began to feel anxious, as we had failed to sight the lightship, when happily a spidery red form was vaguely discerned in the distance ahead. I had made the common mistake of overestimating the distance made good. We passed close to the lightship just before 1500, and her crew turned out to greet us. I had no time to respond to their friendly hails, for a squall struck us and I was busy reefing the mainsail and setting the storm jib.

By then it was blowing half a gale and the barometer had fallen no less than 20 millibars since early morning. From the yacht's position the Sandettie lightship lay about 30 miles to the south. The distance from the Sandettie to Dover is only a matter of another 20, so we determined to carry on and get to Dover or another port on the English side of the North Sea.

On we sailed, with the wind continuing to harden all the time. The wind had backed to a little east of south, and with shoals off the coast of Belgium little over 20 miles to windward the seas, although rough, were not high. The hours slipped by as *Annette II* held on her course, and the glass continued to fall. It dropped another 7 millibars, making a total fall of 27 millibars.

Towards sunset *Annette II* was more than half-way from the North Hinder to the Sandettie lightship. The sun was low and fiercely yellow, and gradually a great bank of purple cloud spread across the sky, covering the whole horizon to the southward. The sun became hidden behind the cloud as it advanced, but a wan light lit up the white-capped sea.

Then it arrived. The first squall was on us. The seas were obscured in the whiteness of pelting rain, and there was a sizzling noise. The yacht heeled far over, the wind whistled in the rigging with every sail, every stay and sheet hardened under the strain. The seas leapt short, steep and breaking. One came aboard heavily and broke over the cabin top and cascaded off to leeward. The yacht was hard pressed with her lee rail under in spite of her tremendous beam, and I let the mainsheet fly. That eased her, and I sheeted the storm jib to weather and belayed it. I sheeted in the reefed main and mizzen hard and with the helm lashed slightly down, the yacht lay hove-to on the port tack.

For a while I sat in the steering well. The force of the wind was stunning. The rain was torrential, ironing out the breaking seas with its pelting drops, leaving only the deep furrows. The squall was accompanied by thunder and lightning.

Annette II lay hove-to very well. The heavy oak hatch to the companionway was closed, and the dinghy was well secured on the foredeck by four lashings. The yacht seemed safe enough, but she might have been better without the mainsail, though I did not lower it. With the onslaught of the gale the wind veered to the south-west, at once dispelling any hope of fetching to Dover. The yacht was slowly drifting towards the middle of the North Sea. There was no danger in that direction for many hours to come, but if the wind veered to west or north-west the Belgian shoals and coast would be under the lee.

There was nothing to be done. Seizing my chance between the seas, I opened the hatch and slipped into the warm interior of the cabin. In the meantime my wife had been busy with the lamps, filling and trimming them: white, red and green. She jammed them between the cabin table and a bunk, where they were ready for use. She had also prepared malted milk, and this, together with some dry biscuits, formed our evening meal.

Night was soon upon us. There was nothing to do but lie as best we could in our bunks. We took watches in turn to open the hatch and look out for the lights of approaching ships and to see that all was well. Down below in the cabin it was wretched. The atmosphere was close and damp, for leaks had developed in the deck and round the coachroof and there was condensation everywhere. The ceaseless hammering and shaking as the seas struck the boat were wearying, and as the sea rose we had to use continual effort to prevent ourselves being thrown out of our bunks. We were both very wet, as the violent rain and spray had penetrated our oilskins, so that when we went below the wet clothes gave a feeling of clammy dampness. Neither of us slept, but we dozed a little from time to time.

When one or other of us went on deck we occasionally saw the lights of ships sufficiently close to warrant exhibiting one or other of the flickering hurricane lamps. On deck the scene was impressive in the extreme. The seas were black, but their formation was outlined by the gleam of the breaking crests. They came in sinister procession. The bows of the boat would rise to meet one, then down it would fall on the sloping mass of water into the trough, before rising again in time to climb the next. Sometimes the top of a sea would break aboard green and fall with a thud on the top of the cabin roof to stream aft in a cascade of water over the closed companion doors into the cockpit.

There was tremendous noise: the wind in the mast and rigging, the hiss of advancing breakers and the splash of running water. Above all raged a non-stop thrumming and vibration in the rigging. I believe the sea steadily increased, for the motion became worse and worse. Two of the lamps got smashed, but it did not matter much, as they flickered so much when exposed to the wind that there was little chance of them being seen. Moreover, we grew lethargic and our visits to the deck became less and less frequent; we were quite game to accept the chance of one in a thousand of being run down.

The position was wild in the extreme: we were hove-to in the night of a severe gale, surrounded at a distance of about 20 miles, except to the north, by a broken circle of shoals, on which the seas would break heavily. We were absolutely in the hands of chance. At the time we were profoundly miserable. We just lay in our bunks only half asleep. The position was not imminently dangerous so long as the wind did not shift to the west.

The night went by steadily, time passing neither slowly nor quickly. The hours went by until the moment arrived when in the dark cabin one could discern the outline of the porthole against the dim light of approaching day. We had only a vague idea of our position and my wife very appropriately suggested that I should go on deck again while it was still dark to see whether the lights of any lightships were in sight. On my reckoning there was little prospect of this, and I remained as I was for a few minutes before summoning energy to do my wife's bidding. Then I climbed out and stepped up into the cockpit. It was bitterly cold on deck and a big sea was running, but the wind was less vicious. It was still dark and, to my surprise, I saw at intervals the loom of several flashing lights reflected in the sky to the west. There must have been a clearance in the visibility. My wife's intuition was right. Then, suddenly, away on the starboard beam I saw distinctly the glow of a distant red light. It disappeared. It came again. A long interval and there it was again!

I had been anxious about our position, but in a few seconds the situation was utterly

The ketch Annette II *in the Stint See, Riga, at the start of her cruise to England. The Estonian family is bidding farewell.*

The Tumlare yacht Zara, *showing the extremely high aspect ratio of the sail plan.*

changed. Unfamiliar as I was with the Thames Estuary I could not be misled by two red flashes. On the face of the whole chart it could mean only one thing: the Galloper lightship (the character of the light has since been changed).

The wind, which had shifted from time to time during the night, was SSW. I unlashed the helm, sheeted the jib to leeward and bore away to the NW. As the yacht gathered way with the sheets eased she crashed over the seas, and the wind hurled the spray across the ship in massive sheets. With a roar a big sea flung itself aboard, struck the dinghy, and fell on the cabin top. A few minutes later another reared up and came crashing over the ship, breaking in a solid mass over the companion cover, and struck me a heavy blow across the chest. My wife joined me on deck. All night we had lain in our bunks damp to the skin and, now in the bitter cold before dawn, the driving water constantly penetrated our oilskins and chilled us to the bone. We were both pretty tired after the sleepless night and the effort of keeping ourselves from rolling off our bunks. We were also hungry, for under these conditions cooking had been almost impossible.

We had a few sips of whisky and Riga balsam. The latter was a potent drink which we had bought in Riga. It was rather an unpleasant bitter taste like medicine, but I fancy it has a very high alcoholic content and is warming. The edge of hunger was relieved by eating macaroons which we had bought at Ymuiden; they were wet, for a sea arrived at the exact moment that we opened the tin. It was rather an odd kind of breakfast and an odd time to have it, but basically it was alcohol, sugar and protein—a good mixture when one is tired and cold.

In a wild welter of foam-capped seas *Annette II* sailed on. Time after time she was swept by seas. But the knowledge of the yacht's position gave us confidence. We could see the friendly red flashes of the Galloper lightship and before dawn it came abeam. The sea was then tremendous and very confused, as we were crossing shallower water in the vicinity of the North Falls.

My wife sat beside me, trying to identify the looms of the distant lights. The steering was too heavy for her, but she was cheerful and took a full share in any work to be done. An hour after we had passed the Galloper, the distant lights still remained below the horizon. Before long even the looms could no longer be distinguished against the lightening sky.

Navigation posed problems, as I was totally unfamiliar with the east coast of England and I was unable to leave the deck to go below to lay proper courses. I had charts of the Continental coast, but the only ones I had of the east coast of England were a small-scale chart of the North Sea and an old Blue Back on an equally small scale. Harwich was little over 20 miles away, but the approaches would be tricky without a proper chart and the wind was likely to veer and head us. On the other hand, Lowestoft appeared to have an easy approach. It was nearly 50 miles distant, but even if the wind veered to the west it would still be free and the passage would take less than ten hours' sailing. So we eased the sheets and altered to the estimated new course.

Running before the gale with the wind on the quarter, *Annette II* was very hard on the helm. I was very cold and it was exhausting work. The hours passed slowly. My wife tried to take a spell at steering, but the effort to hold a steady course before the big following seas was beyond her physical strength.

At length the time arrived when land should be in sight, but there was nothing to be seen except sea and more sea in every direction. Nevertheless, the gale had moderated and we passed a few ships.

At last my wife (who has exceptionally good sight) declared she could see something like land on the port bow. I could see nothing of it myself, but I knew she must be right and the end of our difficulties was in sight. This cheered us immensely. My wife went below and got a primus going to heat the cabin. There was a shambles in the cabin from articles thrown adrift during the gale. The water in the bilges had risen to the cabin sole and everything was soaked. But the sea was moderating under the partial lee of the land and presently my wife appeared on deck with a tin of cold baked beans, which between us we devoured hungrily. She then took a watch, for the yacht was becoming easier on the helm, and I went below into the warmth of the cabin, peeled off my oilskins and slowly changed into dry clothes. When I returned on deck to resume steering, with dry clothes on and a shot of whisky inside me, I felt a different man, but my wife, who was also wet through, would not change.

A low coast was abeam, and short brown seas replaced the confused mass of grey. A black buoy was passed, the sun appeared, but the minutes still seemed like hours. At last we saw a town on the port bow and after some consideration came to the conclusion that it was Southwold. It came nearer and, through glasses, we distinguished two long low wooden piers. References to the Cruising Association Handbook gave a few notes on the shallow harbour at Walberswick which were far from encouraging, but on nearer approach we realized that we had a leading wind up the reach between the piers. The yacht was steered more inshore, the entrance came near and we could

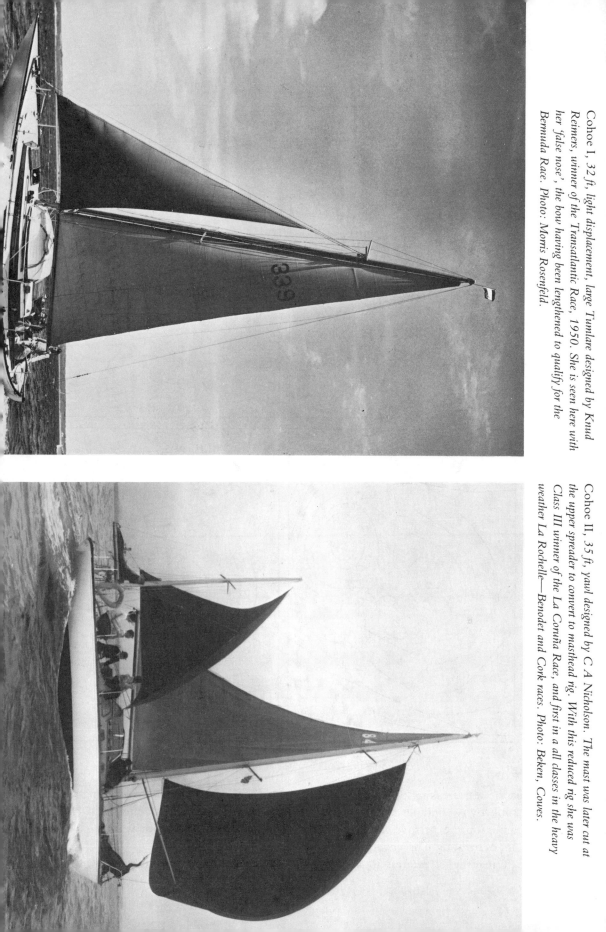

Cohoe I, 32 ft, light displacement, large Tumlare designed by Knud Reimers, winner of the Transatlantic Race, 1950. She is seen here with her 'false nose', the bow having been lengthened to qualify for the Bermuda Race. Photo: Morris Rosenfeld.

Cohoe II, 35 ft, yawl designed by C A Nicholson. The mast was later cut at the upper spreader to convert to masthead rig. With this reduced rig she was Class III winner of the La Coruña Race, and first in all classes in the heavy weather La Rochelle—Benoder and Cork races. Photo: Beken, Cowes.

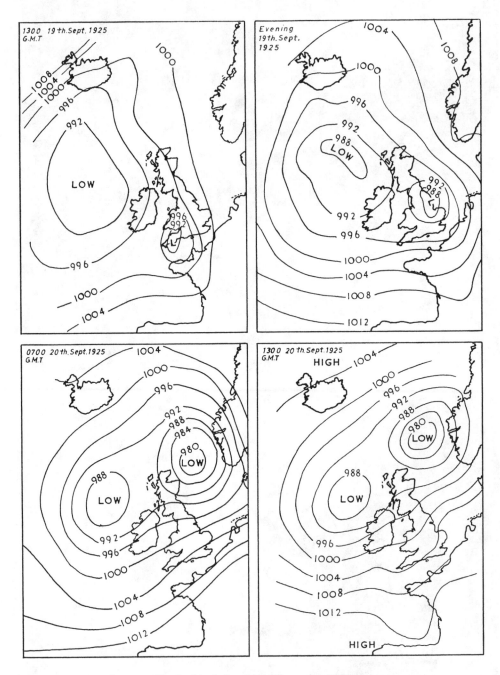

Fig 1.2. Synoptic charts covering the North Sea gale of September 1925.

see the seas breaking on either side. I calculated that it was two hours after high water. I took two soundings with the lead before sailing on straight for the piers. *Annette* foamed up the narrow entrance between the piers and her anchor was let go in the calm of harbour. The voyage was at an end.

It may be of interest to add that the name *Annette II* was adopted by me as a pseudonym in my early book *Close Hauled* (now long out of print) for Arthur Ransome's *Racundra*. When I sold *Annette II* the new owner installed a new engine and reverted to the original name of *Racundra*. Since the war her whereabouts had been something of a mystery until a yachtsman identified her, by means of a photograph, sailing and apparently in good condition in the Mediterranean.

Conclusions

The weather was featured in the Sunday and Monday newspapers, which described the week-end as the worst of the year, with 'violent' gales and fierce rainstorms. The disturbance was reported as developing off the north coast of Spain early on Saturday morning and travelling NE across England at a rate of 40 to 45 knots, but the accompanying synoptic charts show that the cause of the trouble was a deepening secondary. At Dungeness the wind was 35 knots (force 8) sustained from 0900 on Saturday evening till 1700 on Sunday. At Calshot, Hants, and Spurn Head, Yorkshire, it reached about 43 knots (force 9). Many vessels were in distress round the coasts.

From these reports, it seems reasonable to regard the gale as experienced by *Annette II* in the North Sea as mean of force 8 for a few hours, falling on Sunday morning to force 7, as recorded at Calais at 0700. The frontal squalls when *Annette II* hove-to could have been anything up to 50 or 60 knots, possibly bringing the mean to up to force 9 on the Beaufort scale for an hour or so. The following lessons were learnt from our cruise in *Annette II* and the gale on the last lap.

1 *Sail area*. A yacht with a very small sail area, such as *Annette II*, is so slow in light and moderate winds that, without auxiliary power, she is a sitting duck for gales when on long passages.

2 *Time*. Our holiday was drawing to an end when we were caught out, otherwise we would probably have taken shelter long before the gale started. Shortage of time and the need to get a yacht to her home port in a hurry are the most common causes of the cruising man getting caught out.

3 *Heaving-to*. The yacht hove-to well without coming up into the wind or forereaching too much. No doubt the long, straight keel and the sail plan distributed over three low sails helped her performance. Leeway, however, must have been considerable.

4 *Rain*. From later experience I can confirm that torrential rain, so violent that the surface of the sea smokes with it, has the temporary effect of taking some of the viciousness out of the seas.

5 *Tiredness*. On arrival in port we had been fifty-three hours at sea in narrow waters, out of which we had perhaps a total of four to six hours sleep. We felt tired, but we quickly brightened up ashore in front of a blazing fire at the Bell Inn. It is probable that in heavy weather, when sleep is difficult to get because of the din, the lack of it is not exhausting providing that the crew can get a reasonable amount of rest in their bunks. People suffering from insomnia carry on with relatively little sleep.

The passing hardship made no apparent impression on my wife, who had taken it all calmly and uncomplainingly, but years later she told me that for weeks afterwards she suffered from nightmares of huge seas.

2 POOPED FOR THE FIRST TIME

Adlard Coles

The first boat I owned after the war afforded a complete contrast to the heavy *Annette II*. She was a Tumlare yacht named *Zara*. The Tumlare class boats were built to the design of Knud Reimers of Stockholm, and were a kind of 20 sq metre Skerry Cruiser. They were comparable in size and speed to the Dragons, but had a much larger cabin, where, as I once put it, 'one person can be accommodated in comfort, two in tolerance, three in tenseness and four in bitter enmity'. *Zara* had a long low narrow hull with the typical Scandinavian pointed stern, with outhung rudder. She was 27.2 ft LOA, 21.8 ft on the waterline, 6.5 ft beam, and 4.2 ft draught. The total sail area was 215 sq ft, and the sail plan was of very high aspect ratio.

A light-displacement yacht of this kind is a delight to handle. Although inclined to be sluggish in light airs, in any reasonable wind *Zara* was lively and fast, well balanced and light as a feather on the helm. She was good fun for day sailing or racing and week-end cruising, though my wife never really enjoyed more extended cruising in so small a boat. The cramped size of the cabin, the lack of headroom and the absence of a proper galley, or large lockers for food and clothes, took the fun out of it for her.

However, I did put in one cruise down west to Brixham which I made single handed very soon after the end of the war. The following incident occurred during the passage back to the Solent, and I record it as it shows that there is a limit beyond which one must not go in a very small yacht with a large open cockpit.

I had spent the night anchored in Babbacombe Bay, near Torquay. I was up at dawn the following morning, for this, I hoped, was to be the day for the home-ward passage across Lyme Bay. At least, I tried to convince myself of this reason for an early rising, for to tell the truth I had spent an exceedingly uncomfortable night in *Zara*. An unpleasant swell had been running into the bay and all night *Zara* had rolled hideously and kept me awake.

Up early as I was, I did not weigh anchor until 0900, as much had to be done before setting sail. At the start the course was set for the position of Lyme Buoy, a little over 20 miles to the eastward, though this buoy had been removed during the war and had not then been replaced.

There was little doubt that *Zara* was going to get enough wind. It was southwest and fresh; already there was plenty of it. As I sailed out of Babbacombe Bay I brought *Zara* close to a fishing boat and hailed them for a weather report. They shook their heads and one replied: 'Look at the sky. It's going to blow hard. No day for crossing the bay, sir'.

I would normally have postponed my crossing of Lyme Bay, but it was already October, the days were getting short and I wanted to get back to my office before the weather broke. I looked at the sky and it seemed to me that, although a strong wind and a rough passage were to be expected, the wind was not likely to reach gale force in the immediate future. *Zara* seemed to be an able sea boat for her size, well able to hold her own within reasonable limits, her only weakness being the fine lines aft, the low freeboard, and the open cockpit. With me alone on board she would be buoyant aft, and there would be no second person to be asked to sit in the cabin to trim the ship, as is sometimes necessary in a Tumlare yacht, in order to reduce the weight in the cockpit when there is a steep following sea.

So I sailed on. At first the sea was smooth under the lee of the land, and I set the mainsail and the genoa. No sooner was the yacht clear of Babbacombe Bay than the genoa had to come down and the staysail be hoisted in its place. It was a grand day, with a sunny sky and a blue sea, but when the yacht sailed beyond the protection of Hope's Nose and opened up Torbay the seas grew larger, livelier and white crested.

Writing now, I see before me the pencil marks on the big chart showing *Zara's* progress on an almost easterly course over the first half of the 40 miles of open sea between Babbacombe and the Bill of Portland. At 1000 I reckoned she was 6 miles offshore, tearing along at great speed. I suppose that it must have been at about this time that the wind began seriously to freshen. Reefing became necessary even with the following wind and, knowing that conditions would become livelier as each mile of broken water was left astern, I brought the yacht to the wind and hove-to. I did not do the reefing by halves, for I rolled down four complete rolls, bringing the peak of the mainsail down to the upper spreaders. The sail area was reduced from 215 to about 120 sq ft and, more important, the head of the sail was at the same height as the forestay and back stay, so the mast was well supported. Yet under such conditions with even this small spread of canvas (not more than is carried by a 14 ft International racing dinghy) *Zara* was very fast. This was due to the efficient shape of the high, narrow sail, and her easily driven hull. Snugged down like this, *Zara* was a yacht that could take a lot of knocking about, yet still remain lively and close-winded.

Fig 2.1. Course of Zara *across Lyme Bay.*

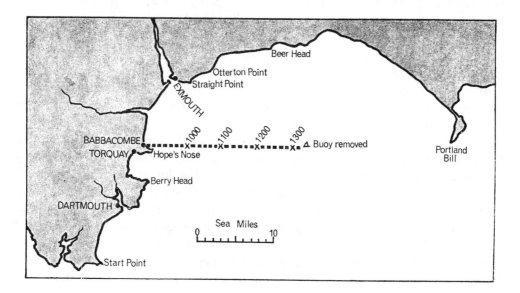

Time passed. On the chart the positions recorded that *Zara* was 11 miles east of Babbacombe Point at 1100 and 16 miles by noon. By then she was well out of the shelter of Start Point, which lay nearly 30 miles to the south-west and the sea was coming with unimpeded fetch from the direction of the Atlantic. The Channel tide was running in the teeth of the wind, though not as yet at its maximum rate, for it is not until a yacht gets well to the east side of Lyme Bay towards Portland Bill that she meets the full strength of the tidal stream.

In an account of a rough passage in a smallish yacht one wants complete accuracy to give a true picture of the conditions and the behaviour of the yacht. Well, this was no gale. It was what the fishermen afterwards described as a 'hard wind', what is often called 'half a gale of wind'. The seas were fairly regular, but inevitably they were large, coming in from the open ocean and meeting the westerly running spring tide. There was a gale warning in operation in the Plymouth area only 40 miles to the westward, so I heard afterwards.

At times I felt a little apprehensive, but the yacht was behaving wonderfully and made light of the following seas. A thing which pleased me was the zest and beauty in the scene. It was a most glorious day, with a blaze of sun on the water and it was really warm. The seas afforded a magnificent picture, a mass of blue and white glinting in the sun, and, on top of all, the dinghy, her wet red enamel shining brightly and spurts of spray flying off each side as she foamed along in the wake of the yacht. From time to time I saw big seas of frightening height rearing up astern, but as they never hit *Zara* I came to the conclusion that they were an optical illusion, as the weight of the wind did not seem to justify anything so vicious. I had never experienced more exhilarating conditions. They were vigorous and lively in the extreme. The seas chasing up astern, the dinghy scudding over their tops, the incessant motion of *Zara*, the hard-pressed sails and the taut sheets, and above all, the strong wind.

Land was out of sight, as there was a slight haze. At 1230 I had lunch on deck. I had heated soup before leaving Babbacombe Bay and put it in a Thermos flask ready for this occasion. It tasted grand and after that I had bread and cheese and apples.

The next position that I marked on the chart was at 1300. The yacht had covered about 20 miles and I made a pencil cross at a position 2 miles west of the normal position of Lyme buoy.

And so *Zara* raced on across the sunlit seas, the dinghy following gamely at her stern. It was, I think, about a quarter of an hour later when something caused me to look back. It must have been the sound of water.

Two enormous seas were bearing down on us. Instantly I put the helm up and ran dead before the wind to take them true on the stern. The towering top came climbing up. The stern of the dinghy lifted high. For a moment the dinghy planed and then, in a flash, I realized that I had never seen a dinghy at such an angle. Her bow was down. It was under water. I could see the water overlapping the forward transom. The stern was rearing up. But there was something higher than the dinghy. A great white crest of foaming water was above her, embracing her aft transom in its curling top. The whole thing, water and dinghy, was coming at a speed I never want to see again. I caught a glimpse of a sea that was steep as a wall, like a breaker throwing itself on a shore such as at Chale in the Isle of Wight, where the sea comes deep to sharply shelving shingle. The dinghy was in the act of making a complete somersault stern over bow.

I ducked under the cockpit coamings and held on. There was a deluge of water, followed by a resounding crash. It flashed through my mind that the mast had gone.

It is difficult to describe the supreme violence of a breaking wave when it comes cascading over the stern. There is a great noise, the shock of cold water, and an impression (whether true or not, I cannot say, for my eyes were filled with water), of seething sea, as though the cockpit is a bubbling cauldron.

When I rubbed the salt out of my eyes, I found *Zara* was still afloat, and I was astonished to see the mast was still standing. A quick survey for damage showed that there was none at all. The yacht had gybed and broached-to. The crash which I had taken to be the mast going over the side must have been the boom against the runner as the yacht gybed. There *Zara* lay, riding quietly to the big seas, her boom resting against the lee backstay and her jib sheeted loosely on the weather side. She was practically hove-to on the port tack. Her short boom and narrow sail plan had saved her from what, in some yachts, might have resulted in the breakage of the mast or boom, or both.

I hurriedly set up the weather runner and cast off the lee one. I sheeted in the mainsail and the weather jib sheet, and *Zara* responded, heading up into the wind and lying quietly hove-to, facing the turmoil of breaking seas.

The painter of the dinghy had parted, and I spotted the dinghy away to leeward, floating bottom upwards on the waves. There was no possibility of salvaging her. To have tried to come alongside the boat in the sea which was running would have been highly dangerous and useless, too, for it would have been impossible to have righted the boat and bailed her out single-handed. My first attention had to be given to *Zara*, and that urgently, so the last I saw of this faithful companion of many months' cruising was the waves breaking over her white upturned bottom as she bobbed up and down after her abandonment. I heard later that she went ashore near Otterton Point (Budleigh Salterton) and was salvaged on 16 October with, in the words of the Receiver of Wreck, 'five planks shattered, one keel cracked, six ribs broken, and a shattered nose'. It read more like a medical report. I wonder what part of her was left whole!

In *Zara*'s cabin I found trouble. The bilges were full and water was running over the cabin sole and over the lee planks, as she heeled to the gusts. This was not quite such a serious matter as it may sound, for it must be remembered that *Zara* is a yacht of light displacement, with narrow floors, so that the amount of sea required to make the water in the bilges rise above the floorboards is not so great as in a larger yacht. Still, it was more water than I had ever seen in her before.

Before the wave struck her, *Zara* had been a thing of life, riding confidently over the following seas. From a buoyant yacht she became in an instant an irresponsive, partially waterlogged object that could sink under the next wave. I realized that she was like the gaily-coloured metal boats that I played with as a child and which sink in the bath if too much water gets into them.

How my friends would have laughed to have seen me manning the pump. I hate exertion, but I set to work with greater enthusiasm than I had ever shown before to get the water out of her before the next big sea came along. It took quite a long time, but once the hull was clear of water there was no immediate danger. *Zara* was lying hove-to like a duck, her tiller locked in the steering contrivance with which I had fitted her.

She was heading up to the seas and they broke over her bows and forward deck and not into the vulnerable open cockpit.

Although there was no longer any immediate danger, I was averse to running the yacht off again, before the wind, on her course, because I did not want to be pooped again, and I thought the wind might continue to harden and the seas get worse until the tide turned to the eastward. Even then, with a fair tide, Portland Bill is not a place I would choose for so small a boat. Disturbed water extends a long way beyond the confines of the race itself. At the right state of tide, the inner passage offers a safer route, but the overfalls are not entirely avoided.

The alternative was to beat all the way back to the shelter of the Devonshire coast. It was quite a tough prospect, for it meant beating against a strong wind and heavy breaking seas, but there was no doubt that this was the right course of action.

I was drenched to the skin despite my canvas smock, and I now began to feel cold, whipped as I was by the stinging wind and spray. Happily I had some soup left in the Thermos, and this hot drink greatly revived me.

Then I set *Zara* to her task. I let the sails draw and she went off at a great pace, taking the seas magnificently. The seas were so big that she did not slam at all as she had inside Portland Race on the outward passage, or as she usually did in the Solent when beating in the short wind–over–tide seas there. From time to time I would luff to a 'big 'un'; she would lift over the top and slither down the steep slope on the other side. Forward she was like a half-tide rock, with the stem cutting through the tops of the waves which would come streaming over her long foredeck in deluges and run off over the lee side in continuous waterfalls. But aft in the cockpit it was comparatively dry. Nevertheless, for a time sailing was not without anxiety. The seas were undoubtedly severe for the size of the ship, but as I sailed westward her wonderful performance gave me added confidence in her windward ability. Close-sheeted, *Zara* could lay almost due west, but it was better to ease the sheets a fraction and keep her travelling fast heading WNW.

After an hour I sighted land again. A misty white smudge appeared against the sky to the northward. It must have been Beer Head, about 10 miles distant. Then I hove-to again, for the water in *Zara*'s hull was once more rising above the cabin sole. With water flowing in a constant stream over her foredeck it was continually leaking through the hawse pipe and the forehatch. I locked the tiller in position so that the helm could be left while I pumped her out. The seas were already easier. There were fewer really big waves, but it remained a lively scene of endless white horses in every direction as far as the eye could see. The yacht was happy enough, bobbing up and down in the seas, so, as I was feeling cold despite the sun, I took the opportunity to change from my wet clothes, and get a rub down with a towel. The cabin was in a dreadful mess, but I found a pair of dry pyjamas in a sail bag, a pair of flannel trousers which I pulled over them and a tolerably dry reefer jacket. With oilskins over the lot I was well protected and it was wonderful to feel dry again after the unpleasantness of the clammy cold garments. I blessed the sun, for in grey and cold weather it would have been grim indeed.

Then I put the yacht back on her course. The seas became progressively smaller as we got farther west, and in the late afternoon I berthed *Zara* along-side a big yacht moored off the Morgan Giles yard at Teignmouth. Close-reefed she had made good 20 miles in four hours including two tacks which is fast for so small a boat sailing hard on the wind in rough water.

Conclusions

The nearest record of wind forces was from the shore station at Pendennis Castle, Falmouth, but it is pointed out in a letter from the Meteorological Office that according to the synoptic charts there is not likely to have been much difference there between the wind forces at Pendennis and Lyme Bay.

The mean wind speeds were west force 7 (28–33 knots) at 1130 GMT force 6 (22–27 knots) at 1230 and 1330, force 5 (17–21 knots) W by S at 1430. The wind must have been fairly steady, the highest gust recorded being 33 knots (almost force 8) at 1115. The times are GMT, to which an hour must be added to coincide with my own report, assuming summer time was in operation. The strength of the wind in Lyme Bay would be higher than recorded by a shore station, but I think a considered estimate might put it somewhere about 25–30 knots, force 6 to 7. The strong winds were caused by a relatively high 'high' to the NW of a 'low'.

Fig 2.2. Synoptic chart, 10 October 1945.

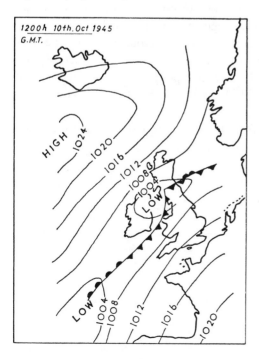

1200h 10th.Oct 1945
G.M.T.

There were several useful lessons to be learnt from this experience.

1 *Pooping*. Ansted's *Dictionary of Sea Terms* states 'when a sea comes over the stern of a vessel it is said to poop her'. If this is so (and many yachtsmen agree with the definition) ocean-racing yachts are frequently pooped when running downwind above force 7, for the heads of waves often break over the stern and half fill the cockpit. To my mind, however, pooping means something much more severe. It means being overwhelmed aft by a breaking following sea. It may result in a broach-to which throws the yacht on her beam ends as she rounds to the wind. This can be a dangerous thing if it leads to dismasting or structural damage to the yacht. Genuine pooping is a very rare thing indeed and I have experienced it only twice in a lifetime.

2 *Danger Downwind*. *Zara* started in

sheltered waters under the lee of the land and running downhill, so to speak, the strengthening of the wind and the increase in the sea was gradual as the fetch increased. With a following wind the temptation is to carry on, whereas with a strong headwind the going is so rough that the temptation is to turn back. Once, when my wife and I were at Keyhaven in a westerly gale, three canoeists set out for Lymington. The water was smooth when they started, as they had a weather shore, but gradually the seas increased and somewhere off Lymington two of the canoes capsized and their occcupants were drowned. Running downwind is the cause of many accidents in open boats and dinghies, since the increase in wind and sea is so gradual as to be almost imperceptible until it is too late.

3 *Freak Waves*. In rough weather there are occasional waves, larger and more unstable than the others. The exceptional wave which may be called a 'freak' wave can be caused by the synchronization or form of wave trains or by tidal overfalls or by shoals or obstructions on the bottom. In this particular case, I think *Zara* had passed over one of the wrecks just west of the normal position of Lyme Buoy. The depth in the vicinity is about 40 m, but it is only 30 m over the wreck a mile WSW of the buoy. One can easily see that the westerly-running stream meeting these big obstructions on the bottom would cause turbulence on the surface which might cause a big sea to break. It is a useful point to remember that such things as big underwater rocks and wrecks, even if deep down, can cause overfalls on the surface if the tidal stream is strong and the sea rough.

4 *Buoyancy Aft*. The Norwegian pilot cutters had pointed sterns and were noted for their ability in heavy weather. *Annette II*, which I described in the previous chapter, ran through the North Sea gale without giving any trouble. *Zara*, however, was too narrow in beam and too fine aft to give sufficient buoyancy for her to rise to an exceptional breaking sea. It is said that an advantage of a sharp-ended stern is that it divides following seas as they run harmlessly past. But the dangerous waves are the ones that do not divide but rear up over the stern like a breaker on the beach. Such seas are rare, but for a boat to rise quickly enough to avoid pooping it must have great buoyancy aft. Certainly the combination of lack of buoyancy aft and a large open cockpit is dangerous.

5 *Time*. The incident demonstrated again that the most frequent reason for getting 'caught out' is shortage of time towards the end of a holiday.

3 THE GALE OFF THE CASQUETS

Adlard Coles

We only kept *Zara* for one season, as she was too small for our family. In her place I bought *Mary Aidan*, a new 7.5-ton sloop designed by Fred Parker and built at the Dorset Yacht Company's yard at Hamworthy, near Poole. She was a cruising yacht with a good turn of speed, somewhat the ocean-racing type, although too short on the waterline to be eligible, even for Class III. Her dimensions were 34 ft overall length, 23 ft waterline, 8 ft 1 in beam and 5.2 ft draught. *Mary Aidan* was sloop rigged with a total area of 444 sq ft, consisting of 286 sq ft in the mainsail and 158 sq ft in the staysail. She was fitted with an 8 hp Stuart Turner engine, and was the first yacht that I owned which had a practical auxiliary.

We had a good season of cruising and handicap racing in *Mary Aidan* during 1946, in the course of which we experienced one gale. I was cruising with my daughter Arnaud, and a friend named George. We had sailed down west in mixed weather to Brixham. From there we took a departure from Berry Head at 1115 on Saturday, 28 July, bound for Guernsey. The weather forecast was good except for 'general rain spreading from the west tonight'.

This was my first cruise to the Channel Islands and French side of the English Channel. Although I have sailed on the French coast nearly every year since and visited practically every harbour and anchorage between Barfleur and La Rochelle, I was at that time a stranger to it. I regarded it as a rather hazardous cruising ground, with its innumerable rocks and strong tides.

After eight hours of fast reaching with a sunny sky above and a blue sparkling sea around us, we made a landfall on a gaunt, craggy little islet above which rose two lighthouses, and in the distance beyond was a bigger island. We were lucky to make this landfall, because even as a stranger I soon identified it as the Casquets, which I still think have a somewhat sinister appearance. Arnaud and George were in high spirits, but my own feelings were tempered by a slight uneasiness about the weather. During the morning we could not have asked for better conditions, but some ominous signs had developed: there was a halo round the sun, the swell was higher than the wind justified, though this had slowly but steadily increased. At the start of the passage *Mary Aidan* had logged little over 5 knots; in mid-Channel she was doing 6.5 and for the last two or three hours she had reeled off a steady 7. The glass was falling slowly.

At about 2000 the bad omens took more definite shape. A great bank of dark cloud was gathering to the westward, the sea was getting lumpier, and the wind began to pipe up in earnest. It was time to lower the genoa and replace it with the staysail. George went forward to do this, but as he was grappling with the big sail he slipped and, with his foot caught in a loop of the sheet, he a made spectacular dive overboard, complete with oilskins and his Army oilskin cap.

The leaden, tide-troubled swell off the Casquets, the gloom of the advancing clouds and approaching night provided a grim background for an accident such as this, but by

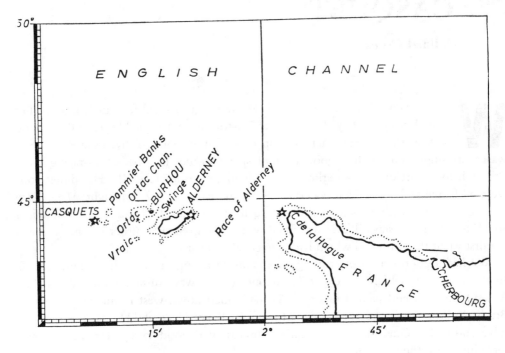

Fig 3.1. Casquets to Cherbourg.

the mercy of Providence, George had grabbed a rope as he fell. He reappeared by the shrouds and was quickly back on deck, where he stood like a great glistening fish, as the water poured off his yellow oilskins. I was glad to note that Arnaud, who was steering, was so quick with a lifebuoy that she had it out ready to cast before George would have been swept past. In *Mary Aidan* the lifebuoys are carried loosely, standing vertically between brass strips and the sides of the cockpit coamings, so they can be thrown in an instant. This is a better practice than having a buoy on the counter, so often recklessly lashed down. Seconds lost cutting a lashing or freeing a slip-knot may make the difference between life and death with a man overboard in heavy garments.

George went forward again to complete his job, and soon the staysail was set and the genoa stowed below out of the way. He came aft cheerfully, declaring he was warm, and I had to be very firm before he would consent to go below to change into dry clothes and have a glass of rum.

When the Casquets lighthouse lay about 2 miles ahead, the wind was rising so rapidly and the general conditions were so threatening that I decided that more reefing was desirable as a precaution, though not yet strictly necessary. We hove-to and tied down two reefs. This took some little time, for we had no tackle on the cringle and the weight of the wind made it difficult to haul it in hard enough. When all was done and the *Mary Aidan* was nicely snugged down, we sailed on towards the Casquets.

If the weather had been reasonably settled, I would have let the yacht sail herself temporarily to the westward close-hauled, or hove-to on the port tackle with the yacht slowly jogging ahead until the tide turned and we could sail to St Peter Port. But the wind was backing and conditions looked so ominous that it appeared wiser to run for shelter to leeward. I did not fancy piloting *Mary Aidan* into Alderney harbour (with which I was then unfamiliar), owing to the submerged breakwater marked on the chart, and the strong tidal streams which would make the approach difficult in the dark. Cherbourg, 30 miles to the east ward, seemed the best bet, with a fair wind and fair tide.

So *Mary Aidan* squared away and ran to the eastward into the gathering night. We gybed her to get a good offing from the Pommier Banks and the dangers WNW of Alderney. The sea grew increasingly rough as *Mary Aidan* ran before the wind through the overfalls of the roaring tide.

By the time it was really dark, the wind was blowing so hard and the seas had become so rough that, in order to take things comfortably, we lowered the mainsail and ran under staysail, but the yacht's speed was hardly lessened. I navigated by the Casquets light astern and Alderney light to starboard. We gybed again, as I had let the yacht run unnecessarily far to the northward owing to a kind of rock-shyness, and my reluctance to bring her too close to the gap between Alderney and Cap de la Hague, where Alderney Race thrusts into the English Channel.

George took the first night watch and I went below, being occupied with calculations of tidal streams and other points of navigation. The tide was due to turn just after 2300 and, as it was springs, there would be a spectacular sea when the westerly ebb stream began to run against the wind.

Gradually Alderney light changed position and we came into the track of ships. The masthead light of one looked exactly like an occulting light, so regularly did it disappear in the troughs of the seas. George was finding it difficult to steer, for even under jib alone it needed a lot of concentration to keep the yacht running straight before the breaking seas striding up astern. Presently the patter of rain was heard and the lights of Alderney disappeared in a squall. It was blowing great guns.

A little later George shouted to me that the weather was getting worse. He asked what was to be done. The night was pitch black, but one could hear the breaking seas above the wind. I went forward and lowered the staysail, lashing it down hard against the forestay. *Mary Aidan* was now under bare pole. Even so, she was running at great speed, but George steered her carefully, always meeting the seas stern on, judging them by sound and by the motion of the ship. I went back to the cabin, closing the companion doors after me, and opened the cocks under the bridge deck controlling the pipes of the self-draining cockpit.

The plan of running to Cherbourg had now to be abandoned, as between the yacht and her port of shelter lay a piece of water marked on the chart '7-knot springs, strongest part of the tide'. This note on the chart, which worried me at the time, is not repeated on recent charts. The rate is probably lower, about 5 knots at spring tides. With a wind then estimated at over force 7, I did not fancy experiments, even with wind and tide together. Moreover, owing to our detour to the northward it was clear by then that there was no chance of the yacht arriving at Cherbourg before the tide turned, when a wicked sea would be running.

At some time after 2300 there was a heavy crash, followed by a strange sound of rushing water. In the cabin Arnaud and I were thrown across the ship. I opened the cabin doors and looked out. The head of a sea had broken on board. It had thrown George from the tiller across the bridgedeck and half filled the cockpit. The unfamiliar sound was that of the streaming water as it gushed through the outlet drains of the self-emptying cockpit. On deck I found there was pandemonium with the howling wind and the din of the seas and rain.

It seemed to me that some water might have got into the bilges as well as into the cockpit. I tried the pump, but it was no use, for the motion was so violent that the water, of which there cannot have been very much, was sluicing across the bilges first to one and then to the other side of the hull, and the pump only worked intermittently as the flood rushed across the strum box.

In the cockpit the exit pipes had dealt with the flood of water at first quickly, but later at reducing speed. Some minutes after the influx there were still some inches of water over the gratings, and it would have been nasty if a second sea had broken aboard before the yacht had rid herself of the first.

Mary Aidan was running too fast even under bare pole. George reckoned she was doing a good 5 knots. I hurried below again, carefully closing the cabin doors after me. Crawling forward into the peak, I extracted a 45 m length of warp. Then, pushing it through the cabin door again, I got out into the cockpit and made fast one end of the rope to a great big cleat that has often been laughed at on account of its massiveness. I paid the other end out, over the starboard quarter. That steadied her immediately. Then I took the tiller from George and lashed it down slightly to starboard and the yacht headed north with the seas on her port quarter.

'All right,' I said to George, 'it's time to turn in now', and we went below to the cabin, switched on the electric navigation lights (hitherto only used when in proximity to ships) and shut the companionway doors firmly after us against the gale.

Down below in the cabin it was difficult to realize what a wild scene had been left behind on deck. The electric cabin light lit up the interior and added a touch of domestic bliss which contrasted oddly with the grimness of the conditions outside. The motion was not so severe as might have been expected. The wind in the mast steadied the boat, and although she rose and fell great distances she did so less violently than before. Certainly one had to hold on to steady oneself, but here is the advantage of a small boat, for it is possible to put one hand against the cabin top, and there is never the risk of being thrown across a big open space. It was possible to lie down on the berths with the canvas leeboards tied up to hold one in. Probably the worst thing was the din – things breaking loose, cooking utensils clanking, and the mess as things broke adrift. A packet of cleaning powder appeared from nowhere and spread itself in a slippery mess at the foot of the companion steps. A tin of blackcurrant puree overturned, followed soon afterwards by a tin of condensed milk.

Above the hubbub came the sibilant sound of rushing water as the seas bore down on the yacht out of the night. At intervals there would be a plonk as heavy spray broke into the cockpit, followed by the gurgle as it trickled down the outlets to the sea.

George was soaked to the skin (although I had not realized it at the time), but kept tolerably warm in blankets. Arnaud asked: 'When shall we get into Cherbourg, Daddy?'

There was some point in this question, for a little while earlier she had been lifted bodily in the air and flung from her bunk against the cabin table. But she took the news that the night was to be spent at sea with cheerfulness, and even managed to get some sleep towards morning.

A major objection to the cabin was the smell of sodden clothes and another very strange smell. Thinking that the latter must have been a leak of Calor gas, I went forward to the fo'c'sle to investigate, but I found the cock there was turned off. As I was returning to my bunk I saw a saucepan which had got adrift and contained three fine chops bought the previous day, which were now chasing each other along the cabin floor. They were the most handsome chops we had seen for a long time, and we had wondered at the butcher's preference in parting with them to strangers at a time when meat was still in short supply after the war. We now realized the reason why. They were high.

At midnight I marked *Mary Aidan*'s approximate position on the chart. I must confess I was apprehensive. Although a good 7-tonner should be able to live through any gale, there is the proviso that she must be in deep water, clear of shoals and tide races. Here, although we were north-east of Alderney, the tidal stream runs very strong at springs. The stream had already turned and would soon be ebbing at full strength directly in the teeth of a WSW gale. *Mary Aidan* was in for a dusting.

None of us felt sea-sick. It was fortunate that such was the case, for with the fore-hatch closed, the portholes screwed down tightly, and the cabin doors shut in case a big sea should break aboard, sea-sickness would have created a fetid atmosphere.

From time to time I went on deck, pushing open the companion doors against the wind and shining my torch on the compass. Then I would take a look round for the lights of ships. Close at hand there was nothing to be seen except for the phosphorescent white of breaking crests. In the squalls the sea seemed to be boiling and the lisp and hissing noise was uncanny.

So passed the night, a night of violence and racket, the roar of the wind, the noise of the breaking seas, and the regular gurgle of the water trickling out of the cockpit outfalls. I switched off the cabin lights. At about 0400 the sky lightened. The portholes admitted a grey light and the outline of the mast and table could be seen. When dawn came it lit up a wet little yacht with a tired crew. At about this time I noticed that the wind had moderated. There was less noise in the rigging, but the sea was higher and more irregular, and spray seemed to be breaking aboard as much as ever. A big sea was still running, but by 0600 the tide had eased, the waves had lengthened, and were no longer breaking. I turned in again until 0730 and then got the yacht under way.

Conclusions

First one must consider whether there was a gale at all. Here are all the ingredients which contribute to exaggeration. First, the landfall on an unfamiliar coast. Next a man overboard off the Casquets. Then the run before an increasing wind at night during spring tides in an unfamiliar area, notorious for the strength of the tidal streams.

It would be all too easy to paint too colourful a picture. Yet when I reread the words which I wrote many years ago there were certain facts which stand out: 'Doing a good 5 knots.' You do not get such a speed under bare pole with a wind of less than force 7. Then again you do not normally, under force 8, experience heads of seas which 'half fill the cockpit', especially when running under bare pole.

I must, however, plead guilty to one element of exaggeration which appeared in an earlier account of the gale. When, twenty years ago, I obtained the wind speed at Thorney by telephone, I took it to be the mean speed in knots, but it must have been the velocity of the highest gust expressed in statute miles per hour and not in knots, because, now that I have received figures from the Meteorological Office, I see that the mean hourly value of the wind at Thorney was only force 5, though gusts of 36 knots and 32 knots were recorded at 2350 and 0020 GMT, which indicates that the wind was exceptionally turbulent. Thorney is a sheltered coastal station and Portland, where the hourly value to midnight was force 7, is a better guide. I think this is a fair assessment, but the wind, funnelling through the Alderney Race gap between Cap de la Hague and Alderney, might well have attained force 8 locally, accompanied by violent squalls and gusts, while the front (which moved from the Scillies to Spurn Head) was going through. I make the following comments:

1 *Tidal disturbances.* When sailing in strong tidal streams these will appreciably add to or reduce the strength of the apparent wind. If the true wind is say 31 knots (force 7), a contrary stream of 4 knots running against it will give an apparent wind of 35 knots (force 8) on the surface of the water. If it is a lee-going stream, it will have the effect of reducing it to 27 knots (force 6). This is an exaggerated example, but often there is one whole grade in the Beaufort scale between the strength of the apparent wind when the stream is weather-going and when it is lee-going.

There is more in it than this. A tidal stream or current causes an increase in the wave height and steepness when the stream is running against the waves or the wind and a decrease when it is running in the other direction. The height may be increased by 50 to 100 per cent, and breaking seas may occur even without much local wind.

This explains why the navigator in coastal waters or in currents such as the Gulf Stream meets seas which may be out of all proportion to the strength of the wind.

2 *Harbours of refuge.* Alderney is a good harbour, but we were right in not attempting it at night for the first time in rough weather. My instinct in a gale to get away from the land was a sound one. It is better to lie a-hull and go to bed in the relative safety of deep water than to attempt to reach an unfamiliar harbour.

4 *Bare poles. Mary Aidan* lay well under bare pole. We discovered the technique of lying a-hull by experiment, found it to be good and adopted it later in stronger gales. But it should be noted that, although the exit pipes of the self-emptying cockpit coped with the heads of seas, they would need to be of greater diameter to be effective if bigger seas had broken aboard. Cockpit drains in yachts are nearly always too small.

5 *Morale.* The morale of the crew was high. Granted that the gale was a short-lived one, no more than force 7 or 8 possibly, we thought at the time that it was blowing harder. It makes a great difference in heavy weather if everybody keeps cheerful.

6 *Stowage.* In a gale anything loose, such as cooking utensils, provisions, etc., will take to flight. Everything should be secured on the *approach* of bad weather.

4 STARTING OCEAN RACING

Adlard Coles

In prewar years I always took an armchair interest in ocean racing. It seemed to me the most genuine form of sport, for the races, if not all on the ocean, were at least offshore over open water, to the Bay of Biscay, Spain and the Baltic, not to mention the 635-mile Fastnet Race, the Grand National of ocean racing. Racing such as this in any weather, gales, calm or fog, impressed me as the ultimate in sailing, providing a test of boats, crews, navigation and racing skill.

After the war I thought I would like to have a shot at it myself, before it was too late to start. The minimum length of yacht eligible for ocean races was then 25 ft in waterline length. *Mary Aidan* was only 23 ft LWL, so I sold her in 1946 and bought instead a yacht named *Cohoe*, which had just been built by A H Moody & Son Ltd, at Bursledon, and was virtually new, as she had only been sailed for two week-ends in the Solent. *Cohoe* (the name is that of a species of small fast Canadian salmon) was an enlarged version of a Tumlare, designed by Knud Reimers, with a 30 sq metre sail plan instead of the 20 sq metre of *Zest* and *Zara*. The shape of the hull was very similar, with moderate overhang forward, wine-glass sections and a Scandinavian stern with outhung rudder, but the stern was much fuller than in the smaller design, giving greater buoyancy aft, and she was thus less liable to be pooped.

Her design dimensions were 32.1 ft length overall, 25 ft 4 in length waterline, 7 ft 4 in beam and 5 ft 2 in draught. She worked out at 7 tons Thames measurement, but she was only 3.5 tons designed displacement, and in reality a considerably smaller boat than *Mary Aidan*.

I stress the word 'designed' displacement, because modifications had been made to the design of her former owner which added considerably to her actual displacement. The height of the topsides had been increased and a long coach-roof and a doghouse had been added. This greatly improved the space below. There was a full-length cabin with a berth on each side, and aft under the doghouse there was a galley with a primus stove to port and a quarter berth with chart table over it to starboard. Forward of the saloon was a compartment with cupboards and storage space separated by doors from the forecastle, in which there was a pipe cot, sail and anchor stowage and a Baby Blake WC. Although *Cohoe* was narrow (Humphrey Barton described her accommodation as 'like living in a tunnel'), the arrangements below were comfortable, and there was fair headroom with over 6 ft under the doghouse and over 5 ft at the forward end of the saloon.

The cockpit was a comfortable size and self-emptying, and situated below it was an 8 hp Stuart Turner engine driving a three-bladed propellor. To compensate for this extra weight aft, 5 cwt of lead had been added to the forward end of the keel. Many of our friends, especially cruising men of the older school, were very critical of *Cohoe*, saying that a yacht of such light displacement was unfit to go outside the Solent. I even overheard her described contemptuously as 'a toy yacht'. Nevertheless, she was a well-built boat and a thoroughbred, and I still look back on her with affection.

We bought *Cohoe* in August 1946, and my wife and I managed to get away in September for our first cruise, which was a delightful one in the Channel Isles and on the coast of Brittany. During the winter I prepared *Cohoe* for ocean racing the following season. The trouble was that the extra weight resulting from the alterations to the basic design and the addition of an auxiliary engine had spoilt her racing performance. She was not nearly so fast as her sister ship *Josephine*, owned by a friend of ours, Mr C Smallpeice, who had kept to the original design. Light displacement is all right as long as it is kept light, but *Cohoe* floated inches below her designed LWL. This made her longer on the waterline, which in theory provided a higher theoretical maximum speed, but, as all racing men know, superfluous weight in the wrong place makes a yacht tender and slower under all conditions except when running in strong winds. At a guess, her real displacement in final racing trim may have been as much as 4.5 tons, but the ballast keel even with the additional 5 cwt was only 1.75 tons, giving a ballast ratio of under 40 per cent, which is low for a boat lacking beam and hull shape stability. When she was measured in 1950 in America for the CCA rules the official measurer remarked that the only thing about her which surprised him was 'how she floated upright'.

The only way I could reduce weight was by having the engine removed. I had this done, much to my wife's annoyance, as it was a good one, and substituted a light 6 hp engine with a feathering propellor. Another alteration I made was to cut down the foot of the genoa, which was too long for the RORC rule and attracted a penalty. I also shortened the foot of the mainsail. The reduction in sail area reduced *Cohoe*'s rating, but made her sluggish in light airs, for to compensate for extra weight a yacht needs more sail area, not less. There is no advantage in a low rating if it is achieved only at a corresponding loss of speed. This lesson I learnt many seasons later.

In our first year of ocean racing we had a good introduction to heavy weather sailing, with wind forces of 6 and 7 and occasionally higher, such as are experienced in most seasons, but we encountered none of the major gales of the kind I refer to in later chapters.

Force Seven

Our first ocean race was the Southsea to Brixham Race, over a course of about 200 miles from Southsea to Le Havre lightvessel and thence to Brixham. We raced with a total of three, our complement consisting of Roger Heron (at that time a partner in the yacht designers, Laurent Giles and Partners) and Jim Hackforth Jones, the son of Commander Gilbert Hackforth Jones, the well-known author. Our routine was for each man to take two hours at the helm, then to take two hours stand-by below (available instantly at call) and two hours off. A boat of *Cohoe*'s size needed only two men to handle her, except when setting or lowering the spinnaker. I did the navigation as well, while cooking was divided between us. It was a good system, as all had full responsibility in watches and the result of the racing depended upon teamwork, which makes for a happy ship. It was a system we followed in all our early offshore races.

There were twenty-eight entries in the Brixham Race, of which eight were in Class

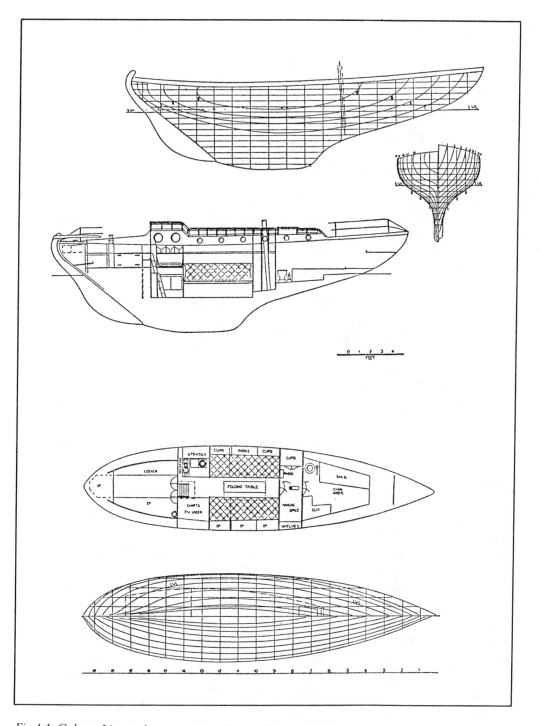

Fig 4.1. Cohoe. *Lines and accommodation. Designed by Knud Reimers.*
LOA 32.1 ft, LWL 25.4 ft, Draft 5.2 ft,
Displacement 3.5 tons; sail area 362 sq ft.

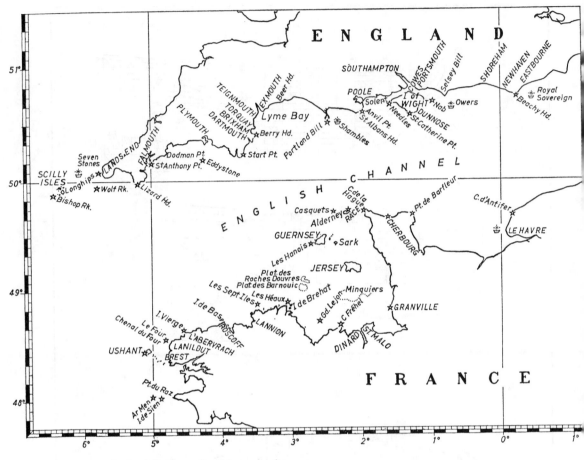

Fig 4.2. The English Channel and its approaches.

III, which is the small yacht class which included *Cohoe*, the smallest of all. The race started on Friday, 27 June, at 1845 and the first leg of the course to Havre LV was eventless, as there was a light free wind, though it was foggy during the night.

After rounding Le Havre lightship on Saturday forenoon, the next leg was diagonally across the English Channel, WNW to Brixham, a distance of 120 miles, which provided genuine offshore sailing which would normally take about twenty four hours. It was on this leg that we were to meet our first gale or near gale when racing.

The wind had been easterly during the night and after rounding the lightship in the morning the course provided a broad reach, but the barometer was falling and during the Saturday evening the wind quickly veered to dead ahead. It freshened steadily and by midnight, when *Cohoe* was in the middle of the English Channel, all yachts were reefed. At dawn on Sunday morning it was blowing harder still and conditions were dismal, as they always are in a small yacht turning to windward in rough water.

In *Cohoe* the angle of heel was extreme. She was a tender yacht, heeling readily, but once heeled to a certain point, the leverage of her keel stiffened her and she went over no farther. However, it is tiring work when a yacht is sailing on her ear. The motion was lively and on deck, sheets of spray were flying aft into the helmsman's eyes. We were close-reefed and I thought the wind was about force 6 to 7. *Cohoe* lay on the port tack. Several sails were in sight but these were not our immediate competitors. The best course she could lay would take her into the middle of Bournemouth Bay. The sun came out when we made our landfall, sighting the cliffs of the Isle of Wight to starboard. I remember that the strong westerly wind against the tide caused a big sea to run and it was rough going.

By the time we reached Swanage Bay we were pretty tired of beating and the tide was due to turn against us. I realized then that if we were to tack in a strong wind against a foul tide off St Alban's Head we should make little progress, so we decided to anchor in Swanage Bay.

There we had the luxury of changing into dry clothes and eating a hot meal. We pumped the bilges dry, kept the primus going, which helped to dry out the cabin, for everything below was wet. We mended a torn sail and carried out a number of minor repairs. Then we turned in for a couple of hours' rest before getting under way to catch the early inshore stream at St Alban's Head.

We had felt a little ashamed of ourselves for sheltering, as when we anchored in Swanage Bay we had seen the RORC club yacht *Griffin* standing out to sea on the starboard tack. *Griffin* was a boat of which we were to see a lot in our future racing. She was a Class II yacht, measuring 24 tons TM. However, when we had tacked from Swanage up to St Alban's we sighted *Griffin* again, and to our surprise we saw that the tide had set her back some 2 or 3 miles. Our self-indulgence inshore had been rewarded. If we had kept at sea, not only would we have made no progress but we would actually have been carried eastward by the tide. It was a useful lesson from which we profited in the future.

The wind by then had moderated and the rest of the race provided another night at sea, but plain sailing. The finish, however, was interesting. We were tired and despondent when we arrived and could see a forest of masts rising above the long breakwater of the Brixham harbour, suggesting all our competitors were ahead of us. We anchored off the yacht club and Roger Heron rowed off in the dinghy to learn the worst.

When he returned he was smiling. 'Cohoe's won,' he announced; 'no other Class III boat finished.' 'And,' he added, 'we've also beaten all Class II on corrected time.'

Thus ended our first ocean race. Sixteen yachts had retired. Some had given up because owner or crew had to be back on Monday morning and others because they had been damaged in the strong winds, among them the Class I yawl *Eostra* of 44 tons which had been dismasted. *Bloodhound* was winner in Class I and *Phemie* in Class II.

Conclusions

I have received data from the Meteorological Office. At the Lizard the wind records showed 27 knots and 26 knots (top of force 6) for two hours to 0030 Sunday morning. For the next hour it rose to 28 knots, which put it at the bottom of force 7, after which it fell to 26 knots. The highest gust was 35 knots at 0100. A front would have arrived later in the Wight area and the strength of the wind at sea might have been a little higher.

1 *Sheltering.* The relative performance of *Griffin* soldiering on at sea and *Cohoe* temporarily sheltering in Swanage Bay shows that, if the sea is really rough, the average small yacht cannot make progress beating to windward against a strong foul spring tide off a headland. It is better to shelter, make such repairs as are necessary and feed and rest the crew. *Griffin* was gaff rigged and *Cohoe* was small, so both were handicapped in windward performance in rough weather. Modern yachts, designed for ocean racing, would not nowadays shelter in winds under 40 knots (force 8) in the absence of special reasons for it.

2 *Historical.* As the modern development of the small ocean racer is indirectly traced in this book, it is interesting to read a cutting from *Yachts and Yachting* on this race. 'Conditions . . . again throw some doubts on the suitability of a long course for the very small boats that are now admitted to RORC events.'

Yachts and Yachting has always been closely associated with ocean racing, so the comments may be taken as a fair expression of the general opinion at the time.

Thunder Storm in the Bay

We experienced heavy weather in two other races in our first year of ocean racing. In the Channel Race there was a strong easterly blow which John Illingworth described as attaining 'over force 7'. It was a hard sail for the three of us on the 70-mile beat to windward, but after rounding the Royal Sovereign lightvessel the wind moderated and with full sail set *Cohoe* headed across Channel towards Le Havre.

We were all feeling happy and Gerald Harding, the mate, got the primus going and hotted up a large saucepan of soup. Then followed a nasty accident, for the swell was still running and the saucepan was flung off the gimbals and Gerald got badly scalded by splashes of soup. I gave first aid, according to the directions for burns in *Reed's Almanac*, but he was in severe pain for two hours. We sailed back to the English coast, and anchored for the night off Eastbourne, where an aunt of his was living. He was out of pain and slept well until the following morning, when a shore boat landed him on the beach and a diligent Customs officer conducted him to his aunt's, but the burns proved worse than I had realized at the time.

I mentioned this race because two things can be learnt from it. The possibility of burns or scalds is a real risk in heavy weather, owing to the violent motion of a yacht. When the gale was over, tension was relaxed and it was then that the accident happened. A considerable swell was still running. A sea cook should never wear shorts, because there is always a risk of things being thrown off a stove. Trousers afford some degree of protection. The second point is more important. An amateur with no medical knowledge may be inclined to judge the seriousness of burns by the condition of the patient and think once the pain is over and he appears cheerful there is nothing to cause anxiety. On the contrary, as I have learned since, only a doctor can judge how serious a burn may be. It is the shock following it that matters, and in the event of an accident of this kind the patient should be got ashore for medical attention as quickly as possible. In the meantime he may be given hot tea and sugar, but no alcohol.

The final race of 1947 was the Plymouth to La Rochelle, and it was in this race we experienced a thunderstorm which at the time we thought was the beginning of a heavy gale. The Rochelle Race was an exciting event for us, as it was *Cohoe's* first venture into the Bay of Biscay. The distance was 355 miles, and at the conclusion of the race the yachts sailed through the battlemented entrance between the towers into the basin of the medieval port.

The thunderstorm occurred in the Bay of Biscay about 20 miles south of Belle Ile and about the same distance off St Nazaire, and I cannot do better than to quote the description of it which appeared in the *Yachtsman*, spring 1948, by Sir Ernest Harston, who then owned *Amokura*, the beautiful Class II yawl later owned by George Millar, and described in his books:

'At six I was awakened by hearing Roger at the chart table and asked him what was happening. He said he was just marking up the DR because there was the most awful sunrise he had ever seen, a dipsomanic's dream of hell, red, green, purple, yellow and black. As Roger is not conspicuously addicted to hyperbole, I knew I had to get up, so I groaned out of bed and went on deck, where covers were being lashed on the hatches and skylight and the genoa changed for the No 1 foresail, a doughty piece of stuff that has stood some hard blows in its time. The reefing handle was brought out, but I was watching the approaching squall and called out, 'To hell with reefing – down with the mainsail!' It was about half-way down when the blast hit us and what a draught it was! It was a tough job getting the rest of the sail down and lashed to the boom, but the crew did splendidly.

How glad I was for our permanent gallows; I kept heading up into the wind as close as I dared to see how she took it. We were heading in the wrong direction, but no one worried about that. The wind freshened and the rain came down in sheets, but the sea was as flat as glass. I spun her round and ran before it for awhile; she handled perfectly under jib and mizzen. Finally, as I got more used to things, I came back on to the course with the wind abeam and watched the speed variation indicator creep steadily round until it registered 9.5 knots. The boat felt like a flying armchair and I wished I had kept to our course from the start. However, we've all got to learn. The thunder and lightning was most impressive. By 0745 hours, however, we had the full mainsail on again – the squall had lasted an hour and a quarter. There have been various estimates of its strength ranging up to force 10 or hurricane – but at any rate force 8 is not far out either way. We had been lucky and nothing carried away, but other yachts reported burst jibs, parted halliards and other troubles of that kind.'

Conclusion

Alan Watts has kindly contributed the following meteorological comments:

Thundery troughs – this was probably a cold front – are usually narrow and stretch across the wind. They may be sometimes referred to as line squalls. One would be most unlucky if a troughful of thunder lasted two hours, providing one did not foolishly run before it, so making the time of its passing slower. Theoretically, the best course would seem to be the one chosen – a broad fast reach with apparent wind on the beam.

It should not be difficult to distinguish between a gale due to a depression and a gale-force wind from a thundery trough, because, although both are preceded by a falling barometer, the former has a long build-up of cloud, while the latter normally has not. The same sort of sky as described preceded the Channel gale referred to in Chapter II, but in this case it was a sunset and not a sunrise which was the portent of hell.

The Ladies' Race

To the end of this chapter I am adding a description of a gale which occurred in the Dinard Race the following year (1948) and it rounds off my account of the different sorts of ordinary heavy weather sailing experienced in our first year of offshore racing, in which *Cohoe* won the class RORC Points Championship.

The Dinard Race is popularly known as a Ladies' Race, as it was then the only RORC event of under 200 miles in length, but it is a misnomer in these days, when women are often as able at sea as their menfolk, and sometimes better. However, there are more wives, sweethearts and daughters taking part in this race than in any other RORC event. The course (See page 28) is a very interesting one, starting from Cowes and taking the fleet past the Casquets to Les Hanois lighthouse on the SW corner of Guernsey, and thence into the Gulf of St Malo, skirting the Minquiers group of rocks, and so to the finish at the fairway buoy off the entrance to the River Rance and Dinard. Although shorter than other RORC races, the course is a tricky one, for in bad weather there are heavy overfalls off the Casquets and along the coast of Guernsey. In fog, the combination of strong tides and rocks provide a navigator's nightmare. Once across the English Channel one is sailing in narrow waters the whole time. As always in this race, there was strong international competition and in our class alone there were eighteen competitors from England, the Channel Isles, France, Belgium and Canada.

For this race I had as crew my son Ross and Gerald Harding, who, after he had recovered from his accident in the Channel Race, had become a regular crew member of *Cohoe* and a much-valued one, owing to his determination and cheerfulness under all conditions of weather.

The race started during the morning of Friday, 16 July, in light airs and easy conditions which held while the fleet was crossing the Channel close-hauled during the afternoon and night. Landfalls were made at various points on the Cherbourg peninsula, depending on the windward ability of the individual boats. Here the wind freshened and there was a forecast of strong winds backing to SW, accompanied by heavy rain and moderate to poor visibility. Bad visibility is the one thing above all others that is not wanted in the rock-strewn Channel Island waters.

At the Casquets the following morning, Saturday, we had to reef the mainsail and set a staysail in place of the genoa. The sea was beginning to build up, as it always does very quickly in those waters. Conditions were soon gloomy in the extreme. There was dense rain which cut down visibility, and when we were beating off the Guernsey coast it was barely possible to sight land on the inshore tack before entering the danger area of the outlying submerged rocks.

It was not until late in the evening that *Cohoe* fetched Les Hanois lighthouse SW of Guernsey; it took a long time to round the lighthouse against a foul tide.

At about 2200 the wind strengthened rapidly. The official forecast was of strong winds (indicating force 6 and 7 on the Beaufort scale), and I logged the wind in the lower region of force 7. Three reefs were taken in the mainsail and the storm jib was set. The rain was fierce, the night was black and the visibility was so poor that no lights could be seen. There was then a weather-going tide and a wild sea, so I decided to heave-to for three hours, between Les Hanois and the dangerous group of rocks known as the Roches Douvres, before entering narrow waters in the vicinity of the Minquiers plateau of rocks.

With the storm jib backed, *Cohoe* lay quietly rising and falling on the seas. All the tumult was over, the slamming, the noise, the spray and the motion. I took the quarter berth, as I always do on these occasions, for a skipper has to be handy to go on deck, and in the quarter berth he can hear what is going on, put his head on deck from time to time and be ready for instant action if occasion demands it. But in rough conditions it is an uncomfortable position, as so much water descends upon the occupant; its sole merit is that there is little risk of falling asleep. It was ironical to be wasting a fair tide and to be waiting a foul one, but in Channel Island waters there is an exceptionally steep breaking sea when wind is over the tide. Then it is like sailing in overfalls.

About three hours later (between 0100 and 0200 on Sunday) the tide turned and became lee-going, so the apparent wind and seas moderated as a result. The seas were higher, but they broke less and the viciousness was taken out of them. The visibility must have still been poor, as no lights were in sight. We let draw and sailed on, still seeing nothing, though we must have passed very close to the Roches Douvres without sighting the light. We gave a wide berth to the Minquiers, setting a course between this plateau of rocks and the French mainland on the west. Here the deep water is 17 miles wide, which should provide easy navigation, but for the strong tidal sets. When dawn turned into day, the weather gradually improved and the sun came out, but we sighted no land until we came off Cap Frehel and altered course for the remaining few miles to Dinard. We had sailed 50 miles and rounded the Minquiers blindfold.

Oddly enough, the end of the race followed the precedent of the Brixham Race. After crossing the finish line we sailed up the river, feeling tired and depressed. Then as we approached Dinard Roads we saw yachts at anchor. As we passed the nearest we hailed her, inquiring what other Class III boats were in. The answer came clearly: 'None'. *Cohoe* had won boat for boat without calling on the handicap allowance due to her small size and despite her loss of time when hove-to. The only other boat in her class to complete the course was *Alethea III*, a 9-ton cruising yacht owned by Vernon Sainsbury, who later made a great reputation as Commodore of the RORC. The winners in Class I and Class II were *Latifa* and *Golden Dragon*.

Conclusions

Only thirteen out of the forty-two starters in the three classes completed the course. *Seafalke* had a man swept out of the cockpit and *Seahorse* was reported to have had two

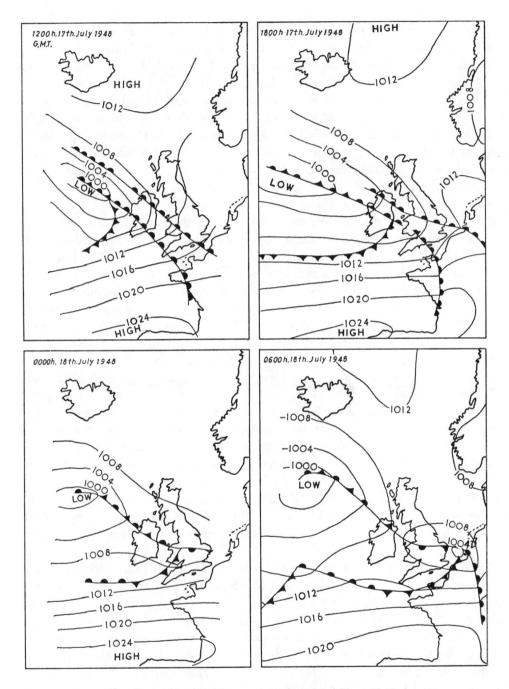

Fig 4.3. Synoptic charts covering the Dinard Race, 1948.

overboard. Happily, all were recovered. Many of the twenty-nine yachts which retired suffered damage to sails (which in 1948 were often of prewar vintage), parting of stays and other breakages. Other retirements were due to the dirty weather, involving a plug to windward in the very rough tidal seas peculiar to the Channel Islands and Gulf of St Malo, and having to round the Minquiers rocks in bad visibility, which was recorded at Guernsey meteorological station at 0600 on Sunday morning (18 July) as 'thick fog'. Nevertheless, the weather conditions were spoken of at the time as a gale or near gale, with general agreement that the wind attained force 7 (28-33 knots).

I was therefore astonished when, nearly twenty years later, I received reports from the Meteorological Office, and found that only the top force 5 (21 knots had been recorded at the shore station Guernsey (only about 10 miles to leeward of *Cohoe*) and from the met. station at the Lizard. From this, cruising men may infer that ocean racing is very sloppy, but I would point out that cruising men, including some distinguished ocean voyagers were among skippers of the twenty-nine yachts which retired. There was something more to it than this.

Referring to the synoptic charts, it will be seen that there was a slow-moving shallow depression to the north-west of Ireland, and that the Channel Isles were under the stable winds of a warm sector. This is amply confirmed by the poor visibility, which is a trademark of maritime air originally from the tropics. The isobar spacing over the area at 1800 on Sunday suggests winds of 20 to 30 knots (force 5 to 7) at the surface under these stable conditions. The tightness of the isobars over the Channel Islands as the cold front moved over Guernsey around 0600 on Sunday, 18 July, is consistent with perhaps force 7 (28-33 knots) sustained temporarily. Taking the recorded wind velocity at Guernsey and adjusting for height of anemometer and multiplying by the factor for increase at sea compared with a shore station (as explained in Chapter 21), we again arrived at force 7. This in turn agrees with my log and I think is a true assessment and has since been confirmed by other evidence.

1 *Windforces.* That wind force tends to be overestimated in rough going if accompanied by dismal weather, driving rain and bad visibility. It is under such conditions (though most had retired to St Peter Port before the fog) that retirements attain their maximum. Given the same strength of wind after a cold front, with sun and good visibility, retirements would be fewer.

2 *Tidal waters.* That a small yacht will have a rougher time in strong tidal waters at force 7 with a weather-going tide or current (such as the Gulf Stream) than in a genuine gale.

3 *Light displacement.* These early ocean races proved that a small light-displacement yacht is safe at sea in the sort of heavy weather conditions she may occasionally have to face when cruising or ocean racing. It showed that a small modern Bermuda-rigged yacht can stand up to and beat the traditional powerful yachts such as Bristol pilot cutters and Brixham trawlers of many times her size, boat for boat without handicap, provided the wind is strong enough and right on the nose, where the gaff rig is at a disadvantage when reefed. *Cohoe* was not the first light-displacement yacht to prove her sea-going ability, as H G Hasler had won the class championship the previous year

in *Tre-sang*, a 30 sq metre even lighter than *Cohoe*. Neither *Tre-Sang* nor *Cohoe* ever won an ocean race except in heavy going.

4 *Heaving-to*. The experience also proved that short-keel yachts could heave-to. It had been thought at the time that this was a tactic available only to long straight-keeled yachts, but the fact is that the ability to heave-to depends on hull and sail balance, though a yacht with a long straight keel still retains an advantage in this respect.

5 *Despondency*. Beginners in ocean racing may feel despondent towards the end of a race. It is a condition of mind to which even experienced crews are prone, particularly if no other yachts have been sighted. It is induced by tiredness, but as the results of the Brixham and Dinard Races showed, a race is never lost before it is over. Time after time we have done well when we thought we had fared badly. One sees one's own mistakes, but forgets that competitors make mistakes too.

5 SANTANDER RACE STORM

Adlard Coles

Our first experience of being 'caught out' in *Cohoe* in open water in a severe gale occurred in the famous Santander Race gale of 1948. The last time I had been properly caught out was as far back as 1925, which goes to show that a real gale is a rare bird indeed.

When I first entered *Cohoe* in this long race, I must admit that I did not know the precise position of Santander. I knew, of course, that it was on the coast of Spain, but only when I turned out the charts did I learn that it lies about midway on the south coast of the Bay of Biscay, approximately 440 miles on a direct course from the starting-point at Brixham. Of British ocean races, the Santander is one of the most genuine. Only once does one approach land, and that is when rounding Ushant, the most westerly point of France and an island of ill repute because of its strong tidal streams and off-lying reefs and rocks and the frequency of fog. For the rest, some 300 miles across the Bay, if the wind is westerly, one is for all intents and purposes racing in the Atlantic, in soundings of well over 36,500 m.

As usual, *Cohoe* had a crew of three in total. Geoff Budden, a schoolmaster, came as astro-navigator, having learnt his craft as an instructor during the war. He was then new to offshore racing yachts, but settled down quickly, as he was an experienced dinghy sailor. As mate and cook I had my son, Ross, also a dinghy sailor, who had sailed with me in several previous ocean races. Watches were arranged in the usual manner, but in practice, with so small a crew, the duties and responsibilities overlap and we raced as a team, each ready to take over anything needed. There was precious little sleep for any of us.

In the three days before the race we were kept pretty busy. *Cohoe* was not in good racing trim, for in the Dinard Race she had bent the fitting at the foot of the forestay, and we had also found that the bolt through the mast at the head of the forestay had cut its way like a blunt knife downwards through the wood, so that the forestays were slack. The rigging screws of the forestays had been taken up as far as they would go, which was not enough to compensate for the distance which the bolt had been drawn down the mast. In the intervals between the races there had been no time to have the mast out for complete repairs, but thanks to Moody's yard a temporary fitting for a new forestay had been provided, together with a new mast-band, since most of the cleats had been broken or torn off. I also had a temporary preventer stay set to the mast above the lower spreader, which, when required, could be set up by a tackle to any convenient cleat.

There were many other things to see to, including provisioning. A small yacht may not make good more than an average of 100 miles a day, and less if she meets prolonged calms or gales. I reckoned we needed rations which could be stretched out to cover ten days in the event of the yacht being disabled or dismasted. Allowing half a gallon of water a day per man, we required about 15 gallons. The main tank held 11

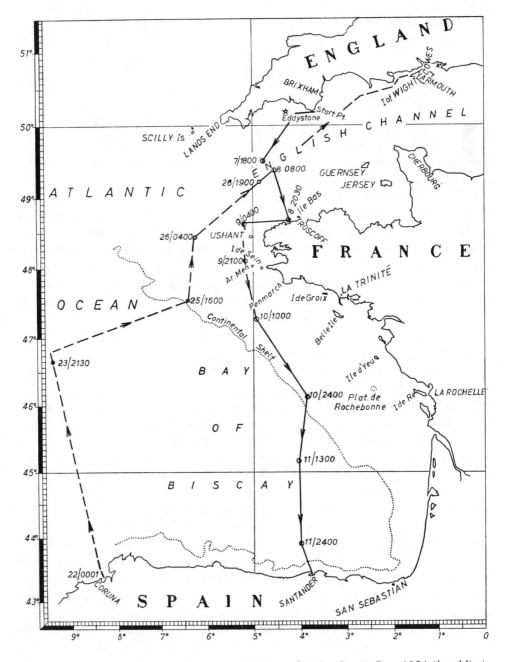

Fig 5.1. The Santander course (solid line) and the return from La Coruña Race 1954 (dotted line)

gallons and we had a 3.5 gallon portable tank made to fit under the cabin sole.

The race was to start from the line of the Brixham Yacht Club at 1530 on Friday, 6 August. Thirty-one yachts were entered, including many of the well-known ocean racers. The morning's weather report was not good. A depression was moving in from the west and there was a southerly gale in the Bay of Biscay. Strong winds were probable, so nobody was under any illusions at to what might be expected.

The wind was light at the start, but soon freshened. For the first twenty-four hours the racing fleet beat westwards off the English coast. The night was thundery and squally. Several yachts retired, reporting 'Friday's gale', but if it was a gale it must have been very local, as we experienced nothing over force 6, but at 0300 on Saturday morning 7 August it was blowing hard. *Cohoe* was going like a train under genoa only, surprisingly carrying weather helm under this single sail. However, by 0730 the wind had moderated temporarily and we were once more under full sail. Later in the morning the wind backed and I noted the barometer had fallen two-tenths of an inch (7 millibars) since the start of the race. We stood close-hauled on the port tack and the best course we could lay was about SW, which would take us some 30 miles west of Ushant.

During the Saturday afternoon the wind freshened again. In view of the forecast and the long swell which was building up, I set about reefing again. At 1500 I double-reefed the main and at 1600 set the brown staysail in place of the genoa. A little later the small white jib was hoisted in place of the staysail and the storm jib hanked on the spare forestay and halyard. At 1730 we lowered the mainsail, ran it off its groove and lashed it down, and set the storm trysail and storm jib. At this time the wind was about force 6 and a fairly rough sea was running. The reefing and shifting of sails had been precautionary and in anticipation of events to come.

It is of interest here to look at the barograph readings of *Golden Dragon* (see page 40), which are reproduced by courtesy of her owner, the late H J Rouse. *Cohoe* was far astern of the Class II *Golden Dragon*, so the depression would have reached *Cohoe* an hour or two later. It will be seen that the slope steepens rapidly, falling three-tenths (10 millibars) in three hours, followed by an almost vertical half an inch (17 millibars) within 2 hours. A secondary depression had formed NE of Spain in the morning and had deepened and moved rapidly across the Bay of Biscay, though we were not to know it at the time.

It was shortly before 1800 that the wind increased to well above force 6, and this, coupled with the falling glass, the driving rain and the menacing conditions, decided me on reefing the trysail and bringing *Cohoe* down to storm canvas. At this time Geoff was at the helm, while Ross was below off watch. I was stand-by, and I pulled down the forward earring of the trysail and made fast. Then in order to take the strain off the clew pendant I checked off the main sheet. Then I returned to the cabin top to swig in the after reefing tackle. The next moment I found myself overboard. Fortunately as I fell I held on to the reefing tackle. I suddenly realized that Geoff was overboard, too, for I saw him floating aft, but hanging on to something. Meanwhile the yacht was well balanced between her storm jib and the trysail, which was checked right off. She was sailing herself on a 5-knot reach to the west.

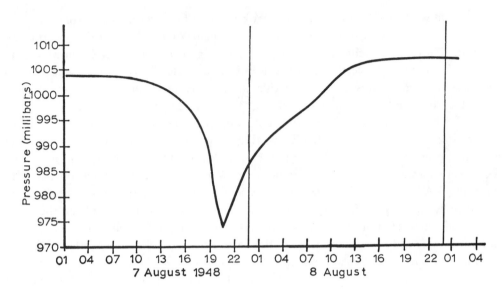

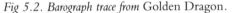

Fig 5.2. Barograph trace from Golden Dragon.

At that instant, it being the 1800 change of watch, Ross had put his head on deck. His appearance was providential, for he was in time to haul aboard Geoff—whose position was precarious, as he had fallen in head first, and his head was towed under water by the speed of the yacht. In some miraculous way he had managed to retain his grip on a lifeline which had gone over the side with the stanchions and was trailing in the water.

My circumstances were less pressing. I had managed to get a foot aboard and wedged it into some wire rope, presumably the runner. But, suspended between the reefing tackle at the outer end of the boom, which was checked off, and my foot at the other, my body was in the water and I was unable to extricate myself. Instead of putting my foot aboard I ought to have pulled myself forward along the tackle to the inboard end of the boom, but one does not think of these niceties at the time—so I was glad when I was pulled on deck again.

It had been a narrow shave, for both Geoff and I were wearing heavy oilskins. If we had lost our hold, there would have been but a slim chance of survival, as with the trysail half-reefed it is doubtful whether Ross could have manoeuvred single-handed in such a sea to find and pick up survivors (not wearing lifejackets). It is easy to make light of such incidents after all is over, but it was something little short of a miracle that there was no casualty. The incident resulted from an extra large sea catapulting both Geoff and me against the lifelines simultaneously. These collapsed under the impact of our combined weight as the after securing eye was torn from the deck, and one stanchion rooted out, taking with it part of the toerail.

All in all the damage was light. Geoff had lost his glasses and for some while he feared he had broken a bone in his hand. I had some nasty-looking scratches on my shin and ankle, which bled a bit. Ross probably came off worst, as it must have been a shock to come on deck and find he was alone with not a single man left on board, and a second shock when he saw us both in the water. Once aboard I finished reefing the trysail (apart from the reef points which were not urgent) and went below with Geoff to change into dry things, and to bandage my foot, leaving Ross at the helm, for it was now his watch.

It is easy enough to talk of dry clothes, but after racing for twenty-four hours to windward there is often precious little left that is dry. Geoff had run out of clothes, but fortunately I had a reserve kitbag forward and Ross had a few spare dry garments. I donned winter pyjamas (what warm, comfortable things these are at sea), sailing clothes on top and oilskins over that. I felt like the well-known Mr Michelin of the inflated tyres.

Meanwhile the barometer was starting on its spectacular plunge and the wind increasing all the time. I decided to prepare for the worst and went back to join Ross on deck. We lowered the reefed trysail and storm jib and secured them.

For this race I had brought a sea anchor. To be of real value at sea one wants a sea anchor of very large dimensions, too cumbersome for stowage, but *Cohoe*'s was a small one. It was sufficient only to satisfy the qualms of those who thought that *Cohoe* was too small to look after herself. However, I decided to experiment with it. To the sea anchor we bent 55m of nylon, and at the inboard end we parcelled the rope with cloth where it would lead through the fairlead before belaying it to a cleat and also to the mast for extra security.

First we experimented with the anchor in the conventional position forward. Here it did no good. The yacht continued to lie broadside to the seas and the anchor lay away on the windward quarter. So we passed the nylon warp aft and made it fast. At the stern it had much the same result as is achieved by towing warps. It seemed to have a steadying effect, and the yacht lay with the seas broad on her quarter. She was perfectly happy, just like a cork, though, of course, the motion without steadying sails was wild in the extreme. At this time there was violent rain and hail. Then we went below, for there was nothing more we could do but reserve our energy for the race, when the gale had blown itself out.

Not long afterwards a Blue Funnel liner approached us closely, evidently desiring to offer assistance. Boats such as *Cohoe* are small from the point of view of a liner, and when lying under bare poles are often the object of investigation. During *Cohoe*'s first two seasons ships stood by on four different occasions, and, although their assistance was not required, the courtesy of the sea was appreciated.

The liner steamed off, and soon *Cohoe* was alone again on the wind-swept seas. I looked at the barometer. It was falling faster than ever.

At nightfall I was confronted by the problem of lights. My big fisherman's oil light had been smashed in the Dinard Race, and it had not been returned from repair by the makers in time for the Santander Race. In any case I doubt whether it would have kept alight under such conditions. The same reasoning applied to the small anchor light I had brought as a reserve. It might not blow out, but, as I have pointed out before, oil lights rarely stand up to the shaking of a gale; it is the vibration rather than the wind which extinguishes them. Beside this I had a small electric all-round light. This I put up, but in the dense rain and spray I felt it was not enough, so I switched on the navigation lights as well, and exhibited a Powerlite electric lamp over the stern. The chances of being run down were small, for the English Channel approaches comprise a large area and ships tend to keep to the north or south.

I have often been asked what it is like in the cabin of a small yacht in a gale. I say unhesitatingly that it is beastly. It is an experience nobody desires, but from time to time is unavoidable. To some extent it is frightening, for a severe gale is such a very violent thing that there is an instinctive apprehension. On the other hand, a yacht's cabin is an extraordinary haven of peace in hard weather compared with what it is like in the cockpit. The cabin lights swing in their gimbals, flooding the little compartment with light and warmth. One has companions to talk to.

When Geoff came off watch he expected to be on stand-by for two hours and had taken the quarter berth so as to be within call from the cockpit. Quarter berths are the wettest in the ship, so Ross and I were lucky to have the cabin berths, one on each side of the saloon. The canvas leeboards were lashed up, so we could not be thrown out. Our principal discomfort was damp. Geoff had got wet again in the quarter berth when the companion door was opened and I went on deck to deal with the lamps. Ross and I got soaked when lowering the sails, and no more dry clothes were available. The cabin was wet, too, not from leaks, but from the water carried below by our clothes when Geoff and I changed after being thrown overboard, and from general condensation. Forward there was an occasional deluge of spray through the forehatch, despite its canvas storm cover. Our cumbersome clothes added to the discomfort, especially the oilskins which we had to wear in case we should suddenly be needed on deck. Ventilation was not good either, as both ventilator cowls had been caught, lifted off the Dorade boxes, and flung overboard by the flogging jib sheets when tacking. We were, however, able to keep a lee porthole open, while the companion hatch was also kept ajar until the head of another sea descended upon the protesting Geoff.

The glass was still falling. It had fallen an inch (34 millibars) since we left Brixham, and the barograph record, which I was shown after the race was over, indicated a fall of over $\frac{5}{8}$ inch (22 millibars) in 3 hours. At midnight the greatest fury of the gale fell on us. The wind and rain flattened the sea. The yacht heeled as though under full sail. On the deck above our heads, blasts of rain hammered fiercely. I think then the wind shifted. How heavy the sea was I do not know, for the sea cannot be seen at night except for the flashing white phosphorescent crests. It was certainly violent and large. One exceptional wave woke us up later as it threw the yacht almost on her beam ends. Some said afterwards that the seas ran 'mountains high'. But as for the wind; it was the highest I had known up to then. If the reader examines the barograph readings he can judge for himself. With a gradient so steep as this, winds far exceeding gale force 8 are inevitable and the gusts may be anything.

In all this *Cohoe* behaved very well. The yacht was lifted bodily great heights in the air and then down she would go. Down, down and one tensed oneself for the crash, but at the bottom she never slammed, but gently came to rest before rising once more to the next. No seas broke over the coachroof or came aboard solid, but Ross said the self-emptying cockpit was often a foot or two under water from heads of seas and driving spray.

By dawn on Sunday (8 August) the wind had moderated, but in *Cohoe*'s cabin, conditions did not seem much quieter. For a while after a gale, the sea is often more truculent than during its height. The yacht is no longer steadied by the wind and the motion is worse in consequence. At 0700 I got up. I felt amazingly well and refreshed, for I had had a good deal of sleep – more than we got at any other time in the race. I went forward through the cabin and tried the radio, as we wanted a weather report. It would not work. I tried various ways, but the only thing to do was to take it into the cabin for drying out and testing. I unscrewed it and carried it aft. At that moment I suddenly began to feel seasick. I handed the radio to Ross and lay down, feeling fit again at once. Ross played with its innards and inserted a new valve, but after sitting up he, too, suddenly felt ill. All of us were fit when lying down, but each felt seedy the moment he sat up and tried to do anything. Ross, however, dealt with the radio, alternately working on it and lying down when he felt sick. I decided not to make sail again until we got the 0800 weather report. The glass was rising, with a jump which might presage more bad weather. By 0800 Ross got the report, though feebly, having to press his ears to the instrument. The report was fair, but the past twenty-four hours was described as a 'vigorous depression with severe local gales in Biscay and the West Channel'. A severe gale is force 9, with a mean wind velocity from 41 to 44 knots.

Once the weather report had been received we immediately made sail. Geoff and Ross got *Cohoe* under way. The two of them heaved on the long warp leading to the sea anchor. Aware of the great strain imposed by a sea anchor without a tripping line, they heaved hard, and with remarkable success, for they got the nylon rope in hand-over-hand as if nothing were at its end. With the final heave they discovered the truth: there *was* nothing at its end. The sea anchor had gone in the night, and none noticed any difference.

Before we had been long under way we sighted a sail. At one moment it would appear high in the sky and at the next it would be lost in a trough. The sea had gone down, but a lumpy swell was still running. As *Cohoe* drew near we fetched cameras and as we passed close under the yacht's stern we exchanged photographs; she was *Mehalah*, a Class III rival. Unfortunately, our arrival on the scene was the sign for feverish sail-making in *Mehalah* and she was quickly after us. We responded by setting more sail. As the morning drew on the wind softened and eventually *Cohoe* was under full sail again.

During the afternoon I took the opportunity afforded by the light weather to check the water position. I was shocked to find only about 3 gallons left in the tank, and that water had been leaking from the cap when the yacht was severely heeled. At first I took a poor view of this, but investigation proved it was not the fault of the crew, for although the cap appeared screwed down tight, the screw thread was broken and when the yacht heeled the water leaked out in the bilges. It must have leaked a lot during the gale. So here we were with 3 gallons left of the main supply and a reserve of only $3\frac{1}{2}$ gallons in the portable tank. Over 300 miles of the course still lay ahead, with possibility of calms or gales. The crew took the news more happily than I did. Geoff calculated we could survive for some days if need be on the juice out of tins of soup and other provisions. In the end we compromised by rationing water to $1\frac{1}{2}$ pints a day per man, which compares with the normal usage of 4 pints per man per day. If this could be maintained, we had enough for ten days. To encourage thrift, tea and all hot drinks and orangeade were cut out. Ross drew a little diagram which he pinned to the bulkhead showing three half-pints per man per day. When one of us had a drink out of a half-pint mug he put a stroke through one of his rations on the communal card. For the benefit of others who may run short of water at sea it can be recorded that we not merely kept to our ration but we saved on it. When we arrived at Santander we had the reserve $3\frac{1}{2}$ gallons untouched.

We had been about mid-Channel when we hove-to and as the wind did not veer as forecast we made a landfall on the French coast some 50 miles east of Ushant, which we did not round until Monday afternoon (9 August) as the wind was light. It was not until nightfall that we entered the Bay of Biscay, leaving to port the Ar Men buoy which marks the end of the Saints, the finger of reefs and ledges which extend from France nearly 10 miles into the Atlantic.

Tuesday (10 August) was a day of moderate winds in which the ship's company settled down to the routine of deep-water sailing as the yacht drew out far away from land into the Bay of Biscay. The only incident occurred when I was below off watch and heard a bang. *Cohoe* had struck a glancing blow on a floating log. There were numbers of these floating about and for a while the helmsman was kept busy avoiding them. We heard afterwards that a timber ship had lost her deck cargo of timber during the gale; it brought a rich harvest to French fishing vessels.

Early in the afternoon there was a sharp fall of rain, and we were able to collect fresh water for washing from the pools in the corner of the lee cockpit seats.

In the early hours of Wednesday morning (11 August) *Cohoe* had left the Continental shelf with its soundings of 180 m or so, and passed into the 1000 m depths of the Atlantic. By 0200 it was blowing force 4 and there was quite a vicious sea and by 0300 we took a double reef in the mainsail, and the wind settled down to a good hard steady blow.

Shortly before 0500 we heard a sharp 'ping'. Geoff and I rushed on deck, but could find no breakage. As soon as it became light we discovered that the bolts of the track for the weather runner had sheared. I made temporary repairs by putting a long wire seizing on the runner and set it up again farther aft on an undamaged part of the track.

At 0700 the wind moderated so we shook out the reefs, but an hour later we had to pull them down again. At 1100 we lowered the genoa and set the intermediate jib. The runner threatened to lift the remaining track off the deck. This was a serious worry, and an unanticipated one, as I had had the track through-bolted during the previous winter, after the same trouble in the La Rochelle Race.

At 1400 we lowered the working staysail and set the small white 6-metre jib, but two hours later this had to be lowered and the tiny storm jib was set in its place. With the runner in such doubtful condition we could not afford to risk the loss of the mast, which was subject to great strains as the ship plunged. *Cohoe*'s speed fell to $6\frac{1}{2}$ knots, a very serious loss of time in a race such as this. Meanwhile the sea was steadily increasing.

Geoff had managed to get a noon latitude by lashing himself to the mast while he took the sight, but in the afternoon the sky was overcast and we sailed through a series of squalls, so no further sights were possible. At 1800 we passed a floating gin bottle, which gave us a clue to our longitude, as we guessed some other yacht ahead of us had passed that way.

During the evening the sea was very rough indeed and the weather stormy. We heard later that even the Class I *Latifa* was at this time down to storm canvas. The wind had veered and freed; under double-reefed mainsail and storm jib we were soon logging 7 knots again, despite the weak runner. When Ross was on watch he made frequent remarks about planing, which as a dinghy sailor he should know something about. Besides being large, the seas had very steep breaking tops. It seemed amazing that none broke aboard. *Cohoe* was lifted up and surfed forwards as each breaking, foaming pyramid came up to her stern. Then she would subside down its back before lifting again to the next sea. It became necessary to lash the helmsman in the cockpit, and at nightfall I shortened the watches to hourly spells. One hour was quite enough, for the boat needed a great deal of concentration to steer, and the helmsman was constantly blinded by spray.

At midnight we knew our distance off the Spanish coast, under 30 miles, by dead reckoning and from a noon sight, but apart from the gin-bottle clue we did not know our exact longitude.

Early on Thursday morning (12 August), at 0130, Ross identified the loom of the double flash of Cabo Major light at the entrance to Santander, less than 20 miles distant. Course was altered and with a quartering wind *Cohoe* sailed very fast, so before long the blaze of lights of the town of Santander could be seen. At 0355 we let off two white flares as we crossed the finishing line. The eventful race was over and, with the dawn, the sun rose above the mountains of Spain, 440 miles and five and a half days from our departure in England.

Eilun proved to be the all-over and Class I winner. Hers was a well-merited victory, as she was an old Fife boat adapted for ocean racing and the smallest in her class. Pat Hall, her experienced owner, told me before the race started that his tactics would be to make as much westing as possible when the wind backed early in the gale and southing when the wind veered, and he proved right. *Erivale* won Class II and *Mindy* Class III. *Cohoe* was second in her class with *Mehalah* third. The ocean racing fleet survived the gale well.

Out of the thirty-two starters only eleven yachts retired owing to damage of one kind or another, and the only serious accident occurred in *Benbow*, one of whose crew broke his arm, but the same night that Geoff and I went over the side from *Cohoe*, a man went overboard from *Erivale*. This happened when it began to blow. It was very dark and raining hard. The headsail was being changed down and Peter Padwick, then a medical student and one of her crew, was winching in the sheet whilst her owner, Dr. Greville, was 'tailing on'. A sea seemed to catch the ship on her wrong foot, so that Padwick was catapulted over the lifelines, but fortunately, lying on his back, he managed to get his arm round a stanchion. The owner grabbed his ankle and, coinciding with the reverse lurch of the ship, got him back on board, when he continued to winch as though nothing had happened. The stanchion, of stainless steel, was bent over at an angle 45 degrees from the vertical.

Conclusions

Let us now assess the gale. A depression was moving west towards Ireland and a vigorous secondary established itself in the early hours of Saturday morning NW of Spain and crossed the Bay of Biscay, arriving off Brest at 1800, and continued across the English Channel approaches towards Plymouth. The centre of the low (which had deepened to 976 millibars) thus passed a little to the northward of the leaders of the ocean-racing fleet and directly across those in mid Channel, which included *Cohoe*, *Mehalah* and others hove-to near by.

In the Bay of Biscay a number of fishing vessels were lost when running for shelter and there were casualties along the coast of Brittany. The gale was reported as one of the worst known in the Channel Islands. The *Isle of Sark*, with 750 passengers, took shelter at St Peter Port, where even in the harbour six pleasure vessels broke from their moorings and were sunk. On the English coast, many yachts and other vessels were in distress. Cross-Channel ferries reported their worst crossing of the year. As I have remarked, a severe gale is always recorded in the newspapers, but this particular gale featured as the leader on the front page of the *Daily Telegraph*, which devoted three columns to it, so great was the damage and loss.

The gale was reported in the newspapers as a 70 mph (61 knot) gale. Mr J S A Rendell (a former officer in the Clan Line), in a letter to *Yachting Monthly* in September 1966, stated that he had passed the yachts during the gale in MV *Stirlingshire* when homeward bound from Australia. If he remembered correctly, it was logged at force 11 (56-63 knots) at one stage.

I thought perhaps there might be an error in the recollection of the *Stirlingshire*'s log or that the entry related to gusts, but Alan Watts, as a meteorologist, commented that 'it seems inevitable that the report of force 11 from the Clan liner was correct, even though . . . such wind strength could not have been sustained for long', and he gives barometric graphs in support of his opinion. The only evidence in the ocean-racing fleet came from *Theodora*, who recorded gusts of 60 knots on her anemometer, which were not necessarily the highest and were probably taken at deck level, which gives a lower reading than 33 ft aloft, which is the correct position for judging the wind on the Beaufort scale.

The only records that I have received from the Meteorological Office show hourly values of force 8 at Portland Bill and force 9 at Guernsey. These are coastal stations where winds are usually lower than at sea, and what is more, they were not in the direct track of the centre of the low.

Looking at the tight isobars on the synoptic charts, I think a fair assessment of the gale when it passed over the yachts was possibly force 10 (48-55 knots) or even force 11 during squalls. If this is true, the gusts might have been about hurricane force. I write with caution, because I frankly admit that I cannot judge wind strengths over force 8 or 9, as they are so few and far between, and in the general hubbub it is difficult to guess the mean speed on the Beaufort scale or to distinguish between gusts of 65, 60 or 55 knots. The latter is quite enough for me.

1 *Severe gales.* A severe gale or storm is a very rare bird in summer months. Eight years elapsed before ocean racers were caught out again in home waters in a storm to match

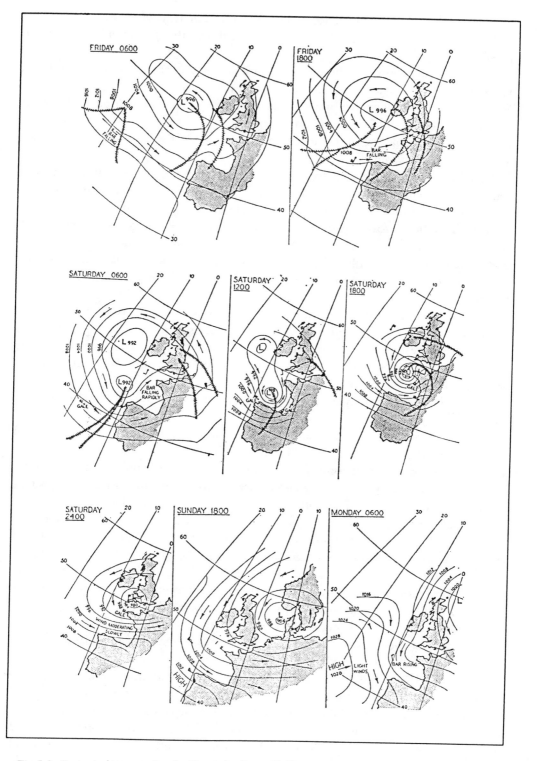

Fig 5.3. Synoptic charts covering the Santander Race, 1948.

it. This was the Channel gale of 1956. Force 10 is so rare (except on weather forecasts) that it need hardly be reckoned with when cruising, though it remains a remote possibility.

Loss of life occurred in fishing vessels running for shelter, and coastal vessels and yachts in narrow waters were in distress, but all the ocean racers in deep water in the West Channel, including the small *Cohoe* and *Persephone*, came through the storm safely, apart from loss of sails and minor damage.

This rubs in the old lesson that in a severe gale or storm the safest place is out at sea, as far from land as possible.

2 *Secondaries*. A secondary depression is sometimes more intense than its parent. It is something a yachtsman should look out for, especially when a major depression is passing to the northward. The development and course of secondaries are not so easily predictable as those of their parents, and hence there may be shorter warning of them on the forecasts. They are of comparatively short duration (the gale referred to lasted under twelve hours) and although the wind forces are high the seas may not attain their maximum height in relation to the wind strength, as the gale may not last long enough for them to develop fully.

3 *Sea anchors*. The experiment with the sea anchor was inconclusive, as it was too small. Nevertheless, it confirmed experiences of other sailing men that a short-keeled yacht will not readily lie head to wind to a sea anchor without riding sail aft. The important fact emerges that *Cohoe*, a very small yacht, was reasonably safe lying a-hull and left to herself under bare pole.

4 *Drift downwind*. When lying a-hull under bare pole the yacht's drift to leeward appeared to have been not less than $1\frac{1}{2}$ knots

5 *Height of waves*. The owner of an 18-ton cruising yacht whose crew were rescued reported 'seas that must have been 40 ft high'. Yachtsmen are inclined to measure a wave by its apparent height compared with the mast, but in my opinion it is an optical illusion and the true height is only about three-fifths or perhaps half of the apparent height. After all, a 24 ft wave ($\frac{3}{5}$ x 40 ft) is immense for the West Channel.

6 *Stanchions*. Stanchions and lifelines must be very strong. In *Cohoe* the stanchions were screwed to the deck and bolted to the toerail, but the combined weight of Geoff and myself not only uprooted the stanchions but tore the toerail out. The RORC has brought in strict regulations about securing stanchions, but I think it unwise to rely entirely upon these safety measures unless the fittings are immensely strong, which is not the case in many yachts. Stanchions seem liable to bend or break, as is confirmed by other incidents recorded later in this book.

7 *Lifejackets and safety harness*. Because of their bulk we never wore lifejackets before 1950. If this is criticised, I may add that the Brixham trawler crews with whom I sailed after the 1914-18 war did not either, even in gales. Nor could they swim.

The thing to avoid is going overboard at all, and it is here that personal lifelines and harness are useful, but they did not come into general use before the 1950 Transatlantic Race to which I refer later. Personal safety harness (if strong enough) would have provided a link with the ship when Geoff and I went overboard. Nevertheless, I still think safety at sea depends primarily upon self-reliance, and that a safety harness is only a secondary aid which should not be overdone.

8 *Water.* Where there is only one water tank, a reserve of water should always be carried in separate containers during a long-distance race or cruise.

9 *Seasickness.* Seasickness is the greatest handicap in a gale. Many sailing men will experience nausea if they do work below which involves bending down, such as looking for the right tin lost among a mass of provisions in a locker, or, as occurred in this race, bending down trying to repair a wireless. Seasickness can sometimes be avoided by keeping on deck or lying down when below. Nowadays, seasick pills may give immunity against nausea, but are not always successful against extreme seasickness.

10 *Anxiety ashore.* Anxiety to relatives was caused by the late arrival of several yachts, which were reported as unaccounted for, including *Cohoe.* A small yacht averages only about 100 miles a day. In a gale she may be long delayed when hove-to or sheltering, possibly where there is no means of communication with the shore. A gale is often followed by a calm which slows her passage. Minor damage such as loss of sails may cause further delay, and serious damage such as a rudder breakage or even dismasting may make arrival several days late. This applies whether racing or cruising, so anxiety ashore is fortunately often premature.

11 *Confidence.* The most valuable part of our experience in the Santander Race gale was that, following successes in lesser gales, it gave us complete confidence in the hitherto much-disputed sea-keeping qualities of a small light-displacement yacht such as *Cohoe.*

6 THREE MORE GALES

Adlard Coles

From the last chapters it may be thought that *Cohoe* was used entirely for ocean racing. This was far from the case, as apart from racing, we cruised 1,000 to 1,500 miles or more each season, and in her first three seasons she made the crossing of the English Channel nearly forty times. Many offshore races finished in a foreign port, which affords a good start for cruising, and we did a great deal of sailing besides this. Sometimes *Cohoe* was manned by crews who had holiday time remaining after the races, but mostly I cruised with my wife or family.

Constantly on the move as we were, we had plenty of heavy weather sailing, but we never really got caught out in the open water in the same way as we did when ocean racing. Thus, I hope the reader will forgive me if I continue what seems to be a catalogue of ocean racing, for it was from these experiences that I acquired most of the knowledge I possess of gales.

Hove-to off Belle Ile

It was a strange coincidence that *Cohoe* should be involved in another depression in the very next race after the Santander, and have the dubious honour of having it all to herself.

The race from Santander to Belle Ile was started on the afternoon of 15 August. The course was a little east of north, 235 miles direct across the south-east of the Bay of Biscay (see page 38, Fig 5.1). In the meantime Geoff's leave had run out and he had returned to England. His place in *Cohoe* was taken by Dick Trafford, a Cambridge friend of Ross.

We had been royally entertained in Spain during our three days' visit and few of us had had much sleep. We were sorry when the time arrived to exchange the sun and the warmth of the hospitality for the cruel sea and tinned provisions.

However, we started on a brilliant Sunday afternoon with a fine fresh breeze to shake us down. The wind headed in the early hours of Monday morning and by 0830 it had freshened so much that we had to take in a reef. We kept the genoa standing, but the bolt holding the runner plate sheared under the strain and repairs had to be effected. At 1000 the mainsail tore right across under the headboard. The sail came down with a run and the headboard and halyard went aloft to the masthead.

Cohoe's mast was solid, high and thin, and above the jumpers it was little more than the thickness of a big walking stick. A considerable swell was running and a fairly rough sea, so there was no prospect of swarming up the mast to retrieve the halyard. The burgee sheave was a strong one, so we tried to lead a wire rope through it by means of the burgee halyard, but the attempts were unsuccessful and finally the burgee halyard broke.

We then shackled the bosun's chair to the fore halyard and hoisted Dick Trafford, the lightest of the three of us, up to the forestay block. But the motion was so wild aloft and he was thrown about so much that he could not reach the peak and retrieve the main halyard. When he was lowered to the deck he was violently sick.

Deprived of her mainsail, the yacht was rolling tremendously in the seas. It was difficult to retain one's foothold even on deck, and aloft, even when secured to the mast, it was like being at the end of a pendulum, as it swung first to one side and then the other over the sea. It was enough to make anybody seasick.

We abandoned the attempts to retrieve the main halyard and lashed the mainsail to the boom. Next we tried setting the trysail by means of the spinnaker halyard, but this failed because we could not get the lead right.

Only one thing was left that could be done and that was to reeve a new halyard through a block and lash the block to the mast above the upper spreader. When the block and halyard were ready Ross volunteered to go aloft with them, but Dick and I did not relish the idea of pulling over 13 stone in weight up the mast by means of the fore halyard; nor would it have been fair on the small mast winch. Accordingly, I, as the next lightest, went up in the bosun's chair. This was quickly done with the aid of Ross's and Dick's beef at the winch. Like Dick, I found the motion aloft made things difficult. I really needed both hands to hold on by, but I got a temporary grip with my knees round the mast when lashing the block in position. The repair was a strong one and I was soon on deck again, when I, too, was promptly sick. All three of us had now been sick, which after all was a compliment to the tremendous parties we had enjoyed in Spain.

As the burgee halyard had broken and hence our racing flag was down, I got a sail needle and cotton and sewed the racing flag to the peak of the trysail before setting it. It was an act of bravado, but it made us feel better.

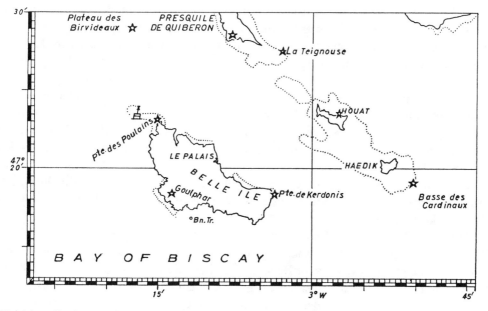

Fig 6.1. Belle Ile to Quiberon.

51

This attempt to retrieve the halyard and the other activities which take only a few lines to describe wasted no less than four hours. Happily, the wind was about force 6, so, despite the loss of the use of the mainsail, reasonable progress was made under genoa and trysail.

However, the wind moderated early on Tuesday (17 August) and remained light. Under reduced sail *Cohoe* was very slow. The loss of the mainsail had already put us out of the race so far as winning a place was concerned.

It was not until 0400 in the morning of the Wednesday, 18 August, that we sighted the loom in the sky of the powerful Goulphar light. on Belle Ile bearing N 15°E, distant 22 miles by dead reckoning. The barometer was falling and the wind had backed and freshened, so we were making over 5 knots.

At 0700 there was dense rain and the visibility closed down to about three-quarters of a mile and often less.

Cohoe sailed on in a grey world of her own. Belle Ile is not difficult to approach when one knows it, as most of the outlying dangers are off the Goulphar lighthouse and at the northern end. The tidal streams, however, are fairly strong and rather unpredictable close in to the island, and less than 5 miles to the eastward lies a particularly dangerous area of submerged rocks between the Cardinals and Quiberon.

At 0900, having sighted nothing and being then totally unfamiliar with Belle Ile, I decided it was best to alter course and stand out into deep water to the north-west of the island. Visibility was very bad owing to the heavy rain, and it was blowing so hard that we had to set the storm jib, which meant that, with the trysail already set, we were down to storm canvas. We found ourselves in company with a big tunnyman. For a short time we followed. We could hold but not overhaul her and she kept eluding us in the rain squalls, so we came back to our course to keep clear of the land.

At 1030 we hove-to. We had been doing 6$\frac{1}{2}$ knots under storm canvas, but now the wind was near to gale force and a big sea was making. The yacht lay heading WSW on the port tack and we lay-to under trysail only, with helm up instead of down, as the storm jib was rather much for her. It was an experiment, but it worked perfectly and left to herself, *Cohoe* lay steady, riding the waves and forereaching slightly.

Our position was to the NW of Belle Ile, but as I had seen no land since leaving Spain, except for the loom of the Goulphar lighthouse, a deck watch had to be kept all the time just in case there was any error in navigation. A motor trawler appeared out of the dense rain, and ranged up close alongside to inquire whether assistance was needed and, when satisfied that it was not, disappeared in the murk. At about 1400 we refreshed ourselves with a bottle of champagne we had bought in Spain. A little later, in better spirits, we tacked under trysail, set the storm jib, and reached back towards Belle Ile, keeping a sharp look-out.

Between 1700 and 1800 the visibility was worse than ever and there was no sight of Belle Ile. Although casts of the lead showed we were in deep water, the bottom shelves rapidly off the rocky NW coast of Belle Ile and there are off lying rocks, so we did not like to stand on any longer. It was blowing harder than ever, so we tacked and hove-to again on the offshore tack. I decided not to attempt to get nearer the land until after dark, when we ought to be able to see Goulphar lighthouse, which has a range of 23

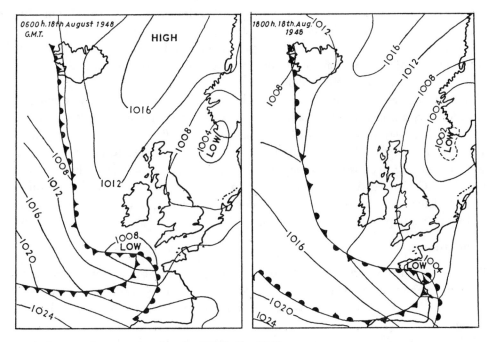

Fig 6.2. Synoptic charts covering the gale off Belle Ile, 1948.

miles in clear weather. We decided to get some rest, as we should have to sail through the night, but at 1930 we were roused by shouting. It was another French tunnyman that was offering assistance.

None of her crew spoke English, but I remembered the words which a governess or schoolmaster used to apply: 'Allez vous en' – and we were able to wave them away with the empty champagne bottle. The tunnyman bore away quickly and disappeared in the rain. We remained hove-to, had dinner and washed up.

By 2100 it was dark and we sighted Pointe des Poulains light, which is at the extreme north of Belle Ile, so we were not far out in estimated position. The light was to the south of us and the harbour of Le Palais was only 5 miles SE of the lighthouse and the wind was free as it had veered. It would have taken little over an hour to gain shelter and get a good night's sleep, but the rules of the race provided that the Pointe de Kerdonis on the SE of Belle Ile had to be left to port. This meant we had almost to circle Belle Ile, which involved a long beat to windward and a distance of 25 miles to cross the finishing line off Le Palais from the correct direction. It says a lot for Ross and Dick that giving up never crossed their minds. Perhaps it was because the racing flag was sewn to the sail, but this attitude was common to all my crews in *Cohoe*.

It was a slow passage. First we had to beat round the north end of Belle Ile. The wind had moderated and we set the genoa. No sooner was it set than the wind freshened and it had to be replaced by the staysail. Then it came on hard again and we had to set the storm jib.

Meanwhile, the sky had cleared and the moon shone. Under the lee of the land the big swell had subsided. At midnight there was a violent squall and it blew so fiercely that the boat was almost knocked down by it. We lowered the storm jib quickly and carried on under trysail only. It was very slow work, but fortunately the tide was fair. The barometer was rising sharply and presently the wind moderated, enabling first the storm jib and later the staysail to be set. At dawn next morning the Goulphar lighthouse and Belle Ile were in sight.

At 0900 we set the genoa. It was a lovely sunny day, but the wind was dying away and under trysail, instead of a mainsail, progress was desperately slow. It was not until 1320 on Thursday that we crossed the finishing line off Le Palais under trysail on a windless summer day. We were the last in the race, but *Cohoe* received a cheer from the ocean-racing fleet as she limped into the harbour, some thirty-six hours overdue. The RORC is not given to sentiment, and in twenty years of racing this was the only occasion I have known such a warm welcome accorded to any yacht. We appreciated it more than any prize.

Conclusions

There is nothing to corroborate this gale, as we received no weather reports and saw no French newspapers. The British meteorological stations were too far away to be much help, but the synoptic charts suggest about force 6. Yet the wind must have been stronger.

I have chanced to come across a cutting from *Yachting World* reporting the race. Referring to Monday night (long before the gale) it reads: 'freshened to about force 6 with occasional heavy squalls and a very awkward sea. ..' 'Nearly everybody agreed that it was one of the most unpleasant nights they had ever experienced at sea.' '*Myth* . . . hard on the wind at an average speed of over 6 knots, some of it in the vilest conditions.'

What, then, was the force of the wind in the gale on Wednesday when the storm jib was set at 0900 and in the squall at night when even the storm jib had to be lowered ? Link this with the fact that a number of large ships sought shelter in Belle Ile roads, and the yachts in harbour were so uncomfortable that many of them went through the locks to find security in the inner basin.

This evidence suggests force 8, but remember that *Cohoe* was an exceptionally tender boat. With the hindsight conferred by long experience I should say the wind was probably force 7, and the squall on Wednesday night probably 50 knots or so, as this is not uncommon when a cold front is going through.

The experience added to my apprenticeship in heavy weather sailing and may be of interest to owners of small yachts of 3 to 4 tons displacement, but the performance would now be outclassed owing to the improvements in the design of small ocean racers. My present boat, *Cohoe III,* for example, has about three times the stability of the *Cohoe*, and with the aid of electronic instruments, the handling is more scientific. At force 7 by anemometer she would carry a small genoa and a mainsail two rolls reefed. A minimum speed of 6 knots to windward, or over 7 knots downwind, would be maintained as measured by Harrier speedometer. DF bearings would render approach to Belle Ile easy. Likewise with a crew of five instead of three, much of the wear and tear of ocean racing would be eliminated.

1 *Damage aloft*. The loss of the main halyard was the primary cause of the trouble, as otherwise *Cohoe* would have been in harbour long before the gale, in company with her competitors. Going aloft in a rough sea proved a very different task from doing it in harbour. Maybe we made too much of it, as wonderful repair jobs have been done at sea in other yachts.

2 *Errors*. When, after the race was over, I checked the course on the chart I found I had made an error of 10 miles in pricking off the distance run. The loom of Goulphar light had been sighted at a distance of 32 miles and not 22 miles. Hence it was not surprising that we failed to sight Belle Ile five hours later, for we still were some miles north of the island when we altered course away from the land.

This stupid mistake was no doubt due to being rather tired. A racing skipper in a small yacht crewed by a total of three has to do navigation as well as his full share of watch-keeping and seamanship, and I had been short of sleep for the whole of the twelve days (allowing for festivities in Spain) between Brixham and Belle Ile.

The moral is to be particularly careful when tired and to double check everything, for it is then all too easy to make a silly slip such as I did. Better still is the independent check by a member of the crew.

3 *Heaving-to*. The method of heaving-to under trysail only was original and worked well. The helm was lashed up instead of down, just sufficiently to prevent the yacht from luffing head to wind and getting in irons. The best position was found by trial and error.

4 *Halyards*. A spare halyard is desirable. A masthead topping lift can be used if there is one, or the burgee sheave can be made large enough and strong enough to accept a halyard.

5 *Visibility*. When cruising it is good seamanship to stand out to sea rather than approach an unfamiliar coast in bad visibility, unless one is certain of position. This is a principle I still maintain, where the coast is rock-strewn, as in North Brittany, and one can get into trouble before land is even sighted. Belle Ile, however, is not a bad landfall and maybe I was rather overcautious by racing standards.

Gusting over 60 Knots

The first occasion on which I have ever been able to get precise confirmation of wind strength in the exact position of the yacht was in the Solent in 1947 or 1948. *Cohoe* was lying at Cowes and there was one of those exceptional gales which occur only once every few years. We made the trip back to Bursledon under storm jib only and after our return I telephoned to Calshot, who told me that it was gusting 70 mph (61 knots, equivalent to the top of force 11) at the precise time *Cohoe* was off Calshot. This was useful, for although *Cohoe* had been at sea under such conditions she had been lying a-hull in open water, whereas on this occasion she was under sail, though in the sheltered water of the Solent.

The feature of this wind force when under sail is the vibration in the rigging if the stays and shrouds are not set up hard. We had to gybe off the West Bramble buoy. The jib was tiny, but to avoid damage we had to sheet it across using both winches, even so when it gybed the force shook the mast from truck to deck.

Although there is only a relatively short fetch in the Solent, there were real seas in the vicinity of Calshot lightship, and I closed the cabin doors in case one should board us. Even off Hamble Spit buoy it was rough, but the seas were more regular, wind and tide being in the same direction. We ran up the river at tremendous speed, passing dinghies sunk at their moorings and yachts which had broken adrift and gone ashore. As we approached Moody's yard I got the engine going, because we had to alter course under the lee of Land's End marshes. Here the wind is always fluky. I did not want to risk getting out control with only the storm jib set, so the sail was lowered and we proceeded up the reach under power. Under the lee of the high land by the Jolly Sailor I brought *Cohoe* round, motored back down river and, with two men forward with the boat hook ready, I luffed for the mooring.

Did she luff, though? With no protection except the low marshes, the gale was too strong for her. She lost way, her rudder became useless, and a gust took charge. With the bodies of the crew acting as a sail the bows were blown off by the wind. In a matter of seconds down she went to leeward straight on to the putty to join the other yachts which had broken away from their moorings.

So, when I read advice to beginners, recommending that if caught out in a gale they should use the auxiliary engine, I wonder to myself what sort of gale and what sort of engine? I cannot imagine any 6 hp or even 10 hp engine in a small auxiliary yacht would have been the least use getting to windward in *open water* in any of the major gales I have described. Windage and seas would be too much, but in Chapter 20 I give an example of where an engine was used by a yacht caught out in a hurricane.

The Wolf Rock Race
(See page 28, Fig 4.2)

Few new yachts were designed and constructed expressly for ocean racing immediately after the war. Of the few, Captain John Illingworth's *Myth of Malham*, built 1947, was the most outstanding and remained one of the best ocean racers for a record number of years. In 1949 the RNSA (Royal Naval Sailing Association) 24 ft waterline class was introduced. These boats were designed by Laurent Giles and Partners in collaboration with John Illingworth, and were commonly known as the RNSA 24's. These were short-ended boats with high freeboard and transom sterns. Compared with *Cohoe*, the RNSA 24 was $1\frac{1}{4}$ ft shorter in overall length, but not having such fine sections as *Cohoe* was of rather higher displacement. In addition, the RNSA 24 set almost 100 sq ft more canvas and therefore carried a higher rating.

The RNSA 24s proved the most successful class designed up to then and swept the board during 1949 and remained near the top of Class III ocean racing for several years. The principal boats in the class were *Minx of Malham* owned by John Illingworth, *Blue Disa* owned by Colonel Dick Schofield, and *Samuel Pepys*, which was the RNSA club boat, later to become one of the most famous small ocean racers.

At that time, Class III yachts were not allowed to enter in the Fastnet Race, as they were considered too small. The Wolf Rock event, which was run at the same time as the Fastnet, was a kind of consolation race for Class III. The course was a good one, starting from Cowes to CH 1 buoy off Cherbourg, thence round the Wolf Rock off

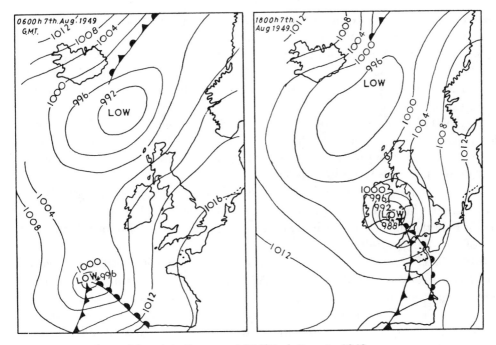

Fig 6.3. Synoptic charts of the gale in Fastnet and Wolf Rock Races in 1949.

Land's End and back to finish at Plymouth, a distance of 305 miles with plenty of windward work.

I do not propose to describe this race, as it was much like many others and I will limit myself to the gale which occurred.

The race started on a Saturday (6 August) at 1000 in summery conditions and it was not until 0800 on Sunday morning, when we were off the Casquets, that we received a gale warning. At noon this was repeated. A deep depression was moving in a NE direction which would give rise to 'severe' gales in the West Channel and Irish Sea. The wind backed to SE and the barometer fell slowly. It was not until the Sunday afternoon (7 August) that the sky darkened and the wind began to rise. The yachts raced on over a dismal sea through a belt of heavy rain, and sail was reduced by stages.

The wind reached gale force SSW in the early evening, when we were south of Plymouth. We adopted our usual tactics of lowering all sail and lying a-hull. The seas seemed to be running as high as they did in the Santander Race gale, but perhaps this was because it was daylight and we could see them, whereas the worst of the gale in the former race occurred during the night, when we were below either sleeping or trying to do so. On the other hand, the wind was not so strong, at least where we were lying. *Cohoe* lay comfortably, but she heeled at a greater angle than she had the previous year. She had been fitted with a stronger mast, which was similar to those in the RNSA class. It was of considerably greater section and both mast and rigging were designed to be strong enough to withstand a gust that would lay the yacht flat on her beam ends. The spar was hollow and supposed to be lighter than the old one, but when it was built I think it must have been made with thicker walls than specified. As a result, the extra windage and weight (remember that the yacht's ballast keel was only $1\frac{3}{4}$ tons) made the yacht more tender than ever.

Gale force conditions lasted little more than six hours. Among the competitors in the Wolf Rock Race and the Fastnet Race there were altogether twenty-five retirements. Two yachts were dismasted, one broke her rudder and many suffered minor damage.

The wind force recorded at the Lizard on Sunday, 7 August, was SW by S force 8 in the morning, rising to force 9 SW at 1330 and 1430. The highest gust (58 knots) occurred at 1340. The gale then moderated to force 8 WSW at 1530, force 7 W by N at 1730, and force 6 NW by W at 1930.

The competitors in the Fastnet Race bore the full brunt of this gale, as they encountered it between the Lizard and Land's End. *Myth of Malham* was the only yacht to beat right through it without heaving-to and was the winner of what I think was the best of John Illingworth's many ocean racing victories.

In the Wolf Rock Race the RNSA 24's put up an equally notable performance. *Blue Disa* was driven right through the gale, under close-reefed mainsail and storm jib, never logging less than 3 knots, and deservedly won the Wolf Rock Bowl. *Samuel Pepys* and *Minx of Malham* (skippered by Commander Erroll Bruce) in second and third place did almost equally well, and if they hove-to in the gale at all it must have been for only a short while.

Mindy, Cohoe and *Mehalah*, the Class III leaders in the Santander Race, all either hove-to or lay a-hull and none of them was even placed.

Conclusions

1 *Plugging through gales.* The Wolf Rock Race of 1949 was one of the most important in the annals of ocean racing. Hitherto, small ocean racers hove-to in gales and victory went to the ones which were quickest to make sail again, after the worst was over. In this race three of the smallest boats (under 5 tons displacement) carried on through all or most of the gale. Their tactics introduced a new technique as far as ocean racing in Class III was concerned, and the days of heaving to, lying a-hull or to a sea anchor were over, at any rate in the ordinary gale, unless special circumstances warranted it.

2 *Wind at shore stations.* The strange thing was that the highest hourly value of the wind recorded at Plymouth was only force 5 as compared with force 9 at the Lizard, less than 50 miles to the SW. From the synoptic charts it is clear that the gale must have attained at least force 8 at sea south of Plymouth. Indeed, it is said that a gust of force 11 (56-63 knots) was recorded at the Eddystone lighthouse only about 10 miles to the south.

This confirms that the records of some shore stations cannot always be relied upon to afford an indication of the wind force at sea, as they are subject to local influences and turbulence depending upon the direction of the wind in relation to the position of the anemometer.

7 GULF STREAM STORM

Adlard Coles

In 1950 the Royal Ocean Racing Club planned a Transatlantic Race from Bermuda to Plymouth to follow after the Bermuda Race organized by the Cruising Club of America. The Class III ocean racers, hitherto excluded as being too small for the Fastnet Race, were to be admitted to the Transatlantic event, as the sea-going ability of very small yachts had been proved over the previous three seasons. The Transatlantic Race was largely inspired by John Illingworth, then Commodore of both the RORC and the Royal Naval Sailing Association. The challenge appealed to me enormously, so I entered *Cohoe* immediately both for the Bermuda and Transatlantic Races. The yachts were to be shipped to Bermuda and would then sail to Newport, Rhode Island, from where they would race back to Bermuda and then cover the 3,000 miles to Plymouth.

I will not go into details of the long preparations, or of the Bermuda Race itself, beyond mentioning two alterations which were made to the yacht. In order to improve stability that had been so poor the previous year (owing to the new mast) I arranged during the winter for a 6-in false keel to be inserted between the existing wood keel and the lead-ballast keel. This would have little effect on initial stability, but once the yacht began to heel, the extra leverage would take effect. The other alteration was the now famous false nose which was added to the bow to lengthen the yacht up to the minimum 35 ft overall length required for eligibility under the rules of the Cruising Club of America. This was constructed of aluminium sheeting over a honeycomb of timber laths and it was a miraculous operation performed by the builders, A H Moody & Sons Ltd, in twenty-four hours.

The false bow proved a success. It had a distinctly steadying effect on the helm and gave *Cohoe* the feel of a longer, faster and bigger boat, more like an 8-metre to steer. I was not entirely easy in my mind about it, because an extension added to the bow is not the same thing as a long forward overhang structurally designed for the job in the first place, with suitable stem and frames. However, it was to be tested sooner than I had anticipated.

After arrival at Bermuda as deck cargo on the 5,000-ton freighter *Araby*, the three small English yachts, *Samuel Pepys, Cohoe* and the rather larger *Mokoia* were unloaded. After a refit and provisioning they set out under sail on their 630-mile passage to Newport, RI, from where the Bermuda Race was to start.

On this passage I had with me Lieutenant-Commander Basil Smith as navigator and Group Captain Jack Keary, who was to be mate in the Bermuda and Transatlantic Races. Basil Smith was on short leave from HMS *Glasgow,* the mother ship for the three small yachts at Bermuda and, before he joined the Royal Navy, had served in the Merchant Service. *Mokoia* was skippered by her owner, the late Major James Murray, with Wing Commander Marwood Elton as navigator and Major Murray's daughter Jean as crew. *Samuel Pepys* had her full racing crew and was skippered by Commander (then Lieutenant-Commander) Erroll Bruce, RN, who was Captain of the RNSA team in the Bermuda Race.

The three yachts set sail from St George's, Bermuda, on Wednesday, 24 May, in fine settled weather and with a good weather report. We had the prospect of a pleasant, easy passage of 630 miles, with perhaps a little rough going when passing through the Gulf Stream, as the strong current often creates a steep, tumultuous sea.

All went according to plan at first, and we hove-to the first night while Jack cooked dinner, and *Mokoia* likewise took it easily. *Samuel Pepys* was ahead and none of the three yachts regained contact until arrival at Newport, RI.

At noon on Thursday, the following day, the wind backed to the ESE and freshened to force 4. By 1630 the wind had backed farther to due east and freshened to force 5. The sea had got up and half an hour later we close-reefed the mainsail and set the small jib, as the wind was hardening. Conditions steadily worsened. The barometer was falling and there was heavy rain. By 2000 the wind had reached gale force and the sea was really rough. We lowered all sail and lay a-hull under bare pole. I remember that when I went forward to lower the storm jib, I crawled on hands and knees, so the wind must have been very strong, as I had never had to do this before.

Before nightfall I went on deck to measure the force of wind by the cup-anemometer. It was extremely difficult to get a proper reading, because the waves were already so high that the yacht was under their lee in the troughs most of the time. It was only on the crests that the full force of the wind could be felt. The anemometer readings varied between 33 and 38 knots and accordingly Basil Smith logged it as force 7 to force 8 easterly.

Wind speeds of 33 to 38 knots give an average of 35 knots. I did not know then that a third had to be added to arrive at the *gradient* wind, 33 ft above surface-level, which is correct for the Beaufort scale. The height of the seas accounted for a good part of the 33 ft, but if you add only a fifth instead of a third to the readings, that makes a mean of 42 knots, which is force 9. Gusts of 50 to 60 knots might occur at this wind force, so the reports of all the yachts tallied approximately for the first night of the gale.

The tactics adopted by *Cohoe* were by now almost routine. With helm down she lay a-hull almost broadside to the sea. The wind was so strong that she heeled over as if under sail, which steadied her and also increased her freeboard on the windward side, so that she presented her greatest buoyancy to the seas. Her keel prevented excessive leeway, but being tender she gave way and heeled over enough to avoid opposition to the seas. Her moderate drift to leeward may have created an eddy to windward to take some of the vice out of the breaking seas. Hers was the strength of the sapling bowing to the storm.

On deck it was difficult to distinguish between air and water, on account of the spindrift and torrential rain. I have little recollection of the sea except that it was steep and white, for, I must confess, it was very wet on deck and I quickly got below .

Later it was a black night outside and an almost tropical downpour rattled on deck like hail. Nevertheless, down below things were comparatively peaceful. Jack and Basil were tucked in their berths, held in by canvas lee boards. I slept in the quarter berth in oilskins, as I had to be ready for any emergency and occasionally put my head on deck to ensure that all was well. Beyond that there was nothing to do. I had warps ready if needed and proper oil bags and two tins of heavy oil in the cockpit locker.

I say that the cabin was *comparatively* peaceful, and so it was in contrast with conditions on deck. But there was tremendous noise all the time: the violent rain beating down on deck, the howl of the wind in the rigging and the seas breaking on the windward side, rushing over the cabin top and running away to leeward. Occasionally a sea would deal the yacht a pretty hard smack, and one sometimes wondered whether the hull and cabin top would stand up to a harder one, but apart from mild apprehension the night passed tolerably. The worst feature was the wetness below. The cockpit was constantly being half filled by breaking seas and the whole boat was virtually covered by flying spray. Under such conditions, a yacht is pretty well under solid water, which penetrates any weaknesses of deck or joinery and seems to come from nowhere. All my boats, even when new, have required regular pumping in gales. A lot used to get below through the cockpit lockers, though this defect was partially cured in later years.

Cohoe lay under bare pole all night. I think the wind strengthened in the night, for I recorded it in my book *North Atlantic* (describing the expedition and the races) as a fresh gale, by which I meant 34-40 knots mean, a fair force 8. Nevertheless, we managed to get a good deal of rest.

Friday, 26 May, dawned as a beastly day, with a dark stormy sky heavy with driving rain. By 0730 the wind had veered to SE and moderated sufficiently to allow the storm jib to be set. The barometer was 993 millibars and still recording its slow fall. The weather was so ominous that Basil as navigator decided to treat it as a tropical storm and we ran off with the wind on the starboard quarter.

During the morning the wind continued to moderate and by the afternoon (1545) we were able to set the full mainsail to a southerly wind, temporarily only force 3. However, the barometer had dropped another 3 millibars.

That evening Jack made one of his splendid stews (he cooked in any weather) and that put new life into us. However, the lull in the gale was short-lived. By 2030 the wind was increasing again and under reefed mainsail and small jib I logged a speed of 8 knots, almost certainly an exaggeration, as it was above *Cohoe*'s maximum theoretical speed. Anyway it was faster than she had ever run before.

Unknown to us, Humphrey Barton's *Vertue XXXV* was away to the NE of *Cohoe* when at 1930 she was struck by an immense sea and nearly foundered. We noticed a deterioration of the weather and at 2140 *Cohoe* was once more down to storm jib. It was about midnight that she entered a very confused sea which I described in *North Atlantic* as 'a huge swell from two directions, and the seas heaped in complete confusion'.

On Saturday (27 May) the gale continued at varying force. In the morning at 0600 the double-reefed mainsail was set. By 0930 the barometer was rising very rapidly, but the wind had backed to ENE, force 6 to 7. At 1100 the barometer had risen to 998 millibars and the wind had increased so much that the mainsail had to be lowered. Half an hour later the jib had to be lowered and replaced by the storm jib. Basil logged a speed of 6 knots under 30 sq ft of canvas, which was as fast as she ran in the gale gusting over 60 knots off Calshot. At 1230 she was under bare pole doing 3 to 4 knots and the entry in the log reads 'gale'. Five minutes later she was hove-to once more under bare pole.

It was not until 1955 on Saturday that the gale moderated sufficiently to allow the small jib to be set and the voyage continued in gradually improving conditions, finally ending in dense fog off the American coast. On arrival at Newport it was found that *Cohoe*'s false nose was still intact, but much of the paint had been washed off it by the seas during the storm.

So far I have written of the gales as we saw them from *Cohoe*, but other yachts were involved and their experiences contribute to form the picture of the gale as a whole.

Mokoia appears to have been the nearest to *Cohoe* and she hove-to in the gale at 1630 on Thursday evening, 25 May. Her experiences were much the same as *Cohoe*'s, but, close as she must have been, her barometer did not fall to the same extent and on 26 May she had different winds: 0915 light airs S, 1800 W force 1 to 3, 2000 new winds from S.

Samuel Pepys was about 60 miles to the northward on Thursday, 25 May. She had been reaching under spinnaker. She lowered this at tea-time and continued under genoa only, the wind freshening to force 6. The gale did not reach her until just before midnight (nearly four hours later than *Cohoe*). The genoa was lowered when, to quote from Erroll Bruce's book *Deep Sea Sailing*, 'before any other sail could be set the sky was black all round, rain torrential and wind gusting at whole gale force 10 from the east. Ran under bare pole. In view of the sudden increase of wind and ominous conditions, decided to treat this as a tropical revolving storm . . .'

At 0530 on Friday morning, 26 May, gusts were of 60 mph (about 54 knots) and the seas were breaking down their whole slope. Shortly afterwards the wind suddenly fell dead, the rain stopped and a spot of blue sky appeared. *Samuel Pepys* may then have been in the eye of the storm, but the lull lasted only seven minutes before the wind pounced again at full blast from the east (the same direction as before) and the barometer, instead of rising, continued its slow fall. Except for a brief respite in the afternoon at 1600, when the wind backed NE and moderated to a fresh gale, the storm continued unabated until early the following morning and was logged at force 10. The seas steadily built up with nearly thirty hours of easterly or NE winds. The height of the biggest waves was estimated at 35 ft.

On Saturday *Samuel Pepys* appeared to have met much the same weather as *Cohoe*. She reported force 6 at 0530; force 7 at 0930 and an hour later heavy squalls required the mainsail to be replaced by the trysail and she was once more down to storm canvas. Later she proceeded on passage and reached Newport RI, a day or two ahead of us.

Our first intimation that *Vertue XXXV* was in the vicinity of the three yachts was early on the following Monday morning (28 May), when we were hailed by the US Coastguard cutter *Castlerock*, who was searching for her. We were astonished, for, although we knew that *Vertue XXXV* had started on her famous east to west crossing of the Atlantic on 15 April, it had never occurred to us that she might be near to us. It was a shock to learn that she was storm damaged.

Thursday, 25 May, was the fortieth day out from Falmouth for Humphrey Barton and Kevin O'Riordan. *Vertue XXXV* was running due west and early in the morning the wind backed and freshened. At noon her position was about 180 miles NNE of Bermuda, and about the same distance to the NE of *Cohoe*, on whom she was converging quickly, as both yachts were sailing fast. She reported she entered the Gulf Stream on that day.

The wind was SE and the glass continued to fall slowly. Obeying Buys Ballot's Law, as Humphrey Barton puts it in his book *Vertue XXXV,* 'Face the wind and the centre of the depression is approximately 100° [90° to 135°] on your right—I find that puts the centre at about SW by W.'

This was in the direction of *Cohoe*. At 2200 the wind freshened: 'it is a simply foul night with blinding rain'. 'The yacht was tearing along at a frightening speed'. Humphrey Barton is a hard driver, as I know, for I have sailed with him, and he kept *Vertue* going all the Thursday night, gradually shortening canvas until at midnight the mainsail was down to trysail size. The rain was torrential.

At 0400 on Friday morning, 26 May. Humphrey Barton handed over to Kevin O'Riordan. He reports the wind as 'blowing hard now—force 8 or 9 I would guess. I have no intention of heaving-to.'

At 0605 *Vertue* was reduced to bare pole, and typically of Humphrey, he still kept the yacht running at 3 to 4 knots. 'It is blowing 65 mph (56 knots) now, quite one of the hardest blows I have ever been out in . . . the sea is all white. The crests are fairly being torn off. Barometer dropped nearly one-tenth (3 millibars) in the last hour.' At 1300 conditions were described as absolutely shocking: 'A wind that has reached a state of senseless fury.' 'It became difficult to make out where the surface of the sea began or ended.'

In the afternoon at about 1600 it became no longer possible to steer down-wind. 'For one thing we were 45° off our course and for another the mental and physical strain were pretty severe.' Humphrey Barton is a master of understatement.

A 21 in diameter Admiralty pattern sea anchor was let go over the starboard quarter. It was considered that the yacht lay better thus than wallowing broad side in the trough, but at 1500 a sea fell on *Vertue XXXV* which Humphrey Barton reported as far and away the worst he had seen yet. The barometer was down to 29.26 (994 millibars) and still falling slowly.

It was about 1930 on that grim Friday evening that the accident occurred. I quote Humphrey Barton's words, because they convey in a single paragraph the sudden contrast between security and danger.

'It happened just as we were finishing our supper, about 1930 I suppose, on the 26th. We had had fried sardines and potatoes, tinned peaches and I had just poured the hot water over the Nescafe. The gale was blowing as hard as ever but there we were in our snug, dry little cabin with an oil lamp burning as it was dusk, almost dark, in fact. It came with devastating suddenness; a great fiend of a sea that picked the yacht up, threw her over on her port side and then burst over her. There was an awful splintering of wood, a crash of broken glass and in came a roaring cataract of water.'

Note the contrast. At one moment Humphrey Barton making coffee in the security of the cabin and the next facing disaster and the elements raging on deck during that fearful night. What had happened was that an immense sea had struck *Vertue XXXV*, throwing her on her beam ends so hard that on the lee side the coaming was split at deck-level for nearly the whole length of the cabin, the doghouse window was smashed and the water was cascading into the cabin at every sea.

The yacht was saved by a narrow margin by running her off dead before the seas, and through the fine seamanship and tremendous energy of her crew in manning the pump and effecting repairs in time. Had she broached and been struck by a second sea she would undoubtedly have been lost.

The log is incomplete during the hours when the crew were at their work and when the exhausted men were resting, but it was not until late on Saturday morning the gale had moderated sufficiently to allow the head of the mainsail to be set. The whole dramatic story of the near disaster appears in Humphrey Barton's book *Vertue XXXV*.

Conclusions

This Atlantic storm has always been a puzzle, because of the different experiences of the yachts involved in it. The first theory was that the depression had two centres and the second that it was one intense depression which became almost stationary over the yachts.

It was not until over fifteen years later that I received from the US Weather Bureau a track of the storm which proved to be an extra-tropical cyclone which made an anti-clockwise loop north-west of Bermuda, with the yachts near the centre of its coil.

At my request Captain C Stewart, who is an extra master engaged in hydraulic and wave research, kindly studied the problem, with the aid of all the relative data available from the British Meteorological Office.

He points out that while the broad pattern of any pressure distribution is correct on synoptic charts no more can be done where information is lacking (particularly if it is lacking in the centre of a storm) than to draw in the lowest isobar for which there is definite evidence. The area thus enclosed may be much as 300 miles or more in diameter, in which a deeper centre may exist. It was in such an 'area of uncertainty' that the yachts met their severe storm, but it is impossible to determine its exact track in the absence of regular barometer readings together with wind directions and forces from all the yachts. There were reports from many ships to the westward between Bermuda and Cape Hatteras and elsewhere, but no ships' reports from the area in which the cyclone made its loop.

Captain Stewart arrives at the conclusion that the centre of the low moved rapidly from the north-west from off the Chesapeake towards Bermuda, but the track shown by the American authorities shows it moving in from a south-westerly direction. It is impossible to reconcile the two opinions without the data on which the US track is based, but it is immaterial, as the track from 0200, 26 May, Bermuda time, more or less coincides in the anti-clockwise loop in the area over which the yachts were sailing. This US track is shown on the diagram (page 65), together with the alternative approach from the north-west which is shown in dotted line. Captain Stewart has also

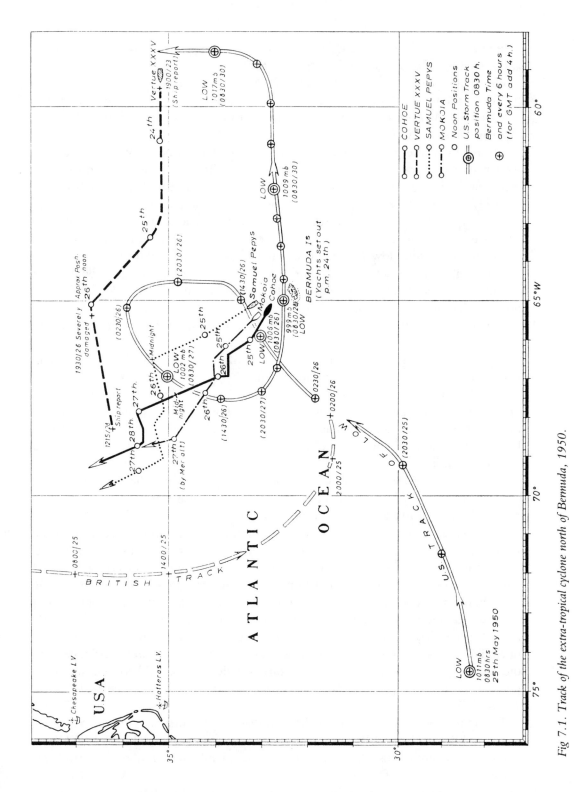

Fig 7.1. Track of the extra-tropical cyclone north of Bermuda, 1950.

plotted the positions of the yachts from such information as was made available to him and altered all times, including the US track, to Bermuda.

On the first day of the gale (Thursday, 25 May) no problem is presented. Whether the low came in from NW or SW, the centre passed first to the south-west and close to *Cohoe* and *Mokoia*, and the following day and night moved round towards *Samuel Pepys* and *Virtue XXXV*.

My report of the gale proved to be rather an understatement, as my opinion was influenced by several factors. I had no idea that a third has to be added to the wind speed recorded by an anemometer used in the cockpit, and hence the gale was probably one grade or more higher on the Beaufort Scale. Secondly there was no spectacular fall in the barometer as occurred in the Santander Race, so I did not anticipate such strong winds. It is only recently that I have learnt that a fall in barometric pressure in lower latitudes has a much greater effect on the strength of wind than in higher. For example, a pressure gradient which at the Scillies, about 50°N, would produce 30-knot winds would produce about 40 knots in latitude 35°N, to the north of Bermuda.

It is on the second day of the gale (26 May) that the experiences of the yachts differ so much.

At 0730 the wind had moderated to force 5, SE, and by 1545 *Cohoe* was running under full sail before a force 3 southerly wind. It was not until 2130 that the wind increased again and it was just after midnight that she entered an area of very confused swell and sea. She had a respite from the gale of about fourteen hours and *Mokoia* to the NW experienced lighter conditions, and had a respite which lasted about thirty-six hours, for she was gaining distance from the centre and did not heave-to again until 1730 on Saturday.

This was the puzzle, because, while *Cohoe* had a respite from the gale until shortly after midnight, *Samuel Pepys* and *Vertue XXXV* to the north were fighting a battle for survival in winds gusting up to hurricane force.

From this I arrived at the conclusion that the first theory was right, and that while *Cohoe* was running before increasing southerly winds a vigorous secondary had developed in which *Samuel Pepys* and *Vertue XXXV* were involved to the northward. However, Captain Stewart does not think that there were two distinct centres of the low, although he does not discount the possibility. He thinks that it followed the loop as shown, although possibly the curve may have been flatter at the top and the loop more lozenge-shaped, with the centre wandering or oscillating and at the same time intensifying, probably due to 'backbending' of the fronts that had already occluded near the centre. It seems fairly clear that the centre rounded and passed between the yachts, somewhere to the north of *Cohoe* on 26 May and to the south of *Samuel Pepys* and *Vertue XXXV*.

From the approximate track it will be seen that after the first night the centre moved round to NE, north and SW in such a way that *Samuel Pepys* and *Vertue XXXV* were involved in it continuously for a much longer time, while the centre of the disturbance circled round south of them, which gave longer for ocean seas to build up. As previously mentioned, Captain Stewart believes the depression intensified as it moved and its isobars were much tighter on the north side of the centre, which would account for storm to hurricane gusts for *Samuel Pepys* and *Vertue XXXV*. In the second place I attach significance to the dramatic change in the character of the seas encountered by *Cohoe* at midnight on 26 May. This could have been due to the seas left in the wake of the centre of the low after it had passed, but it could have been that *Cohoe* had entered the edge of the Gulf Stream and that *Vertue XXXV* and *Samuel Pepys* could have been in a meander of the Gulf Stream during the whole of the storm. This would cause more dangerous seas, and the sharply warmer temperature of the Stream could account for an intensification of the storm and squalls of hurricane force.

It was certainly a survival storm for *Vertue XXXV* and *Samuel Pepys* and a near miss for *Cohoe* and *Mokoia*. The only positive certainty is that the disturbance was a very complex one of near-hurricane violence.

1 *Sea anchors.* The sea anchor used in *Cohoe* during the Santander Race gale and in *Vertue XXXV* in the Atlantic appear to have been of identical type. *Cohoe* lost hers because the ring broke, whereas the failure in *Vertue XXXV* was due to a 2-in manilla warp chafing through at the taffrail. The loss of the sea anchors confirm the experience of other deep-sea sailing men that tremendous strains are imposed on a sea anchor and its gear, which should be very strong.

The question remains whether the sea anchors did any good? I doubt it, and Humphrey Barton goes further in believing that the trouble was due to *Vertue XXXV* being tethered to a sea anchor. But it must be noted that neither yacht set a riding sail.

2 *Running before the gale streaming warps.* This tactic was adopted by *Samuel Pepys* with complete success. In exceptionally heavy weather this throws a strain on a small crew, as steering dead before the wind requires concentration and the cockpit is often filled with water. Nevertheless, the tactic seems to have been the best one.

3 *Lying a-hull.* This method of dealing with gales was almost routine in *Cohoe,* but it must be remembered she had hitherto only encountered force 9 or possibly force 10 (with gusts of over 60 knots) in the Santander Storm. Gales in the English Channel of short duration are different from prolonged gales with the longer fetch of the Atlantic which produce more formidable seas.

My opinion is that *Cohoe*'s tactics of lying a-hull in the Bermuda gale were adequate for a yacht with less buoyancy aft than is afforded by a transom stern. I attributed her immunity partly to her light displacement and the considerable angle of heel at which she lay, giving way rather than resisting the seas. Humphrey Barton, on the other hand, thinks lying a-hull is safe only up to about force 9. Above that he recommends running off, streaming warps as *Samuel Pepys* did, because of the risk when lying a-hull of being knocked down by a sea, or even rolled over, as sometimes occurs with yachts in exceptional storms.

4 *Freak waves.* It appears that *Vertue XXXV* was struck by one of the freak waves which ride high in most gales. I describe a parallel experience, in Chapter 16, where a yacht was nearly sunk, though only in a force 7 to 8 Biscay gale. Turbulence of wind has much to do with sea formation and in both cases the wind must have been gusting far above the mean speed of the Beaufort scale.

5 *Doghouse and superstructure.* Note that *Vertue XXXV* was not damaged by the weight of the invading wave, but was thrown on her beam ends so violently that she was split open on the lee side as though she had fallen on a pavement. As will be confirmed later, when damage is suffered it is usually on the lee side.

6 *Barometer.* During gales there is a tendency to take barometer readings only occasionally, such as first thing in the morning, at noon, and at 1800 in the evening. This makes it impossible to assemble a true picture of a depression, especially if the lowest reading is not recorded. For many years now I have carried a barograph in my boats. The graph produced on this instrument shows the relative steepness of the gradients and provides the only really complete permanent record.

8 POOPED IN THE ATLANTIC

Adlard Coles

For our next gale we jump to the second week in the Transatlantic Race. In this event the three small yachts *Mokoia, Samuel Pepys* and *Cohoe* were entered together with two larger yachts, Jack Rawlings's new Class I *Gulvain* a Lieutenant-Commander G C L Payne's Scandinavian double-ender *Karin III*. Before the start of this Transatlantic Race I had *Cohoe*'s false bow removed. It had given no trouble, but I thought the boat would be better without it. I need not have worried, for it proved so strong that it took us a tremendous time to dismantle it, even with the aid of two shipwrights from the dockyard. There were hundreds of screws to undo and we were thankful when at last the task was done and it lay by the roadside in pieces. It would have survived a hurricane.

With me in *Cohoe* I had Jack Keary as mate, Tom Tothill as navigator and John Halstead, a young American ex-marine who wanted to work his passage to take up a vacation job in France. They were a tough crew and I needed them, for the race was to prove something of an endurance test. With a total of four, watch-keeping arrangements were altered. In day-time, watches of three hours were kept and at night six hours, in order that the watch below could in theory get six hours' continuous sleep, though in practice they rarely did. There were always two on watch at a time, although the one not steering was free to go below if he was not needed for spinnaker handling or other duties.

The race started on Sunday, 2 July, in an almost complete calm, and it was not until the fourth day, with *Cohoe* tailing (as she always did in light airs) over 100 miles behind the fleet, that the wind freshened, bringing with it almost trade wind conditions, with a smart following breeze, a big blue sea under a sunny sky and zest and movement in hull, spars and sails. These conditions gradually gave place to grey skies and rough sailing in the second week as the yachts entered the mid-Atlantic stage of the race, where a series of depressions were moving across.

On Tuesday, 11 July, *Samuel Pepys* and *Cohoe* were very close, had we known it, and *Cohoe* for the first time had temporarily taken the lead. Both yachts reported force 5 to force 6 winds, both crews were beginning to feel fatigue following on ceaseless driving and spinnaker work. On Wednesday, 12 July, conditions were much the same, but the wind backed to the west and the barometer fell 7 millibars to 1009 millibars.

The night had been very tiring for the crews, owing to the size of the seas, and spinnakers had been handed in both yachts.

In the early morning the wind was only force 4, and *Cohoe* was logging 6 knots, but by 0730 it had freshened and following seas seemed so high that we lowered the mainsail and ran under genoa. This brought the speed down to 4 knots, the reduction in sail having been made on account of the seas rather that the weight of the wind.

During the morning the seas continued to build up and the wind freshened, but it was sunny and warm and I took photographs of the steepening seas. By 1500 *Cohoe*'s motion was so uncontrollable that the genoa had to be replaced by the storm jib, the 30 sq ft pocket handkerchief.

The following extracts from my diary carry the story a stage further:

'Half an hour later we lowered even the storm jib and ran under bare pole. The wind is well *under* force 8, but the Atlantic seas are so big that they throw the yacht about and punish her severely, making steering difficult.'

'Under bare pole we only make about 2 knots in the troughs of the waves, but on the crests she runs at 4 knots and seems to skid on the breaking crests.'

'At about 1630 I was called on deck to see the Italian liner *Saturnia* which was tearing close past on a reciprocal course to the westward. She made no sign of seeing us. In fact, in the seas our helmsman only saw the liner when she came close by.'

'I returned below, leaving Jack and Tom on deck. Then all of a sudden there was a roar of breaking water. The yacht lurched violently, then went right over. The cabin went dark, water spurted through the tightly screwed port lights. There was a tremendous noise of rushing water, and a loud report as though the hull was cracked in. *Cohoe* had been pooped by a big sea and broached to.'

'The yacht rose again and came to even keel. John and I tried to open the hatch to see that all on deck were safe, but it took a minute or two (which seemed hours) to get it opened, as somebody was sitting on it.'

'Jack and Tom were safe and told us what had happened. A large sea (but not much larger than the others) with an immense breaking crest had struck the yacht on the quarter. This caused her partly to broach-to, threw her over on her side and half filled the cockpit.'

'While the yacht was staggering under this blow a second sea, larger than the first, came roaring down. It was this second sea which gave her such a tremendous crack. It completely filled the cockpit, ran as high as Tom's arm where he was clinging to the runner, and knocked the yacht on her beam ends right down over the cabin top.'

'The yacht recovered and Jack, at the helm, straightened her out before the seas. No serious damage was done and the pump quickly dealt with the inrush of water.'

'Shortly afterwards it was observed that the *Saturnia* was turning. It was probably her wash that had caused the big wave to break.'

'Round she came in a great circle. Well handled, she passed slowly by. She seemed crowded with passengers. We signalled MIK 'Please report me to Lloyds, London' and exhibited a strip of canvas with the name of the yacht *Cohoe* painted on it.'

'We hoisted our ensign and the liner, acknowledging our signal, made a complete circle around us to satisfy herself that no help was needed.'

The sea puzzled me. The wind was strong, but not force 8, and yet the seas were big enough to make the yacht almost unmanageable. On the tops of the 'big-uns' she would simply be picked up bodily and thrown forward, planing on the crest.

Three times she nearly broached-to and at 2000 at the end of my dog-watch I agreed to the mate's suggestion that we should heave-to.

Waiting for a 'smooth', we put down the helm and she came gently up into the wind and fell off broadside to the seas. As usual, she was happy in this position, riding the seas like a duck.

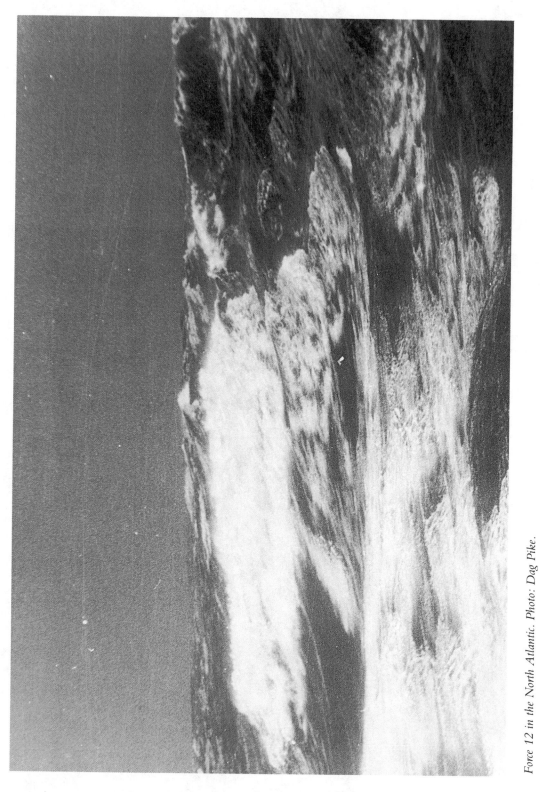

Force 12 in the North Atlantic. Photo: Dag Pike.

Full Gale, North Atlantic. Conditions appear to be much the same as when the British yachts were encircled by the anti-clockwise loop of an extra-tropical cyclone north of Bermuda. Photo: Jan Hahn.

Nevertheless, as a result of this delay *Cohoe*'s noon to noon next day was only 65 miles against *Samuel Pepys*'s run of 158 miles, which was her third best in the whole race.

If I seem unduly severe, I will add that we made good our shortcomings later. We set the spinnaker at force 6 the following evening and carrying it through a ramping tearing night, *Cohoe* achieved a noon to noon run of 177 miles. The yacht surfed on the tops of the big seas, the bow wave each side abreast the mast, a great wedge of foaming water shooting out like a fan 3 ft above the level of the guard rail, and at the stern a high wave, boiling almost to the top of the rudder. Her speed was anybody's guess, possibly 10 to 12 knots for perceptible moments until she slowed as the crests passed. The speed was higher than I have ever experienced before or since, even with a speedometer jammed at the 10 knot maximum. It was followed by another record of 174 miles. Hour by hour, day by day and night by night in ceaseless din, like the roar of a weir or waterfall, *Cohoe* gradually whittled down *Samuel Pepys*'s lead until five days later the pair were almost level. Then *Samuel Pepys* began to draw ahead again and on the final day crossed the finish line at Plymouth in 21 days 9 hours.

Cohoe saved her time by two hours on handicap and won the Transatlantic Race, but *Samuel Pepys* was first home and had the honour of making the fastest elapsed time ever accomplished by so small a vessel.

Conclusions

There is nothing to suggest a gale on the synoptic chart for 12 July at 1200 GMT (about 0900 by ship's time), some six hours before she came down to storm jib. The warm front might give rise to strong winds and squalls, but the isobars are widely spaced and suggest a geostrophic wind force of only 20 knots (force 4 to 5) at the weather ship about 100 miles to the NNE. On another chart, a conspicuous wave low with a sharp kink in the isobars is indicated and the front is shown as occluded, but there is no notable difference in barometric pressures, and according to Captain Stewart's investigations *Cohoe*'s barometer was registering low.

The only suggestion of strong winds come from the synoptic charts covering the whole of the Atlantic Ocean. A deep low giving force 10 winds had crossed the Atlantic and was situated SW of Iceland on 12 July. This would leave moderate to fresh westerly winds for *Cohoe* and her immediate competitors. These winds would be augmented on the north side of the high to the SW and by the low NE of Newfoundland. There was yet another low (996 millibars east of Hudson Bay) with an occlusion between the centres of the two. With a family of depressions, some forming and some filling, the weather situation can rapidly change hour by hour, and can vary locally even in mid-ocean.

Let us look for other evidence. *Mokoia* about 90 miles to the NE of *Cohoe* logged force 8 and hove-to. *Samuel Pepys* about an equal distance to the SW reported force 6 and ran under twin headsails. Both yachts had experienced skippers and crews and, as *Cohoe*'s position was about half-way between her competitors, it seems reasonable to put the wind at force 7, as I logged at the time.

However, the most authoritative evidence comes from the captain of the Italian liner

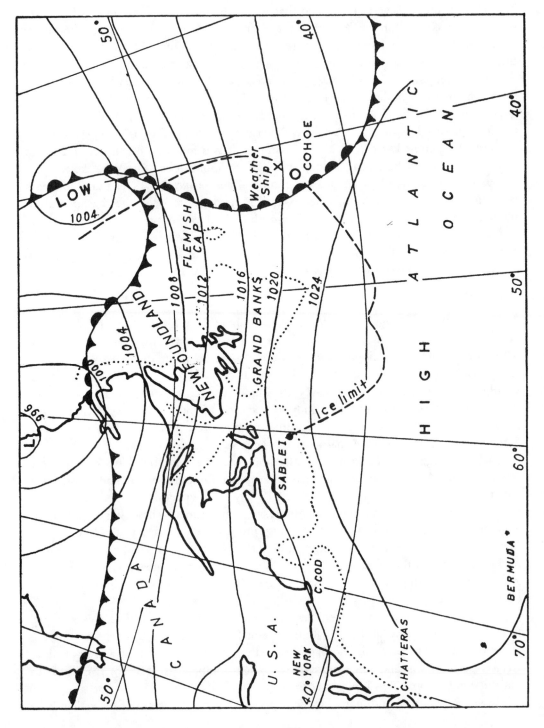

Fig. 8.1 Synoptic chart Western Atlantic 1200 GMT 12 July 1950

Saturnia, who logged the conditions as 'Sea Disturbance No. 6 Beaufort Wind Force 6'. Sea Disturbance No 6 is usually associated with wind force 7, so I think a reasonable guess at the strength of the wind gives a mean of about 25 to 30 knots (force 6 to 7), gusting perhaps up to 40 knots, which is, in Alan Watts's terms, a 'yachtsman's gale'.

I have gone into these assessments of wind force with particular care, because they illustrate the point that the state of the sea, even in the open ocean, cannot be judged entirely by the mean force of the wind on the Beaufort Scale, and that a small yacht may occasionally run into trouble in winds of under gale force in which in the ordinary way she would only be moderately reefed. In another chapter I shall give a converse example where in a vigorous secondary, giving verified winds of force 9-10, the seas never rose high enough to warrant attention.

There is no doubt that the seas on 12 July were out of all proportion to the wind, which had backed from NNW to west in the early morning. The height of the seas during the afternoon was independently judged by each of us at 30 ft in relation to the known height of the mast. Such estimates by eye from the deck are inaccurate, however carefully and coolly judged, and usually should be halved. This gives 15 ft, but as this seemed too conservative at the time I logged 18 ft, being three-fifths of the estimate by eye. This, as it happens, was confirmed by the *Saturnia*, for sea Disturbance No 6 is 'very rough' and gives a mean height of the waves as 19 ft. Tom Tothill, our navigator, estimated the length of the waves as 250 ft, but this was probably on the low side, since oceanographers state that, while height is usually exaggerated, the length is often underestimated. The size of the seas could be accounted for by the long duration and unlimited fetch along the straight isobars to the north of the huge high to the south-west, but the feature was their confused pattern, which must have been due to the shift of wind, frontal gusts and the combination of wave trains caused by the movements of the depressions.

1 *Pooping and speed.* I think the pooping was due to the wash of the liner (passing fairly close at high speed) superimposed on the already high and confused seas. The wash probably combined to form two pyramidical, high, breaking seas which struck *Cohoe* on the quarter and did the damage. Freak waves can be caused by any interference with the normal run of the seas.

This, however, does not account for the difficulty which was experienced by all the helmsmen in *Cohoe* when steering that day. It may well be asked why *Samuel Pepys* was able to carry on efficiently under twin staysails at a time when *Cohoe* was running under bare pole? Why did *Cohoe* make so much better weather of it running before the gale north of Bermuda when the gale was incomparably stronger?

According to the synoptic chart and her log, *Samuel Pepys* possibly experienced less severe wind and sea conditions, but I think her immunity was due to carrying more sail. She averaged 6 ½ knots. Likewise, when running under storm jib in the Bermuda gale, *Cohoe* was averaging about 6 ½ knots, and at the end of the Transatlantic Race she effortlessly maintained the same speed in a depression for which a gale warning had been received.

I think that on the day *Cohoe* was pooped she was not carrying sufficient sail. As the seas got worse, sail was progressively reduced in the traditional manner until she was down to bare pole. If the seas were running 18 ft. high she would be partially under their lee in the troughs except for the upper part of the mast. Her speed would thus fall in the trough and when the next crest rode up it would be inadequate to give the quick response to the tiller, which is necessary in high confused seas. She ought to have carried more sail, not less, because under heavy ocean conditions speed is needed so that a boat is quickly responsive and can be steered with a flick of the helm to take a 'big-un' stern on.

To maintain quick control I suggest a speed of about 5 knots is desirable, but the right speed will only be found by trial and error, as so much depends upon the characteristics and size of the individual yacht in relation to the pattern (or absence of pattern) of the seas. There may, however, come a time when speed becomes dangerous. It is still necessary to keep the stern to the seas and this is accomplished by towing warps to steady the yacht and the use of the helm as well. In other words, if a yacht runs in heavy weather in the ocean she must *either* maintain sufficient speed to keep her lively and responsive (but see Chapters 19, 20 and 22 for running in storms or a hurricane) or, on the contrary, she must adopt gale tactics and tow warps.

2 *Shape of stern*. Only on two occasions in my life have I had the experience of being in a yacht when she has been well and truly pooped by a following sea. In each case, the wind was below gale force and the yacht was a double-ended one with a Scandinavian pointed stern and outhung rudder. The Norwegian pilot cutters and the Colin Archer designs with their pointed sterns are noted for their ability to run well before gales, and Vito Dumas demonstrated the efficiency of the type when he ran round the world in the Roaring Forties in *Lehg II*. But in these vessels the pointed stern was associated with wide beam, which enables the buoyancy to be carried aft. This cannot be done so effectively in narrow light displacement boats such as *Zara* and *Cohoe*. There was little to complain about in *Cohoe*, as the pooping was probably just a coincidence due to the wash of the liner, but I think that a transom or a well-designed counter stern is better, as it provides greater buoyancy aft.

3 *Damage*. In my book *North Atlantic* I referred to a 'loud report as though the hull was cracked in' when the sea struck *Cohoe*. Long after the book was written and published, *Cohoe's* builders, A H Moody & Sons Ltd, made a survey of the yacht and found two oak timbers fractured on the port quarter inside the stern locker in the position exactly where the sea struck.

When a yacht offers resistance to a big breaking sea damage nearly always follows.

4 *Safety harness*. In the Bermuda and Transatlantic Races we used personal lifelines in *Cohoe* for the first time, having on previous occasions merely used the end of a sheet or a short length of rope in the cockpit when steering in exceptionally bad weather.

Commander Erroll Bruce devoted much time to the subject of considering lifelines and safety precautions and later Peter Haward designed and marketed a very efficient type of safety harness which came into general use. The RORC added safety harness to the compulsory equipment under the Clubs regulations and today almost all sea-going yachts, whether racing or cruising, are equipped with them.

9 RETURN FROM LA CORUÑA

Adlard Coles

I n 1952 I had a new yacht built for me, partly because my wife wanted a more comfortable boat and partly because *Cohoe* was becoming rather outdated for racing in home waters.

Cohoe II, as we named her out of affection for the former boat, was a yawl designed by Charles A Nicholson and built at Cowes by A W Souter. In case I should want to enter her for the Bermuda Race without repeating the facial operation of adding a false bow, she was built to comply with the minimum rule length of 35 ft overall, and her length on the waterline was 26 ft. With a beam of 8 ft 6 ins the Thames measurement worked out at $8\frac{1}{4}$ tons. She had a fuller bodied hull than *Cohoe* and displaced $6\frac{2}{3}$ tons. In common with most of Charles A Nicholson's designs at the time, she had broad shoulders and a fine run aft, in the cod's head and mackerel tail tradition. But her bilges were slacker than is commonly found in this designer's work, and the lines were experimental, producing something in the nature of a cruising version of the former International 8-metre class.

The accommodation below in *Cohoe II* was conventional and consisted of a fo'c'sle with one root berth, sail locker, stowage room and the heads. A large locker and wardrobe were arranged between the fo'c'sle and the saloon, which had a settee berth on either side and a folding table in the centre. The mast was stepped on deck, which was reinforced by steel beams and further strengthened under the mast by a steel tube which carried through to the keel. Abaft the saloon, but divided from it by a curtain running on slides in a track, was the galley on the port side and a dresser and cupboards opposite. Right aft there were two quarter berths. The arrangement was simple but effective, except in very heavy weather, when, despite canvas protective curtains, the quarter berths were always wet. A portable chart table fitted over the starboard quarter berth, and a light, horizontally opposed piston petrol engine was situated under the companionway. The interior was spacious and airy.

When we came to race *Cohoe II* in 1952, which was a season of fresh and strong winds, we found her fast in light or moderate breezes, but she proved to be overmasted and overcanvassed in strong winds, and the world's champion rhythmic roller. This was partly due to her being designed to carry a lead keel, but having had an iron one substituted, as lead reached a peak price in the year that she was built.

Accordingly, in consultation with her designer, I had the sail plan reduced during the following winter, by cutting the mast at the jumpers and cutting the mainsail. The reduction in sail area was drastic, being equivalent to two reefs. As it chanced, the two following Fastnet years of 1953 and 1955 (in which, through a change in the rules, the small class of ocean racers were admitted) happened to be light seasons in which *Cohoe II* would have fared better under her original rig. Nevertheless, the alteration greatly improved the yacht. From being a tender boat she became a stiff one, not only because of the reduction in sail area, but because of the reduction in weight and windage aloft.

Gone was the rhythmic rolling, gone was the excessive leeway. And what a good rig for cruising is the masthead yawl! There are no runners to bother about, which is a great advantage when short-handed, and the foot of the mainsail is short, so that there are no terrors in gybing, however hard the conditions may be. *Cohoe II* handled admirably under a small genoa and mizzen alone, even in light winds. In strong winds my wife and I never reefed. We merely lowered the mainsail. Although perhaps more suited to a longer boat, the yawl rig lends character in these days when almost every yacht is a sloop. For seagoing it is safe rig, because the sail area is split into smaller units, and the mizzen staysail is a practical sail which can be set or lowered quickly. Besides which, I prefer two masts (independently stayed) to one, because in the event of losing one there is a sporting chance of being able to make a reasonably efficient jury if the other remains. This argument may be contentious, but the fact remains that I like yawls and would still have one if they were not virtually penalized under the ocean-racing rules.

The experience I now describe occurred in 1954, not in a race but when cruising home afterwards.

We had raced from Cowes to La Coruña, a slow event involving a 200-mile beat to Ushant mostly under rough conditions with rain and poor visibility, which were followed by calms over an immense area of the Bay of Biscay and finally by dense fog when we arrived off the Spanish coast. My crew consisted of Alan Mansley, Jim Kentish, Mike Awty and Barrie Kendall, and after the race, which we won in our class, we relaxed in La Coruña, a sunny, breezy town where we were most hospitably entertained by the Real Club Nautico. We remained in these happy surroundings three days, when peace was disturbed by one of my crew. He had sent a cable to the head of his department asking for an extension of his sailing holiday on account of inclement weather. The reply was prompt and to the point: 'Leave granted. Divorce pending.'

This carefully worded missive seemed to have a more atomic effect than if the reply had merely read, 'Sacked on the spot', and the whole of my crew evinced a sudden enthusiasm to return to work. As I have remarked before, time running short at the end of a holiday is the most frequent cause of the cruising man getting caught out in bad weather, and our homeward passage was to prove no exception.

Our last evening in Spain was spent at the club having dinner at a table on the balcony overlooking the yachts anchored immediately below. The meal was a pleasant one, and we lingered so long over our coffee and brandy that the hour was already late when we paid our bill and jumped into the dinghy to row off to the yacht.

Once aboard we changed from shore-going clothes and donned our seagoing rig for the night's sail. Five men in a tiny cabin – clothes everywhere, everything anywhere. It was not until nearly midnight that all was ready, order had been restored, and the anchor weighed. So it may be said that our passage started at 0001 on Thursday, 22 July (see chart, Fig 9.1, page 83).

When the yacht was 10 miles out, and beyond the sheltering arm of Cabo Priorino we began to feel we were really at sea again. It had been blowing fresh for some days from the north-east, and the midnight weather forecast gave promise of strong winds in the Bay of Biscay and rain. The night was very dark, and to the eastwards over the land the sky was blacker still. Ominous clouds were approaching and soon the rain came in torrents, and the wind hardened. We lowered the mainsail (one of the advantages of the yawl rig) and continued under genoa and mizzen alone, doing a good 6 knots. All turned in, except the man on watch, for we were cruising now, and only one was needed on deck. With five of us it meant two hours on and eight off, apart from sail shifting, cooking and navigation.

What never fails to surprise me, even after many years of sailing, is the sharpness of the contrast between life ashore and life at sea. Only a few hours before we had been part of the land world. The table at which we were seated was steady, the wine and food were excellent and well served, and the brightly lit club was gay. It was a well-ordered and comfortable existence. But once at sea again, the yacht became a compact little world unto herself in which all our activities were centred. Life ashore was as remote to us as life on another planet. On deck the helmsman is alone. As he looks seaward nothing breaks the impenetrable darkness of the night except for the phosphorescent top of a breaking wave and perhaps the flash of a lighthouse. The yacht plunges forward on her course, throwing aft sheets of spray which patter over the cabin roof and drive into the helmsman's face.

Throughout the night the wind hardened. A strong onshore wind nearly always means rough conditions until one gets into deep water. I went on watch at 0200. We were 3 miles west of Cabo Prior and now exposed to the full force of wind. It was blowing about force 7 and raining hard. Watches had been reduced to an hour at a time. It was long enough, for the incessant spray driving in the helmsman's face is quickly tiring. By the end of my watch the salt was burning my eyes, and I was delighted to be relieved and return to the cabin for breakfast.

Although the wind remained strong the seas became more regular as we got into the deep water of the Atlantic. Conditions were not so severe during the second night and at 0400 on the next morning (Friday 23rd July) the wind moderated, so that we were able to set the mainsail again. The day developed into a pleasant one. The sun broke through the clouds in time for a noon sight to be taken and in the afternoon we got our longitude. The sight placed us nearly 30 miles west of our dead reckoning, so we assumed that the strong north-east winds had produced a west-going current off the Spanish coast.

In the afternoon the wind petered out and *Cohoe II* lay becalmed in a long ocean swell. The course combined with the current had taken the yacht out of the Bay of Biscay and well into the Atlantic. At 2100 our position was 46° 40' N, 9° 20' W— over 300 miles from the shores of the Bay, and almost equidistant, at 200 miles, from La Coruña and Ushant, with over 3000 m of water under the keel. The sea remained calm all night, but next morning (Saturday, 24 July) the weather forecast gave south-west winds in the north of the Bay of Biscay. Sure enough to the north the sky was overcast. At 0400 we went on the other tack and an hour later a light breeze got up from west of north, and we were sailing once more, but this time in the direction we wanted to go, to Ushant instead of out into the Atlantic.

On Sunday morning (25 July) the wind freshened. An overcast sky still lay to the north, but we were getting much closer to it. The barometer at 1016 millibars showed a tendency to fall. During the morning the wind steadily increased and we lowered the mainsail. At noon I was just in time to get a sight of the sun through the gathering clouds, and in the early afternoon the spinnaker was handed, as the yacht was rolling heavily, and a small genoa was set in its place. The midday weather forecast was bad. There was a gale warning for the Sole area just north of us, and a forecast of strong winds, gale force locally, coastal fog and otherwise poor visibility, for the north of the Bay of Biscay and Plymouth to the east of us.

The 1800 forecast was even more ominous. A depression was deepening off Ireland and there was a deep depression off Iceland. The yacht was then running fast, doing nearly 8 knots under small genoa and mizzen only. Before nightfall both sails were handed and the storm jib was set. It did not set well, as we found the strain on the genoa halyard had nearly drawn the winch off the mast, so that the storm jib halyard had to be set on a cleat, which also worked loose. The speed under only 50 sq ft of canvas dropped to about $4\frac{1}{2}$ or 5 knots, but the yacht was incomparably easier to steer, and if the moderate gale matured into something serious she would be under the right canvas for it, without the need for sail shifting at night.

Ushant, the most westerly point in France, is a bad corner to round at night in severe weather. The navigator has to anticipate the possibility of the lights being obscured by fog and rain, and owing to the strong tidal streams in the vicinity, the seas in a south-west gale from the Atlantic can be formidable. For these reasons, although I wanted to fix the yacht's position by the lights on the French coast if they could be seen, I decided to give Ushant a wide berth. The yacht was gybed and stood away to the north.

As anticipated, it proved a dirty night. It was exceptionally dark owing to driving rain, with not a glimpse of moon or stars. The only light was from the phosphorescent tops of the waves now breaking in every direction, and the fiery wake. The distant breaking crests were so luminous that they could easily be mistaken for the loom of a distant lighthouse. The barometer was still falling smartly; it fell 24 millibars in twenty-four hours. The boat was behaving beautifully. The long smooth run left a clean, though phosphorescent, wake. She answered the lightest touch of the helm and no heavy water came aboard. Nevertheless, steering was responsible work, and once again the helmsmen were relieved hourly. As the visibility was so bad and we were approaching shipping routes, a radar reflector was hoisted in the rigging, and a white flare was placed in the waterproof bag for the helmsman in case of need.

It was a turbulent scene on deck, and below decks there was the usual discomfort of four men penned up in a small space—wet oilskins, cigarette ends and spent matches. The motion made sleep almost impossible. There was also a slight uncertainty about our position, for we had sailed over 300 miles since leaving sight of land.

However, the night, although rough, proved uneventful. Even the glutinous spaghetti au gratin which we had for dinner was shaken down. At dawn (Monday, 26 July) we reckoned *Cohoe II* was a good 30 miles west of Ushant. Although visibility was still thick, we had plenty of searoom to enter the English Channel, and a course was set for Portland Bill, still 200 miles distant. Now that it was light we could see the seas and judge their potential. For the weight of the wind and the long fetch, they were not excessive. They were, of course, breaking heavily and occasionally forming into white-topped pinnacles of water, rearing against the sombre sky, but they were not unmanageable, so we set the mizzen, as we wanted to get well round the corner into the English Channel as quickly as possible in case the weather should worsen further.

The visibility did not improve in the morning, but we sighted a small tanker making heavy weather of it as she passed close to us, and at noon the sun showed through the clouds conveniently in time to give us a sun sight. We also got a fix from radio beacons which put us well in the English Channel off the French coast some 40 miles north by east of Ile Vierge lighthouse.

However, the glass was still falling. By then it had dropped from 1,016 millibars to 989 millibars. The weather forecast that evening was greeted by the crew with hilarity. By the time the annoucer had finished his gale warnings there was hardly an area left where the weather was normal. A deep depression was moving east across Scotland, and, if my memory serves correctly, gale warnings were given for Rockall, Malin, Shannon, Fastnet, Lundy, Sole, North Biscay, Plymouth, Portland, Wight and some of the North Sea. We were in the Plymouth area and approaching Portland and Wight, so there was no escaping a blow which covered such a vast sea area.

So we prepared for another night of discomfort, though we were becoming acclimatized to it by then. The yacht was again proving herself an able sea boat and we had a strong crew. One hour on watch and four below is light work, even if sleep is difficult. The night passed much as had the previous one, but we were lucky next morning (Tuesday, 27 July), for the sun came out. True, it was an unhealthy sort of sun and there was a haze over the sea (which if anything was running higher than before), but it was sufficient to give us a sight. The yacht was sailing quite fast enough for normal cruising, but divorce by their bosses once again became a topic of conversation among the crew. To hurry things up we reefed the mainsail, and set it. Immediately the yacht picked up to her maximum speed, and although steering became more difficult she still remained under perfect control. It was blowing a moderate gale and no more. We passed a trawler, a tanker and a coaster, so we concluded we were crossing the shipping lane off the south coast of England.

It was our sixth day at sea and, allowing for the headwind in the Bay of Biscay, the yacht had sailed nearly 600 miles through the water, the last part in thick weather, so it was an exciting moment when land was sighted on the port bow at 1030. England showed as distant hills, and then a ray of sun passed over a yellow headland, none other than Golden Cap in Lyme Bay. Not long afterwards Portland could be identified. It had not been far distant, but obscured by low cloud, so our landfall was precise and a credit to Barrie, who did most of the navigation.

The tide was foul, so we gybed to get an offing of 5 miles when passing the Bill. The notorious Race of Portland only extends a couple of miles, but in bad weather the sea is disturbed even 10 miles out and it should be given a wide berth, except when racing. It was a good day for taking photographs of the sea, as the sun was out, and Alan and I seized the opportunity.

The sea, as the yacht drew south of the Bill, became heavier. Alan took a trick at the helm and enjoyed the experience of surf riding, with the speed indicator reading 9 knots, no doubt over registering, but none the less an exhilarating performance. We were lucky again with our tide, for it was fair when we arrived off the Bill. There would have been a really dangerous sea if the tide had been weather-going. As it was, the seas were building up formidably.

It is always difficult to assess the strength of the wind when running before it. Under such conditions it is often underestimated, just as it may be exaggerated when beating into it. The observer's judgement is affected by such things as the height and severity of the sea, and especially by the conditions, whether sunny or overcast, with rain to add to the gloom. The variations are subtle, and it was not until the next watch, with Barrie at the helm, that I guessed the wind had touched at least gale force. We were then off St Alban's Head and crossing the tail of the bank, when Barrie called for relief. The boat was getting unmanageable. It had been blowing hard from the Atlantic for some days, and the waves by then were pretty big. The sea was indeed a mass of foam. It was time to slow down, so I lowered the mainsail instantly. That relieved the boat, and she was happy once more.

After passing Anvil Point the seas grew smaller, and we crossed Bournemouth Bay and safely entered the Solent through the North Channel, which is partially sheltered by the Dolphin Bank and the Shingles Bank.

As we approached Yarmouth in the Isle of Wight we saw the Roads were dotted with coasters at anchor sheltering from the weather, which confirmed that we had not overestimated the strength of the wind when running before it. We tied up in the harbour before 2000 and the Customs came alongside immediately on arrival and cleared us, in time for three of my crew to hurry ashore to return to their jobs. Before they left, the Customs officers gave us a bit of news. Another gale warning!

Conclusions

The synoptic charts are reproduced and it will be seen that a fairly deep depression to the south of Iceland moved rapidly across the north of Scotland, where it slowed down and deepened to 980 millibars. It was accompanied by minor troughs of low pressure which moved across England and the English Channel. The weather affords a typical example of that experienced in the English Channel when a depression is passing to the northward, giving rise to force 6 and force 7 winds and force 8 gales locally. This particular depression caused a longer period of bad weather than usual, as it became almost stationary over Scotland and lasted long enough to allow the seas to build up to maturity.

The nearest shore station of which I have records is at the Lizard. The wind forces recorded there on Monday, 26 July, were a consistent force 6 pretty well all day, rising to force 7 for the two hours to 1230 GMT and force 8 for the hour ending 1530 GMT. On Tuesday, 27 July, the wind was maintained near the top of force 6, but rose to force 7 in the early morning from 0230 to 0530. The notable feature on both days was the squalliness of the wind. On Monday a gust of 43 knots (force 9) was recorded at 2025 GMT, nearly double the mean wind speed for the hour of force 6 (22-27 knots);

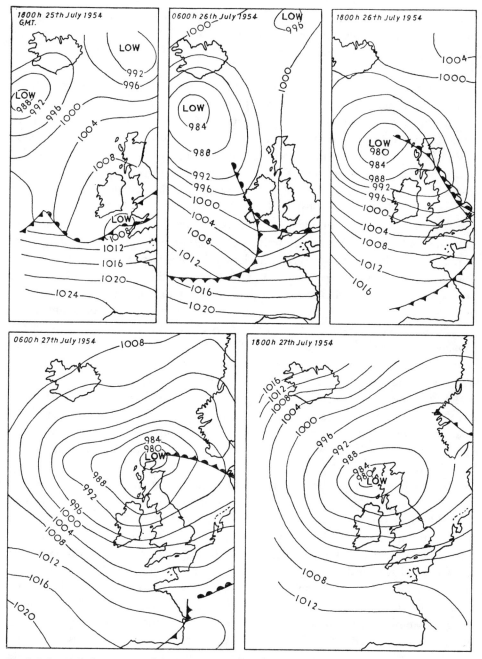

Fig 9.1.Synoptic charts, return from La Coruña, 1954.

on Tuesday a similar 43 knot gust was recorded at 0335 GMT when the wind was mean force 7. At sea, the strength of the wind would have been higher, and undoubtedly there were considerable periods when it attained the 34-40 knots of force 8.

1 *Shoaling water.* High seas were experienced off Portland Bill. I have noticed before and since, during gales in this vicinity, that the waves take the form of a very high steep swell, literally towering much as they do approaching a steep-to shore. With a lee-going tide they are toppling rather than breaking, although white-crested. Such seas are impressive, but harmless provided nothing causes them to break heavily. It would be dangerous if one of them actually did so. With a weather-going spring tide against a south-westerly gale they would break and the whole locality is dangerous. It is safer to keep 10 miles south of Portland Bill.

The roughest seas were found SE of St Alban's Head, where there is an irregular bottom with only 16 m as much as 5 miles seawards, coupled with strong tidal streams. After passing St. Alban's the seas gradually lessened. When approaching the Solent by the North Channel, the seas grew steeper and whiter, but they were much smaller than out at sea, probably due to being under the lee of the Dolphin Bank, which, although it has 10-16 m over it, acts as a brake on size. As a matter of purely local interest, the North Channel is said to be safer than the Needles Channel in a south-westerly gale.

2 *Running in gales.* As I have remarked before, it is extremely difficult to judge the force of a following wind. When a yacht is running at 6 knots or more this alone reduces the apparent wind by about one grade in the Beaufort scale. A force 8 gale will only feel force 7 on the face, when looking astern. A lee-going tide, so that the yacht may really be doing 10 knots over the bottom, even further reduces the apparent wind.

The interesting thing is to compare the performance of *Cohoe I* in the Atlantic near gale (described in the last chapter) with that of *Cohoe II* in stronger winds though not such confused seas.

I reduced sail as *Cohoe I* became more unmanageable; eventually coming down to bare pole, doing perhaps 4 knots on the crests of the seas and only 2 knots in the troughs. The speed of *Cohoe II*, on the contrary, under mainsail, mizzen and storm jib was $7\frac{1}{2}$ knots, and when racing it would not be allowed to fall lower, even if the yacht became difficult to steer. Under mizzen and storm jib from Portland to St Albans (when the wind was probably force 8 locally) it was reduced to 6 knots and she was docile and much easier on the helm. Heads of seas often came aboard and there was continual spray flying across the cockpit, but at no time was there any threat of real pooping. It seems to me that when cruising in strong winds and ordinary gales, each yacht has a natural speed at which she is easy to steer and responsive. It is undesirable when cruising to exceed or fall much below the natural speed unless wind or seas dictate resorting to gale tactics, such as streaming warps. This point might be arrived at somewhere about a genuine force 9, according to the ability of the individual boat and other factors such as turbulence, tidal streams and shoal water.

10 A RACE TO CORK

Adlard Coles

W hen *Cohoe II* was back on her moorings at Bursledon there remained little over a week in which to turn her round, refit and provision for the Cowes to Cork Race due to start on Saturday, 7 August.

For this event I had two of my Coruña crew, Alan Mansley and Jim Kentish, but Mike Awty and Barrie Kendall had no leave left and their places were taken by Dr Aubrey Hudson, who had sailed with me on the previous Fastnet Race, and John Webster, then a Lieutenant RN, making a total complement of five. The course was direct from Cowes through the Needles Channel and thence to Cork, a distance of 330 miles. Only ten yachts entered, three in Class I, six in Class II and *Cohoe II*, which raced with Class II in the absence of any other small boat entries. There was a south-westerly gale on the Saturday, so the start of the race was postponed until Sunday (8 August), owing to the risks involved in beating through the Needles Channel under such conditions. The weather had improved on Sunday and after crossing the starting line, the boats turned to windward against a light SW wind. Good conditions continued throughout the day, until on the evening weather forecast there was mention of another depression coming in from the Atlantic.

By 2200 the wind had freshened and the ocean racers were in rough water off Portland Bill, but the wind backed to the SE, so that they were able to lay across Lyme Bay.

On Monday (9 August) the wind continued to freshen, and at 0700 there was heavy rain, but the wind veered later and it was a pleasant day, though rough going. At 1800 there was a gale warning owing to 'a vigorous depression centred over sea area Thames' and another off the Hebrides. Two Class II yachts with whom we had been cross-tacking all day retired to Plymouth.

At 1930 we brought the Eddystone lighthouse abeam. The sky was sunny, but the wind had hardened, and there was a very steep rough sea, although it had only a 20 mile fetch.

The gale came in then with a violent squall which hove the yacht down until her decks were awash. The mainsail was lowered at once and *Cohoe II* carried on under mizzen and small genoa. Cooking dinner, consisting merely of soup and eggs, was a grim job which Alan tackled with his customary determination. The yacht was sailing at a considerable angle of heel, crashing to windward in a cloud of spray.

Down below it was stuffy, as the Dorade ventilators had to be closed, for the seas breaking forward sometimes filled the boxes and sent occasional spouts of water, solid as from a hose pipe, through the apertures into the cabin.

The scuttles were screwed down and the only air admitted below came from the companionway, accompanied by dollops of water. Two crew members were seasick though neither was incapacitated.

The night was cold and unpleasant, during which we made a long tack shorewards towards Fowey and another seawards. There was a note of disappointment at dawn (10 August) when we found the Lizard still some 10 miles to the west of us. Beating to windward in a gale is punishing to all. The yacht had to be sailed rather free to maintain speed and momentum to smash her way against the weight of the seas, and under such conditions the leeway was considerable. Tedious and slow as it seemed, *Cohoe II* made good 30 miles dead to windward in twelve hours; few small yachts of her size are able to do much better than 6 knots through the water and $2\frac{1}{2}$ to 3 knots made good when beating against a gale.

When we drew clear of the land into the English Channel, the height of the seas increased considerably. Although high cut, the genoa began to take heavy water as the boat plunged through the seas. There was a risk not only of bursting the sail but to the mast, rigging and gear, for there is immense weight and power behind a big sea. It was blowing gale force, and on the morning forecast there was another gale warning, so it was time to alter the sail plan. Alan and I tied down a reef in the mainsail and set it, and the genoa was replaced by the storm jib. There was a large reduction in sail area, but the storm jib was small and set high, so it did not take the seas forward as the genoa did, and the mainsail was a safe sail. Cut down as it was, it was equivalent to double reefed before we started and the extra reef brought it down to storm trysail size.

Sailing now under storm jib, reefed mainsail and mizzen, the yacht was snugly rigged and well balanced. The tiny storm jib took no seas in it, and all strains were distributed between three sails with a low sail plan. The storm jib was, however, of the conventional pattern, roped all round and immensely strong, but cut full, and hence it was no racing sail and the yacht was not so close-winded as she had been under genoa and mizzen.

Under gale conditions, the seas near the land (especially in the fast tidal streams off the Lizard Race), are very high. They run higher than they do farther east towards Portland Bill, so we continued on the starboard tack well into the deep water of the English Channel. The wind was very boisterous, with tremendous squalls followed by comparative lulls. Aubrey Hudson prepared porridge and eggs for breakfast, but part of the porridge was lost in a particularly violent lurch and spattered all over the galley and cooking utensils. Conditions below were scarcely jolly: porridge on the cabin sole near the galley, cigarette-ends, spent matches and fluff off the blankets in every corner, water oozing up from the bilges over the cabin sole as the ship heeled, and damp throughout. Wet clothes, wet blankets, wet everything ! The smell was unpleasant, too, as among us was a rather heavy smoker, and stale tobacco leaves a nasty aroma in a confined wet space.

After sailing some 20 miles south of the Lizard we tacked and were delighted to find we could lay Newlyn.

At 1900 we were off Newlyn and the yacht was in partial shelter of the land, for the wind had veered. We tacked again and lay close under the cliffs with a fair tide. The evening weather report had been comparatively good, a forecast of strong winds instead of gales. While under the lee of the land, Alan got busy with the cooking and we enjoyed hot soup and eggs. Then we lowered the storm jib and set the staysail in its place, which enabled us to sail a point closer to the wind.

By 2300 we had cleared Land's End and were close to the Wolf Rock lighthouse. The beam of the light lit up the sails and we wondered whether any other yachts were in the vicinity, for we had seen none that day. I had expected a heavy sea off the Wolf, as it had been blowing from the west for so long and was still fresh to strong, but the seas proved much easier. They were longer and the boat was able to ride the big seas far better than the shorter ones off the Lizard. At the Wolf we tacked and headed for Ireland, though we could only lay to the east of Cork. It was a clear night. The Longships light was on the starboard bow, the Seven Stones lightship and Round Island lighthouse to port.

That was the end of the heavy weather part of the race and by 1100 the following morning (Wednesday, 11 August) the wind was moderate and the sun had broken through. It is remarkable how quickly spirits rise on a bright morning. In no time wet blankets and wet cushions appeared on deck and wet clothes hung in the rigging. Everybody was hilarious, but in this extraordinary race the weather still had a card to play. By 1730 it started to rain and by midnight it turned to dense fog. With it the wind had backed and freed. Throughout the night we ran fast and at dawn (Thursday, 12 August) we picked up the Daunt radio beacon on which John Webster homed by DF so exactly that when the lightship loomed up like a dark shadow we had to alter course promptly to avoid collision with her. As we approached the land the fog cleared and at 0924 we crossed the finishing line at the entrance to Cork Harbour, second boat home. Only two other yachts (*Jocasta* and *Marabu*, both Class I) completed the course, and *Cohoe II* had put up the best corrected time. It marked the first occasion when a Class III yacht beat Class I in gales, for small yachts had hitherto won mostly in light weather.

If I may digress for a moment, I will add that my wife joined *Cohoe II* at Cork and Ross and my daughter-in-law joined at Dingle. We cruised to the Blasket Islands before sailing home, and I can recommend south-west Ireland as one of the best cruising grounds in Europe. It was one of our happiest family cruises.

Conclusions

The period of heavy weather was a prolonged one. As on the return from La Coruña it was caused by a depression slowing up over the north of Scotland. The fundamental difference between the experiences in the return from La Coruña and those in the Cork Race were that in the former we had a joy ride running before the gale, whereas in the latter we were beating against it, the point of sailing which provides the real test of ship and crew.

On 10 August, force 7 was recorded at the Lizard until 0530, followed by a consistent force 6 for the rest of the day, but it would have been blowing harder at sea, probably force 8 for part of the time. As on the return from La Coruña, the feature was the gustiness of the wind. The highest gust occurred at 0355, when *Cohoe II* was still east of the Lizard. Oddly enough this was precisely the same velocity as in the previous blow – 43 knots (force 9). *Cohoe II* was kept going at between 5 and 6 knots, which was an improvement over the speeds achieved by the small yachts in the Wolf Race, but *Cohoe II* was a bigger and more powerful boat than the RNSA 24's.

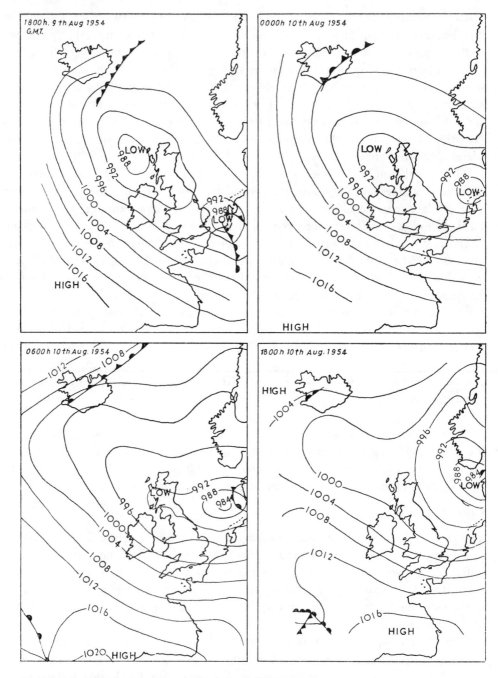

Fig 10.1. *Synoptic charts covering the Cork Race, 1954.*

1 *The yawl rig.* This is a good one when beating to windward in a gale. It enables the sail area to be distributed between storm jib, reefed mainsail (or trysail) and mizzen, all low-setting sails, but high enough in the foot not to be filled by any but a freak wave. The alternative sail plan of mizzen and a heavy genoa is effective up to force 7, but, once the wind rises to force 8 and the seas have had time to build up, any low-cut sail like a genoa becomes dangerous. The head of a big breaking sea may strike it with irresistible force, so that something is bound to go.

2 *Storm jib.* The old-fashioned storm jib of heavy canvas roped all round is adequate for cruising and for heaving-to, but the rope tends to shrink and the sail becomes too baggy for racing. Of recent years I have used terylene and sheeted it hard, so that it is as effective in a gale as a staysail in a strong wind. Terylene is so strong and the sail area is so small that storm jibs no longer need be very heavy. What is most likely to wear is the stitching.

3 *Rounding headlands.* Progress round headlands, when beating to windward in a gale, often seems slow. This is most disheartening to a tired crew, when after hours of tacking, soaked and half blinded by spray, the objective seems no nearer. A small yacht even if hard driven at 5 or 6 knots, will only make good about 3 miles in an hour. Tacking against a strong foul tide, she will make hardly any progress and when the tide comes fair, but contrary to the wind, she will be beating in very rough water in which she has to be sailed rather free. In the English Channel, it is the headlands, Portland Bill, the Lizard and others that provide the endurance tests. Nevertheless, progress is made even if it does not appear to be. Ten hours' gruelling work means a gain of 20 or 30 miles and gales don't last for ever. In some circumstances it may pay, in a small yacht, to shelter during the period of foul tide, of which I shall have more to say in Chapter 12.

11 STORM IN THE ENGLISH CHANNEL

by Adlard Coles

In the Channel Storm of 1956, the wind attained force 11 (violent storm) at the Lizard, and it was the first occasion when an overall picture was obtained of all the yachts involved and how they fared in positions ranging from the open sea to a dangerous lee shore. The Editor of *Yachting World* sent a questionnaire to owners, and a subcommittee of the RORC was appointed to examine their reports. Many lessons were learnt and the findings of the committee provide a valuable contribution to the knowledge of handling of yachts in heavy weather and the necessary preparations for it.

I cannot give a first-hand report of the gale, as I was not racing that year, but the RORC has kindly given me access to all the records, from which I have been able to prepare the following account, aided by information from other sources and my personal acquaintance with the yachts involved and with their owners. In the first part of the chapter I will confine my comments to the yachts which were caught in the worst of the storm on a lee shore, and in the second I shall refer to the experiences of the remaining boats and general points arising from the gale.

Twenty-three yachts divided into the usual three classes started in the Channel Race of 1956, on the 220-mile course from Southsea to the Royal Sovereign lightvessel off Eastbourne, thence to Le Havre lightvessel (for which a buoy had been substituted) and back across the English Channel to finish between the forts at Spithead (see chart, Fig 11.1, page 91).

The race started gently on the Friday evening of 27 July and except for two sharp squalls, nothing of interest was reported on the passage to the Royal Sovereign lightvessel, which was rounded in the early hours of Saturday morning. The wind had freshened by then from the south-west and the next leg of the course to Le Havre buoy at first provided a beat dead to windward, but by the evening the wind had backed to the south.

It was late during Saturday night (28 July) when the wind started to increase. By this time it appears that the big yachts had rounded Le Havre buoy, but the smaller ones were still on the wrong side, and many of them did not round it until 0900 or later on Sunday morning, 29 July.

The midnight forecast had given SW gales for Portland and Plymouth, veering NW. For Dover and Wight it gave no more than strong southerly winds, increasing to force 7 and veering SW. However, a depression was moving into the English Channel and a front of cool air was moving in towards the NW of its centre, which was rapidly deepening to 976 millibars. The low was centred over Wales on the morning of Sunday, 29 July. When the fronts had passed through they were followed by a phenomenally fast rise in the barometer and, with the veer, the winds on the English side of the Channel increased to storm force or over.

Gusts of 100 mph (86 knots) were reported, possibly from ships, as I cannot

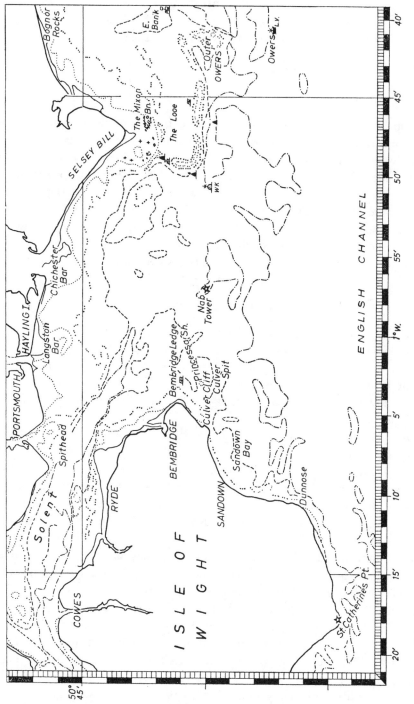

Fig 11.1. St Catherine's Point to Owers LV.

ascertain the source, and the following are the figures which I have obtained from the Meteorological Office.

Station	Wind force	Duration force 8 and over	Highest gust
The Scillies	45 knots (force 9)	7 hours	68 knots
The Lizard	57 knots (force 11)	10 hours	81 knots
Dungeness	42 knots (force 9)	2 hours	70 knots

The principal characteristic of this storm was its extreme turbulence, with gusts well over hurricane force. At the less-exposed shore station of Thorney Island, where the wind was only 37 knots on the Beaufort scale, a gust of 67 knots was experienced, nearly double the mean force. These were the weather conditions with which the leading yachts had to contend, but farther away to the south of the depression the winds were progressively less. Common to both sides of the English Channel was the long duration of severe weather (nearly four days) as the depression slowed down and became almost stationary over the North Sea on Monday, 30 July, without filling in much. Let us now see what occurred to the leading yachts when they approached the lee shore of the English coast between St Catherine's Point and the Owers lightvessel.

The Leading Yachts

Lloyd's Yacht Club's big 70 ft yawl *Lutine* was the first to arrive. She had a very experienced skipper and a non-seasick crew. She reported force 4–5 at 0600 on Sunday, 29 July, which steadily increased to force 8 before 0900, with squalls of 44 to 52 knots. She made her landfall downwind of the Nab Tower, where she set a trysail for the last leg and got under the lee of the Isle of Wight before the worst of the storm and finished the race at 1100.

Bloodhound, likewise a large yacht, was about 10 miles astern of *Lutine* and to the eastward. She reported about the same forces of wind until approximately 1000, when it increased suddenly with squalls of hurricane force. A veer of the wind must have occurred when she arrived east of the Nab Tower for she was beating up westward under working staysail and mizzen when the wind increased. The visibility was reported as nil to windward and only 50 yards to leeward, owing to rain and spray. *Bloodhound* was then making $3\frac{1}{2}$ to 4 knots, 5 points off the wind. At 1030 the track of the mizzen boom started to lift, so she was left under staysail only, lying 7 points off the wind and making little headway. At 1100 the staysail split and the storm jib was set in its place. This lasted only half an hour before the luff hanks broke.

Bereft of all sail, she drifted to leeward towards the Owers Rocks. She let go her 120 lb anchor as a last resort and this ultimately brought her up short of the breakers off Selsey Bill. She was in a position of extreme danger, but her crew were taken off in good order by the Selsey lifeboat, which returned when the weather had moderated next morning to take the yacht herself in tow. Afterwards it was discovered that both flukes of the anchor had broken. By a miracle something held, possibly the stub of one fluke jammed in a rock and held her until the lifeboat arrived to tow her off, or perhaps the flukes may have broken when the anchor was being recovered by means of the lifeboat's powerful winch.

The Class II cutter *Uomie* broke a runner fitting, and next her forestay fitting pulled out of the deck. She ran off under bare pole, but a rigging screw of one of her shrouds then parted. It was a lee one and the damage was due to its flicking, but it must have been blowing great guns for this to happen with the boat under bare pole. Further evidence of the tremendous violence of this storm was that mud and pebbles were thrown on *Uomie*'s deck by the seas when she drifted to leeward of the Owers. A frigate arrived, and as there was a risk that the yacht might drive on the lee shore she had to be abandoned. Her crew were taken off by the frigate, which was manoeuvred to bring the low afterdeck of the ship alongside *Uomie* so that the crew could clamber up the scrambling nets. The frigate must have been very well handled. Happily the

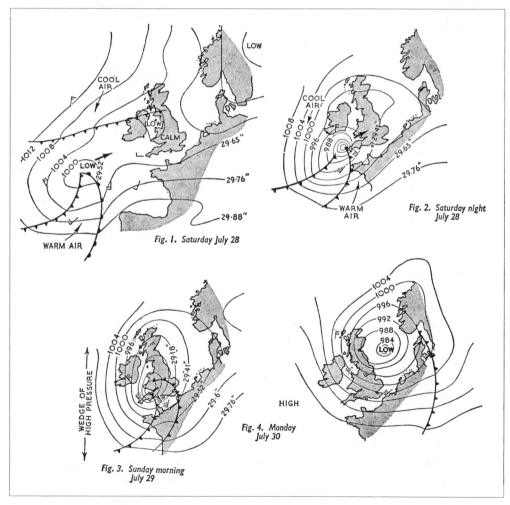

Fig 11.2. *Synoptic charts for the Channel Storm of 1956.*

yacht drifted clear of Selsey Bill and was eventually picked up by a French fishing vessel and towed into Dieppe.

The most remarkable experience in this storm was that of *Tilly Twin*, a Laurent Giles-designed 10-ton TM light-displacement yacht measuring 32 ft on the waterline, with high freeboard, reverse sheer, narrow beam, and deep fin keel, owned by Mr W F Cartwright. When *Tilly Twin* was about 12 miles north-east of Le Havre buoy on Saturday evening, 28 July, the forecast of strong southerly winds and of gales further westward was received. As the owner and crew had to be back on Monday morning, it was decided to retire from the race and to make the best of the southerly wind before it headed. A course was set direct for St Catherines, and it was proposed to use the DF loop to check the yacht's position. They intended on approach to the Isle of Wight to reach or to run off round the east side at the beginning of the easterly running stream which was due to start at about 0900 on Sunday morning. The decision was logical, but owing to a sharp change in the weather, it brought *Tilly Twin* into the height of the storm in company with the big class.

Tilly Twin had been carrying full sail on Saturday, but when the wind increased, the yankee was lowered and at about 0700 on Sunday morning (29 July) the wind veered SW and increased to force 8 (34–40 knots). The yacht was running so fast and so wildly that the mainsail had to be lowered and she continued under staysail alone.

By 0800 *Tilly Twin* had crossed the English Channel at an average speed of 7 knots in a sharply rising sea and a smother of spray. Her position at this time was estimated at 7 or 8 miles south-east of St Catherine's, but may have been less, as visibility was bad and the land above the headland was only seen for a few minutes. The weather forecast at 0740 had been worse: in Dover/Wight southerly force 8 veering SW and for Portland SW up to force 10. She altered course to the northward, intending to sail in about half-way between Bembridge Ledge buoy and the Nab Tower. By that time she was running under staysail only. Her tactics were absolutely orthodox and there appeared to be no danger, for she would arrive east of Dunnose at about slack water and soon have a fair tide.

At 0930 the yacht must have been somewhere to the eastward of the Isle of Wight, off Dunnose. Visibility was so bad that no land could be seen. The wind suddenly veered further and increased to force 10 (48–55 knots) or over. The seas steepened equally sharply. It was decided to lower the staysail and run under bare pole.

Two men were forward lowering the sail when an immense wave struck the yacht. At the time she appeared to be running directly before the wind, but the sea struck her abeam and threw her over to starboard until her mast was pointing below the horizontal.

One man forward broke his wrist and was washed overboard. The dinghy, two lifebelts and the electric flare went over the side at the same time. One of the light alloy stanchions had snapped off like a carrot. The doghouse window framing was cracked on one side and water poured through the ventilators, which were submerged. One man in the cockpit was thrown over the mainboom in its stowed position 5 ft above the cockpit floor and the other two were under water, but personal lifelines saved them from going overboard.

The ship righted herself. The man overboard had managed to preserve a hold on the staysail sheet despite his broken wrist. He was recovered safely, but with great difficulty, by the other foredeck hand.

The wind by then was phenomenally strong, and remained so for at least an hour, with gusts and squalls over hurricane force. The barometer had fallen to what was understood to have been the lowest ever recorded in that area, winter or summer, and, owing to the wedge of high pressure to the westward, the rise was abrupt after the front had passed. The seas were exceptionally high, very steep and breaking heavily with no great distance from crest to crest. They struck from directions varying over as much as 90 degrees, owing no doubt to a shift of wind. The fact that she was struck on her beam when running suggests that the wind may have veered as far as WSW.

Tilly Twin was in a dangerous position. A tanker of about 5,000 tons passed quite close to the yacht and appeared to have seen her, but she was herself having trouble, manoeuvring with difficulty in the high wind, and eventually she proceeded on her way. There was nothing she could have done to help, except perhaps send down some oil. The seas were such that she was invisible every time she was in a trough.

Tilly Twin was run off before the storm and all available warps were streamed to steady her. Her CQR anchor was also let go astern and towed on 50 m of nylon. This slowed her down, but she became very difficult to steer and to keep stern on to the huge breaking seas. Her adventures were not yet over.

Visibility at times was less than 50 yards, with no apparent dividing line between sea and air. In order to keep her stern to the seas, the course of the yacht varied between north and east, for the wind was still veering. Water was breaking over the entire hull as she scudded off before the seas, and the bilge pump was choked. At about 1100 *Tilly Twin* found herself over Selsey Shoals, with the Looe Channel buoy about a cable to port. The anchor had checked the yacht's progress a little as she crossed the shoals, where the water was full of sand and gravel, but it had little effect when she regained deeper water. On the lee side of the Boulder Bank the seas were shorter and very much reduced in height, though equally breaking. There was no respite from the wind.

The visibility must have improved, as soon afterwards the Mixon beacon was sighted, and later the beach at Bognor Regis. *Tilly Twin* ran on until, some 200 yards from the beach edge (at half tide) of Bognor Regis, the CQR anchor on the nylon rope got a hold on rocks, little more than a cable off the breakers.

As the yacht was riding by the stern the seas broke over the cockpit, so that it was full of water to the brim the whole time and there seemed every prospect of her ultimately filling and sinking. It might have been possible to have let go the other anchor, which was stowed in the forepeak, but it seemed dangerous to open the forward sail hatch to get it on deck. Even had this been accomplished and the yacht brought head to wind, she would have been battered to bits on the rocks when the tide fell later in the day. In the meantime it was urgently necessary to get the injured man ashore to hospital. It was learnt later that his wrist was so badly hurt that it took over a year to mend.

The owner-skipper decided to surf up on the beach. Accordingly, the anchor warp

Mid-Atlantic. This photograph was probably taken under spinnaker, with wind force 6, but regular seas. Cohoe surfed on crests at 12 knots or more, but what was gained on the crests was lost in troughs. Noon to noon run 177 miles on 25.5 ft designed LWL overloaded to 26 ft.

was deliberately cut. *Tilly Twin* ran on. The keel bounced on some rocks with a very severe jar, but quickly drove over them. The yacht, with her deep keel acting broadside-on as a drag, surfed gently into the shore on her beam ends, the hull itself at no time touching the ground. Her crew stepped out into about 2 or 3 ft of water. A policeman and people on the beach gave all the help they could. The injured man was rushed to hospital and at 1630, near high water, the yacht was secured fore and aft with her topsides resting on the steep shingle bank. A few days later *Tilly Twin* was lifted by Crane on to a lorry and taken to Emsworth by road.

The worst of the gale seems to have been encountered between St Catherine's Point and the Owers lightvessel. The 10-tonner, *Dancing Ledge*, which was not racing, but cruising, foundered off St Catherine's. She was a boat which my wife and I had sailed and knew well, a powerful heavy-displacement yacht rather like a big *Vertue*, capable of cruising anywhere. At the time she was lost she was under charter to Lieutenant-Colonel H Barry O'Sullivan, MC (an experienced sailing man and a member of the Royal Cruising Club), and crewed by his wife and two friends.

They had sailed out of Salcombe at sunset on Thursday, 26 August, and crossed to Ushant, where they altered course to the eastward for the Channel Islands because of fog. On the evening of 28 August they reefed *Dancing Ledge* at about 1730 and made for Cherbourg, as the forecast told of two shallow depressions. They arrived off the entrance of Cherbourg at approximately 2300, by which time it was blowing so hard that with the wind funnelling out of the harbour it was impossible to get in. Every combination of sail and engine was tried and tried again, but the gusts caught the bows and blew her off. At midnight it was decided to run off under storm jib.

Colonel O'Sullivan took the helm all night, with one or other of the two men helping. He sang most of the time and shouted out occasionally that it was marvellous outside. This was very reassuring to Mrs O'Sullivan, cooped up in the dark cabin in the din of the gale, listening apprehensively to the noise of the seas streaming past close to her ear and separated from them by a mere 1-in thick planking.

The approximate course was for Bembridge Ledge buoy, but the actual course was dictated by wind and sea, as the yacht had to be run off before the seas to avoid being pooped. This was achieved and the ship kept dry; she did not need pumping once.

At dawn on Sunday, 29 July, St Catherine's Point was sighted in the distance, but it was already blowing force 8 and rising. The necessity of keeping stern-on to the seas, coupled with a west-running stream, made it impossible to keep to the east of St Catherine's Race.

The height of the storm with 80-knot gusts was near, because it was at 0930 that *Tilly Twin* (about an hour ahead of *Dancing Ledge*) reported the sudden veer and increase in the wind to force 10 or 11 and shortly afterwards she was struck by the sea which knocked her down over her beam ends.

At 1020 *Dancing Ledge*'s storm jib blew out with a loud report, just as the overfalls of St Catherine's Race made it more than ever necessary to have steerage way to keep to the eastward. The engine was of no real use, as the propeller was out of the water most of the time.

Quoting from a letter from Mrs O'Sullivan, 'we were pooped almost immediately (the alarm clock fell on my knees at 1030). Water broke through the starboard cabin top coaming, which burst inwards. Two or more seas heaping together spilled a few more tons of water on top of us, and *Dancing Ledge* went down very quickly. I was in the cabin, which seemed to fill from every direction. I believe that the fore-stay went, and the mast heel pulled out of the step. *Dancing Ledge* went forwards and downwards, the engine running for what seemed like a long time. We hit the bottom or something hard, and the engine hatch/cabin step dislodged and fell on my feet. In the cabin full of water, and dark, I got free by wriggling my feet out of my shoes and groped out. The boom was in the way, I remember. The lifejacket took charge once I got into the cockpit, and I went up fast for a long, long way.

Once in the fresh air, we saw each other fairly soon, and also saw the dinghy, which must have broken from its chocks on the cabin top. It was upside down, but we hung on to it (aided by lifejackets) for nearly four hours and were carried inshore until we were close enough to see the window panes on the Ventnor houses. We were anxious lest we should be involved in the breakers. But the tide took us eastwards far out into Sandown Bay. There was less rough water there, just a huge swell with the crests blown off to make a thick head of spume on the sea. We knew we were invisible in this. The tide turned (we could feel) at about 1400.

Barry insisted that we should 'bicycle' continuously with our legs to keep warm and avoid stomach cramp, which it proved successful in doing.'

The youngest member of the crew had only a kapok lifejacket. This proved to be totally inadequate and as he became cold and weak his head had no support. Colonel and Mrs O'Sullivan held on to the dinghy with one hand, and supported the crew's head with the other. The fourth member of the crew had already died.

A frigate, HMS *Keppel*, approached at about this time. Colonel O'Sullivan took off his orange jacket to wave it above the spray to attract attention and this was seen as a tiny dot by the naval watch. He then attempted to put it on to the exhausted member of the crew to give him chin support, and he had therefore no jacket for himself.

When the frigate was manoeuvred alongside (no mean feat) a rope was thrown to Mrs O'Sullivan. She let go the dinghy with one hand to grip the rope. Her hand was so cold and rigid that she could not close it round the rope. She let go the dinghy with the other hand to attempt to get a stronger grip, but it was impossible to hold the rope and it ran through her hands as a wave, deflected by the bulk of the frigate, swept her along the length of the ship and she drifted away into the clear, supported by her lifejacket, and became unconscious. The frigate returned and Mrs O'Sullivan was rescued by a man, secured by a lifeline, who went into the water and got her up the scrambling nets. The search was continued for the others, but there was no trace of the dinghy or the men.

The Royal Humane Society made the rare distinction of awarding posthumously the Stanhope Gold Medal to Colonel O'Sullivan. It would be presumptuous of me to say more, but Mrs. O'Sullivan has sent the following information on points to aid survival.

'(a) If caught below, or thrown under the water, take a breath of air while you can, so that you get caught with your lungs full. (b) Take in a breath before any wave looks likely to deluge down on you when you are in the water. (c) Lifejackets must be bright orange or fluorescent to be seen in the water. They must provide support for the head, and lift the chin up, if they are to be effective for more than a short time. (d) When in the water 'bicycle' with your legs continuously to keep warm and avoid stomach cramp. Keep moving .'

Medical opinion has altered of recent years as a result of wide research on both sides of the Atlantic into survival at sea. Doctors now recommend that if an adequate life-jacket is worn and there is no place of refuge survivors should don extra clothes before entering the water if this is possible. They should stay *still* in the water, thus minimising the loss of heat from the body to the cold sea by conduction.

Conclusions

No anemometer recordings are available to me to show the exact force of the wind off St Catherine's Point on Sunday, 29 July 1956. From the records at the Lizard and Dungeness an average is given of 50 knots, which is force 10 on the Beaufort Scale, with gusts of 76 knots, well above hurricane force. It was reported to have been the worst depression since 1922, which was the storm in which *Moyana* was lost in the Sail Training Race.

To the south of St Catherine's, which is a very exposed position, the wind may have been even higher. To the north-eastward the wind aggravated by the high land behind St Catherine's and Dunnose would probably have caused gusts of greater violence when the cold front went through, possibly the 100 knots which has been said. Alan Watts has some most instructive comments to make in Chapter 30 on the meteorological peculiarities of this storm.

It is the turbulence created by the gusts striking like shots out of a gun that cause steep breaking cross seas and the rogue seas which are so dangerous. There is no doubt that the four leading racing yachts and *Dancing Ledge* met phenomenal weather and seas, which might not be encountered in the English Channel during a lifetime's sailing.

1 *Running for shelter.* In the ordinary way there would have been no great danger in running for the shelter of the Isle of Wight, for there was little warning of the sudden deepening of the depression and still less of winds attaining hurricane force in gusts. The weather conditions were unprecedented in an offshore race. The wind suddenly rose to storm force with frontal squalls coinciding with the arrival of the leading yachts on a lee shore. If this could have been anticipated, the leading yachts and *Dancing Ledge* would have adopted storm tactics and remained in mid-Channel, for the experiences confirm the rule that in exceptionally severe weather it is always safer to remain at sea and keep away from the land. But this is hindsight, as none could have foreseen what lay ahead.

2 *No uniformity in gales.* When *Dancing Ledge* had to give up the attempts to gain Cherbourg and at midnight on Saturday ran off under storm jib, only force 7 was registered at Guernsey to the westward and by *Tilly Twin* to the eastward, so it might be inferred that the wind at Cherbourg was the same force. On the contrary, it was

stated by French fishermen that force 10 was recorded locally. There is independent evidence of this, as a new 20-ton sloop, *Wawpeejay*, was caught off Cherbourg in much the same position as *Dancing Ledge*, but an hour later. Her experienced owner told me that it was entirely due to having a very powerful diesel engine that he was able to make up 'submarine fashion' against the weight of the wind. The sea was flattened and had a boiling look. It was blowing so hard that the weight of the wind funnelling down the valley and in Cherbourg Inner Harbour against *Wawpeejay*'s bare mast heeled her over to her gunwale in the gusts, even when she was secured to a mooring.

This illustrates the important point that gales are not uniform in strength over a given area, and wind forces can be much higher locally. This applies particularly when in the vicinity of land, where violent squalls and gusts can funnel down valleys, as they did in this storm at Cherbourg with a southerly wind and off the east side of the Isle of Wight when the wind was SW. The same variations sometimes occur in open water, which explains the references sometimes made in weather forecasts of gales or severe gales 'locally'.

The tragedy of *Dancing Ledge* serves as a warning that in extreme gales, with gusts of hurricane strength funnelling down valleys to the sea, it may prove impossible to beat into shelter even when within a mile or two of the harbour or inlet under the lee of the land.

3 *St Catherine's and Dunnose Races*. When caught out in a southerly or SW gale in the area south of the Isle of Wight the normal course is to make for St Catherine's. The powerful radio beacon aids navigation, if the sea disturbance is not so great that it interferes with reception, and at night the light has a range of 16 miles. On approach to St Catherine's (before reaching the race), it is in theory only necessary to bear away to the eastward to avoid the race and to get under the lee of the Isle of Wight and make harbour safely.

In practice this approach can be dangerous in gales because (a) the tidal disturbances are more extensive than they appear on the chart and rough water stretches in patches the whole way from SW of St Catherine's to the NE of Dunnose and beyond, because the bottom is irregular, and there are underwater ledges with only from 12-16 m over them and off Culver Cliff as little as 9 m. (b) In my opinion the gusts are aggravated by the high ground of the Isle of Wight. (c) It may be impossible to get shelter close under the lee of the land without entering the areas of tidal disturbances. Coming up on Dunnose in *Cohoe III* from Le Havre in 1965, during the early part of a gale which was forecast at Force 8, with thick rain and fog, we got into the overfalls before we had even sighted the land. It is better to approach nearer the Nab Tower, leaving the Princessa Shoal to port.

Much the same difficulties can be experienced at other headlands in the English Channel, where there are strong tidal streams and overfalls, such as between Anvil Point and Swanage.

4 *Pooping and speed*. In the period when *Tilly Twin* was running before the storm trailing 50 m of nylon rope and warps and a CQR anchor her owner comments 'it made it very difficult to steer the vessel and keep her stern-on to the huge breakers'. This links with the difficulty of steering *Vertue XXXV* in the Bermuda gale when she

The ketch Melanie approaching Needles Channel when mean wind velocity at Needles and St Catherine's is 40 knots, gusting 52 knots. Photo: D Ide from Cohoe III.

was running at 1 knot with a sea anchor astern, and *Cohoe*'s experience when running too slowly before a near-gale in the Atlantic. It looks uncommonly as though there is a minimum speed necessary to provide adequate steering control when running before gales, and further reference to this is made in Chapters 19 and 20.

5 *Anchoring in gales*. *Bloodhound*'s experiment in letting go her anchor as a last resort shows that an anchor may hold despite seemingly impossible seas and weather conditions. The anchor in question was a 120 lb. Thomas and Nicholson type, on 80 m of $\frac{1}{2}$ in. short-link galvanized steel chain, Lloyd's tested. When the anchor and chain were recovered it was found some of the links had started to collapse, and others at the top of the chain had stretched so much that they jammed each other and made the chain absolutely rigid for a short distance. It is amazing that they stood up to such a prolonged strain and it proves that to offer any chance of survival off a lee shore the anchor, chain, shackles and leads must be immensely strong. Probably 100 m of nylon rope, with chain at the bottom end, would have been better, owing to the greater elasticity reducing the shocks as the yacht lifted to the crests of the seas.

Tilly Twin's experience was no less interesting. It is to be noted that her anchor held until the warp was cut. When first I heard that she had been deliberately put ashore and suffered no damage I could hardly believe it. I have seen seas breaking on open shores during gales and it seemed to me impossible that the yacht was not broken up immediately. It was amazing that her crew managed to get ashore safely through the undertow. Three points should be made, however.

If the gale was WSW or had veered farther at the time *Tilly Twin* struck, there would be a bit of a lee to the east of Selsey Bill, although there would still be an onshore swell left over from when the wind was southerly. In the second place, I believe the skipper was familiar with the locality and knew precisely what he was doing and where he was putting the yacht. In the ordinary way I cannot believe it is feasible to run a yacht ashore in a gale on an open beach except as a last resort in at least partial shelter. Thirdly, it should be noted that *Tilly Twin* is a light-displacement yacht with reverse sheer and high freeboard. Her deep fin keel would act like an anchor and lying broadside-on tilted over on her side, the buoyancy was such that the hull would float in only 2 ft of water. In fact, when she piled up, the bilges never touched the bottom and consequently the hull (immensely strong by reason of her double diagonal multi-stringer construction) was undamaged. This tactic would only be possible, without damage to the hull, in a light-displacement yacht of exceptional strength with a very deep keel. The owner, who has wide experience of both light- and heavy-displacement hulls, considers that a normal heavy-displacement yacht would have been smashed to atoms when she went ashore.

The Other Yachts

The conditions of wind and sea experienced by the other yachts in the Channel Race varied according to their position on the Sunday morning. Storm force of wind seems to have been limited principally to a strip of water near the English coast. Yachts approaching this area, such as *Maid of Malham* and *Theta*, would probably have experienced force 9 to force 10 (say 40 to 50 knots) and violent gusts, but farther from the centre of the low off the French coast the strength fell to force 8 or 9. As I have stated, at Guernsey, where records often seem to be lower than elsewhere, only force 7 was recorded, but the wind at sea must have been higher.

The reports from the yachts themselves are factual and instructive. Inevitably estimates of the wind forces vary to some extent, as they always do in accounts of major gales, as owners tend to remember the gusts and squalls and to forget the lulls. The lowest estimates come from Commander Robin Foster, skipper of *Seehexe*, who reported force 7 to 8, and from Mr J M Tomlinson, owner of *Rondinella*, who wrote that he could not honestly put the gale over force 8 (34-40 knots). Both are highly experienced skippers and class-winners in this race, but undoubtedly some of the boats experienced stronger winds, at any rate locally, since gales are rarely uniform in force over such a wide area.

The estimates of the height of the seas also vary. The two highest guesses are 30 ft and 35 ft but if these were measured by eye against the height of the mast my formula of three-fifths gives a height of 18 ft to 21 ft. The lowest estimates come from *Seehexe* at 14 ft to 18 ft and *Rondinella* at 15 ft to 18 ft. Most boats reported that near-by ships disappeared in the troughs of the waves. This is, of course, due to the line of sight from an observer in a trough to the crest of the next wave forming a wide angle with the horizontal, so that vision beyond the next crest is masked off to a considerable height. Even so, the reports show that a quite exceptionally high sea was running.

Suprisingly little real damage was sustained by the bulk of the racing fleet in this severe gale. The Royal Engineer Yacht Club's Class II sloop *Right Royal* was dismasted near Le Havre buoy at 0115 on Sunday morning, 29 July, in a southerly force 3 to 4 wind and a moderate sea. The port preventer stay parted above the hook and the shock was such that the mast snapped 11 ft above the deck, and went overboard together with all sails and rigging.

The sea anchor was put out over the bow and by 0300 the top of the mast and all sails and rigging were brought back on board and secured. Bolt cutters were used where necessary to free standing rigging. The sea anchor was then recovered and the yacht was run off before the wind under bare stump, which just gave her steerage way. As the stump did not provide enough windage to enable her to be steered properly, a headsail was lashed in the pulpit to afford the windage to control her. The intention was to stay the stump of the mast and to set a jury rig of staysail and trysail as soon as dawn broke and then attempt to run back to the Nab.

The wind increased and by 0500 had veered to SW and reached gale force, so work on a jury rig was impossible. *Right Royal* ran at from 4 to $4\frac{1}{2}$ knots under the mast stump and windage of hull and the headsail stowed on the pulpit, which suggests that the strength of the wind was well over force 8, at any rate locally. All efforts were concentrated on avoiding being broached by keeping stern-on to the seas, on pumping out and keeping the best DR possible in the circumstances. The cockpit was completely filled on several occasions and the gale reached its height at about noon (an hour or two later than off St Catherine's). At this time the surface of the sea was covered by a large layer of flying spindrift.

Right Royal's danger was that the wind might veer farther to the west or to the NW, which would put her on a lee shore, but during the afternoon and throughout the night she was able to hold a course to take her through the Straits of Dover. A jury staysail had to be rigged for her to weather Cap Gris Nez as dawn broke on Monday (30 July).

Right Royal had spoken to several ships which offered assistance. This had to be refused owing to the risk of collision while attempts were made to pass a towing line. At 0500 on Monday morning the tanker *Caltex Delft* was sighted and called by a pocket torch and asked to summon the Calais lifeboat. *Right Royal* was in narrow waters, but she was manoeuvred through the Straits of Dover and reached the Dyck lightvessel and made fast to her. She was later towed into Dunkerque by the Dunkerque lifeboat. *Right Royal* had run 130 miles with the stump of her mast, aided by the headsail stowed forward. She had had a remarkable passage, but not the sort one would choose for a summer week-end, even when coupled with the fine seamanship shown by her skipper and crew.

Although a considerable number of minor breakages occurred, the only other yacht to receive structural damage was *Maid of Malham*. She was one of the larger Class II yachts and had rounded Le Havre buoy at 0325 on Sunday morning in company with *Theta*. She ran off under mainsail and staysail, logging 36 miles in four hours. When the wind increased she lowered her mainsail and ran under staysail only, but her speed was only reduced from 10 to 7 knots. At 0900 when the storm was approaching its height she was under bare pole running at $3\frac{1}{2}$ to 4 knots. She was badly pooped and about a ton of water must have got into the boat.

After being pooped she lay a-hull approximately 18 miles SSW of the Owers lightvessel. Fifty-five metres of grass warp at the end of 55 m of manilla were streamed, with a canvas bucket at the end, and she lay virtually broadside to the wind. She was less than 20 miles south of the leaders in trouble near the Owers, and the seas were reported as being spectacular.

At 1030 *Maid of Malham* was struck, when she was almost beam-on, by the only heavy sea while she was lying a-hull. It was exceptionally big and landed on the port bow and smashed about 16 ft of the gunwale. As it happened, the damage did not endanger the yacht, but it shows what a single sea can do. If it had broken aboard aft and broken the superstructure or companion doors, matters might have been more serious. When a yacht suffers damage in a gale it is usually due to being struck by a sea and literally thrown down in the trough on her lee side, so that the doghouse or coachroof is stove in on the lee side. In the case of *Maid of Malham* the damage occurred on the weather side.

Three Class II yachts completed the course. *Seehexe* was first and finished at 1835 on Monday, having been kept sailing despite the handicap of old sails. She blew out three headsails (one three times), and for a considerable period was under close-reefed mainsail. *Seehexe* took one sea aboard severely. It was so high that the coachroof disappeared from sight, but the yacht came up straight away, without damage.

Theta, who had laid a-hull in company with *Maid of Malham* at the height of the gale, finished an hour after *Seehexe,* to take second place in her class. Third place was taken by *Joliette.* She is one of only two yachts to have reported taking anemometer readings. The maximum gusts 4 ft above deck were 44 knots, which gives between 50 and 60 knots at masthead and suggests a mean speed of about force 9 on the Beaufort scale.

A remarkable performance was put up by *Rondinella*, the only Class III boat to complete the course, for she sailed throughout the gale and crossed the finishing line second to *Lutine,* beating all other competitors boat for boat without handicap. *Rondinella* is a 9-tonner, designed by Peter Brett and built by Allanson of Freckleton in 1952. She is a short-ended beamy boat with a long keel and the shallow draft of 4 ft 6 ins, which one would have thought to be a disadvantage in really heavy weather. The design is straightforward with no frills and she is very well constructed and was meticulously kept. She had no sail or gear failures other than pulling the clew out of her working jib and stripping a number of rawhide slide ties. Her crew were very experienced and non-seasick. They consisted of Mr J M Tomlinson, (her owner) and his wife, Mr J R Leggate, Miss Bibbington and Dr Hargreaves.

Rondinella rounded Le Havre buoy at 0930 on Sunday. The wind was force 7 plus (a sensible description when one is not sure of the strength, but it is more than one likes) and her mainsail was lowered and she continued the course under storm jib, finally tacking from 2 miles west of the Owers to the finish line under this sail alone. As Tomlinson puts it, 'our only tinge of worry was the size of the breakers', but as it turned out her cockpit was filled by seas only two or three times.

The 6-ton plywood *Aweigh*, one of the two smallest boats in the race, was knocked down by seas on three occasions. Her dinghy was swept off the foredeck despite strong lashings, and the motion was so violent that the Taylor paraffin stove jumped out of its gimbals and hit the deck-head above before landing on the chart table opposite. More serious was the fact she was making water nearly as fast as it could be bucketed out. Her owner traced the source to the exhaust pipe, which had sheared at its mounting in the transom, and when this had been plugged her crew felt happier.

General Conclusions

Although the findings of the RORC sub-committee were circulated many years ago, they still hold good in principle. The most important point is made in clause 2: 'In severe conditions it is far better to be caught out at sea in open water away from land influences, where, provided the vessel is well-found and not hampered by the human element, she has the best chance of coming through without serious trouble.' Two of the boats which survived this storm, *Aweigh* and *Fizzlet*, were smaller even than *Cohoe*, so the experiences of both the Channel Storm of 1956 and the Santander Storm of 1948 confirm that even small yachts can look after themselves in really severe weather.

The report goes on to say that the experiences do not help to indicate 'any preference for heaving-to, lying to a sea anchor from the bow or stern, lying a-hull, or streaming warps ahead or astern'. This opinion, based on one of the worst summer gales, will be heartening to anybody caught out in a Channel gale for the first time, but, as I shall show in later chapters, it requires qualification when it comes to storms in the ocean, and I include the Bay of Biscay during westerly gales within the definition of ocean. The RORC report should be read in detail and here I only make some general comments.

1 *High freeboard*. The yachts with high freeboard and reverse sheer seem to have been the driest. *Tilly Twin* had the worst ordeal, but her owner seems to have been well satisfied with her hull design, and indeed attributes her survival to her combination of light displacement and strength of construction. *Petasus's* skipper reported that, with her high freeboard and buoyant hull shape the yacht was very dry.*Callisto's* skipper stated that no heavy water was taken aboard at all.

2 *Navigation*. It was found in *Tilly Twin* and *Dancing Ledge* that precise navigation was impossible at the height of the storm, because the courses were dictated by the need to take the seas stern-on to avoid pooping.

Several boats reported that visibility was reduced almost to nil to windward at cockpit level by the spray and rain driven almost horizontally by the wind. Another trouble was that the height of the waves interrupted the view of distant objects, which added to the navigator's problems even in daytime and at night made identification of the characteristics of lighthouses difficult. The skipper of *Right Royal* and several others commented on this. The seas were such that the lights could only be seen on the crests of the waves, making it impossible to count the flashes or time the periods.

Most yachts confirmed *Tilly Twin's* difficulty in getting DF bearings on beacons, even on the powerful St Catherine's beacon.

3 *Stanchions and pulpits*. One owner stated that he had two experiences of alloy stanchions breaking. The leverage when a man is thrown against a stanchion, whether of iron tube, alloy or stainless steel, is so great that I think it is a mistake to place too much reliance upon its strength.

The pulpit in *Theta* broke as the yankee was lashed to it, and the seas breaking aboard put so much drag on the sail that the pulpit was carried away. It is quite common practice during heavy weather to lash down sails to the pulpit and stanchions on the foredeck, partly to avoid bringing a big wet sail below into the limited space of the fo'c'sle and partly in order to have it ready to rehoist without delay when the wind moderates. A sail thus stowed on the foredeck offers tremendous resistance to breaking seas and *Theta's* experience suggests that headsails must be stowed below if exceptionally severe gales are expected.

4 *Cockpit lockers*. The RORC reports that in some vessels a great deal of water got below through cockpit lockers. I believe this is the principal cause of the frequent need for pumping in heavy weather. Grooves and drains under the cockpit locker lids are usually useless in really rough going. Locker lids should be watertight, seated on rubber with fasteners that provide a seal; or otherwise made leakproof.

5 *Trysails*. Mr F W Morgan, owner of *Joliette*, gave a useful tip in suggesting that there should be eyes in the luffs of trysails so that luff lines can be laced round the mast below the lower spreaders as a safeguard against tracks and slides being carried away.

6 *Sea anchors*. In the report from *Right Royal* the skipper states as his considered opinion that 'if the sea anchor had been used at the height of the gale the ship would have been overwhelmed'. This is the only opinion given on the subject, but it is expressed by an experienced sailing man.

7 *Wash from ships. Ann Speed* shipped only one really heavy sea which cracked the skylight and one scuttle when the yacht was knocked down on her beam ends. Her skipper considers the exceptional sea may have been caused by the wash of a passing ship. This confirms *Cohoe*'s experience, when she was pooped, that a ship's wash crossing the big seas running in a gale confuses the pattern and causes them to break.

8 *Sail fastenings.* In the RORC report it will be seen that attention is drawn to the weaknesses in sail fastenings. In the Channel Storm a number of yachts broke the piston hanks on the storm jibs, some of which were alleged to have been crystalline. From this experience (and confirmed by my own), it is clear that large strong hanks are essential for storm sails. Indeed, some skippers consider that all storm jibs should have shackles (at any rate at the head and foot) instead of hanks.

Seizings to slides were another common cause of failure and I personally prefer to use shackles for the mainsail slides. Damage to hanks or slides was almost always due to sails being allowed to flog, either in tacking, setting or lowering.

9 *Lifejackets and safety harness.* Mrs O'Sullivan has stated that a kapok lifejacket was incapable of supporting the head of one of *Dancing Ledge*'s crew after a long period in the water. *Tilly Twin*'s owner goes further in stating that a kapok lifejacket was quite useless. Owing to the high velocity of the spray it acquired negative buoyancy after two or three hours.

In 1956 a lifejacket was just a lifejacket and I fancy few owners gave much thought to them. I must admit that I had kapok jackets in my own boats. Since then a great deal of research has been made into the subject of lifejackets and standards have greatly improved. They fall into two groups. In the first are buoyancy aids which can be worn without inconvenience, because they are not bulky. These serve their purpose in helping to keep a man afloat under ordinary weather conditions until he is picked up, provided this can be effected fairly quickly. For survival in rough weather or for a protracted period in the water it is necessary to have lifejackets to British Standards Institution or equivalent requirements which gives the full buoyancy required and support to the head.

Several skippers who were in this storm had a good word for the Hayward safety harness, which may well have saved some lives. It appears to me that when at sea, safety harness is the first requirement, as it prevents a man overboard from being detached from the yacht, so that he can be recovered, whereas with a lifejacket only he must first be found and secondly picked up, which may take time, especially if running under spinnaker. There are many satisfactory makes of harness and lifejackets. As a result of reading the *Dancing Ledge* tragedy I have recently bought sets of a combined harness and lifejacket giving 20 lbs permanent buoyancy and 40 lbs when inflated. The lifejacket is cased in nylon, which is more comfortable to wear, and is better then PVC, which sometimes gets brittle and has a tendency to crack or get punctured.

10 *Courses steered under storm canvas.* The efficiency of sails in heavy weather depends upon the type of yacht and the position of the sail or sails left standing, which varies in sloops, cutters and yawls. The following information throws light on a somewhat neglected subject.

The yawl *Lutine*, under trysail only, sailed 6 to 7 points off the wind and beat to the finish. The yawl *Bloodhound* at about force 8 made about $3\frac{1}{2}$ to 4 knots sailing 5 points off the wind under staysail and mizzen, and under staysail only at the height of the storm 3 knots, 7 points off the wind. When reduced to storm jib, after the staysail had split, she sailed to 8 points off the wind and made very little progress.

Maid of Malham made 7 knots under staysail only before coming down to bare pole, under which she made $3\frac{1}{2}$ knots until she lay a-hull.

The class III sloops *Rondinella* and *Galloper* sailed remarkably well under storm jib. *Galloper* rounded Le Havre buoy at 0500 on Sunday and might have saved her time on *Rondinella* had she not stripped all her mainsail slides and retired. During the height of the gale under spitfire jib she made 3 knots sailing 70 degrees off the wind. Her drift was about 20 degrees so she made good 90 degrees. *Rondinella* under storm jib steered 55 degrees off the wind and her angle of drift was 20 degrees. This is a most useful figure for leeway in a small yacht going to windward in a gale under storm jib, but the angle off the wind and leeway depends upon the individual boat. Much also depends upon the cut of the storm jib, as old-fashioned ones were too baggy for efficient windward work, compared with modern flatter-cut terylene ones.

Seehexe sailed under mainsail alone during the worst of the gale. With six to ten rolls reefed she headed 60 to 70 degrees off the wind and made $1\frac{1}{2}$ knots, so she was virtually hove-to. She was at a disadvantage in having old sails of odd shapes not made for the boat. Her skipper stated that in his opinion a good storm jib is far more valuable than a mainsail for getting to windward in gales. *Ann Speed*, hove-to under mainsail, reefed six rolls, and storm jib, and forereached at about $2\frac{3}{4}$ knots.

Yachts lying a-hull found they had enough steerage way to run off before the wind if desired and could then steer 25 to 30 degrees each side of the wind. *Theta*, when lying a-hull estimated the leeway at 1 knot, but I think the navigator of a yacht lying a-hull in a position where a lee shore has to be considered should be prepared for the possibility of a higher rate of leeway. *Theta* forereached at $\frac{1}{2}$ knot, but the owner of *Maid of Malham* told me that in his opinion his own boat, when lying a-hull, forereached at a higher speed than this. From this it appears that the speed may be a knot, or even more, which means that considerable distance may be covered in a long blow.

12 THE FASTNET GALES OF 1957

Adlard Coles

The Fastnet Race has a reputation for bad weather. Like the races to Spain, it starts on the last Saturday of Cowes week, early in August, and as often as not there is half a gale from the south-west blowing on that particular day.

The Fastnet is essentially a windward race along the South Coast, and when one has beaten as far as Land's End, often against a gale for part of the way, and has passed through the gap between the Longships and the Seven Stones light-vessel into the western approaches to St George's Channel and the Irish Sea the wind may suddenly veer to provide a beat of another 180 miles dead to windward to the Fastnet Rock. Once round this turning mark there is usually a lively reach, or a run, over the remaining 230 miles round the Bishop Rock off the Scilly Isles to finish at Plymouth. As the late American authority, Alfred Loomis, put it: 'If the Fastnet isn't an ocean race, it has all the worst features of such a contest plus mental hazards that have to be experienced to be fully appreciated.'

But the Fastnet is not always rough. It can be so calm that the betting would be on a 14-ft International dinghy if the rules permitted one to enter. So it was in the years when I entered *Cohoe II*. With her cut-down rig she was sluggish in light airs and we were lucky to finish even in the first half of the fleet.

In 1956 (the year of the record Channel gale) I sold *Cohoe II*, and placed an order for a faster all-round boat which we named *Cohoe III*. She was built at Poole by Newman & Sons Ltd, and designed, like her predecessor, by Charles A Nicholson. Her dimensions were 32.6 ft overall, 26 ft LWL, 9.1 ft beam and 6 ft draught. Like *Cohoe II,* she was an example of cod's head, mackerel tail hull with a fine run aft below water. Although 2 ft 5 in shorter overall than *Cohoe II*, she was much bigger, as she ended in a wide transom stern which in some respects is akin to having a larger yacht and chopping the counter off. Compared with the original *Cohoe*, of about equal length overall and on the waterline, she was more than double the size and displaced about 8 tons. She was a sloop with a sail area of 533 sq ft as designed, but when I later converted her to masthead rig set to a tubular stainless steel bowsprit, this was increased to 615 sq ft. The features of the new yacht were her stiffness and sail-carrying ability in hard weather and her roominess and strength. Everything about her construction, whether hull, mast or rigging, was above average strength and her coachroof was stiffened with steel. All this involved some sacrifice of speed, but I had hoped to sail her over for the Bermuda Race, and I remembered Humphrey Barton's advice about deck structures after his *Vertue XXXV* was nearly sunk in the Atlantic. Short-ended and sturdy looking, *Cohoe III* was one of the most powerful and durable small sea-going yachts ever built.

The first Fastnet Race she entered was in 1957 which then was said to be one of the roughest in Fastnet history. I had with me as crew Ross, Alan Mansley, Patrick Madge and Peter Nicholson, the son of the designer and a brilliant helmsman.

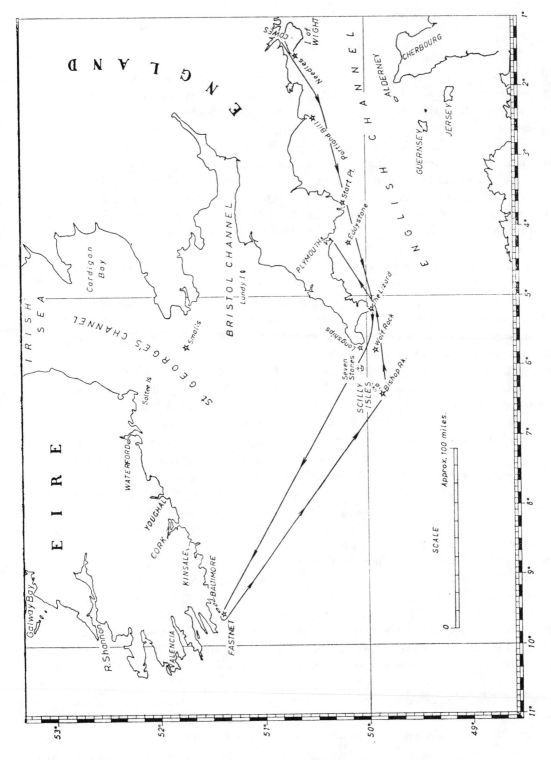

Fig 12.1. The Fastnet Course.

Forty-one yachts crossed the starting line at Cowes on Saturday morning, 10 August, in really dirty weather. It was described as a south-westerly gale. Force 8 was reported at the Scillies and force 6 and 7 at two neighbouring inland weather stations. *Cohoe III* started under staysail and had a few turns rolled in her mainsail, so, although it was said she was carrying relatively more sail than others, I do not think the wind force was more than 25 to 30 knots (force 6 to 7), gusting perhaps 40 knots.

The conditions suited *Cohoe III*, and she quickly took the lead in her class, and by the time she had reached the Needles she had overhauled many of the Class II and Class I yachts. With the ebb stream against the SW wind, the Solent was a mass of short breaking seas and the racing fleet took a dusting. I expected a big sea off the Needles, over the Bridge where the English Channel is entered and the seas build up as the tearing ebb tide meets the full force of a south-westerly. The seas proved big and broken, but not dangerous, and as *Cohoe III* smashed her way into the open Channel the waves lengthened and speed improved. Nevertheless, it was hard going in dirty conditions with rain and flying spray adding to the poor visibility.

Tack by tack and watch by watch *Cohoe III* beat her way westwards throughout the day, with her crew half blinded by flying spray, so that the helmsmen took short spells at the tiller turn by turn.

By the late evening *Cohoe III* in a mass of spray had beaten across Bournemouth Bay, past St Alban's Head, which was shrouded in driving rain, and had arrived off Portland, where she just missed her tide, with *Myth of Malham* and another yacht within sight ahead. The tides were springs, and with 4 knots against us, our speed over the bottom slowed down, and before long night fell upon us. On deck all was dark except the friendly orange glow of the compass and the reflections of the navigation lights. The cold regular four white flashes of Portland Bill lay on our starboard bow. The watch on deck were secured by safety harness, and they needed it, for every sea broke aboard forward, and in the gusts the yacht lay far over.

Down below, the aft end of the cabin was like a half-tide rock. As each sea struck the cabin top forward it came streaming aft, flooding through the aft hatch and a cabin door which had broken, as that year we had no spray hood. Both quarter berths were flooded and the chart table unusable, so I had to spread the sodden chart on the table at the forward end of the saloon.

Navigation was a whole-time job as we were skirting Portland Race, with the spare man on deck taking hand bearings on Portland Lighthouse. It was also physically difficult, as I was thrown about so much. If the chart was left for a moment it would shoot across the cabin together with the parallel rulers and dividers on to the lee berth. Progress was desperately slow, as it always is when rounding a light, but hour by hour the bearing changed and I was able to plot each position a little west of the last one.

Regularly the yacht needed pumping. Masses of water found its way into the bilges. The seas must have been getting through the cracks edging the cockpit lockers, and spray through the cabin hatch, through the broken cabin door, and through the ventilators. It is extraordinary how much water gets below during gales and it has been the same with all my yachts, even with *Cohoe IV*, which had a fibreglass hull which cannot possibly leak. We had two pumps, but the one in use was situated in the cabin with a long hose which was led into the cockpit, so that the water returned to the sea via the self-emptying drains. I took it in turns with the spare man to man the pump. It was tiring work and a joy each time that the water subsided and the pump sucked dry.

Only one man remained on deck at a time except when taking bearings, for it is a mistake to expose two to the flying spray and the cold of the night for longer than necessary. The watches were four-hourly, but the spells at the helm off Portland were fifteen minutes only. On release from the tiller the spare man would hurry below to light a cigarette to add to the fumes of my own cigarettes as I worked on the charts. There was a record fug. As Peter was the only non smoker, he must have suffered a lot, but he never complained.

Cohoe III did not slam when working to windward in rough water as did *Annette* and to a lesser extent *Cohoe I*. But off Portland Bill, thrashing to windward at 6 knots, she would very occasionally fall irregularly in a wave formation and come down on her stem with a most dreadful shock. The whole ship shuddered, the saucepans rattled and the teapot would be thrown off the gimballed stove. This was not a matter of ordinary, regular pounding as is found in many yachts, but it was the effect of nearly 8 tons weight of boat throwing herself at 6 knots over the head of a big sea and falling on her stem at a sharp angle on water that sounded as hard as a pavement. In the saloon one wondered how timber construction could stand such treatment, and after each impact I would lift the cabin floorboards to check whether there was an inrush of water resulting from damage forward, but my anxiety was uncalled for. Peter remarked that he thought for this sort of work an ocean racer should be double-diagonal planked.

As we soldiered on against both wind and tide, progress seemed desperately slow, but inch by inch we edged through off the tail of the race and by the midnight change of watch we had broken through and lay about 2 miles to the SW of Portland.

Shortly after midnight, when the new watch had taken over, I was called on deck, as the helmsman thought the yacht was carrying too much sail. When I took the tiller I found that he was right. The wind had freshened to force 8 (34-40 knots) gusting higher in the squalls, and though the yacht was sailing grandly she was hard pressed and there was a risk of things carrying away.

It was time to shorten sail. Patrick and Peter rolled the mainsail down bringing the peak to the upper spreaders; the staysail was lowered and the storm jib set. All this was done as smartly as if in daylight in the Solent. Alas! The yacht was now overreefed and I had forgotten that the storm jib (roped all round), inherited from *Cohoe II*, was cut too full to be of service as a racing sail. *Cohoe III* sagged badly to leeward and there are unlighted buoys to the west of Portland Bill. Under reduced sail we could no longer clear them and it was improbable that they would be spotted in the dark in time to avoid collision. Thus, we were forced to come on to starboard tack, bringing the weakening foul tide on our beam and thus losing much of the distance that had been so hard earned.

I have clear impressions in my memory of that outward tack. The moon appeared between the clouds flying overhead and at times the yacht and sea were bathed in light. The seas were high, but under reduced sail the boat was an easy match for them. A yacht passed close ahead of us, running east under bare poles, her port light shining

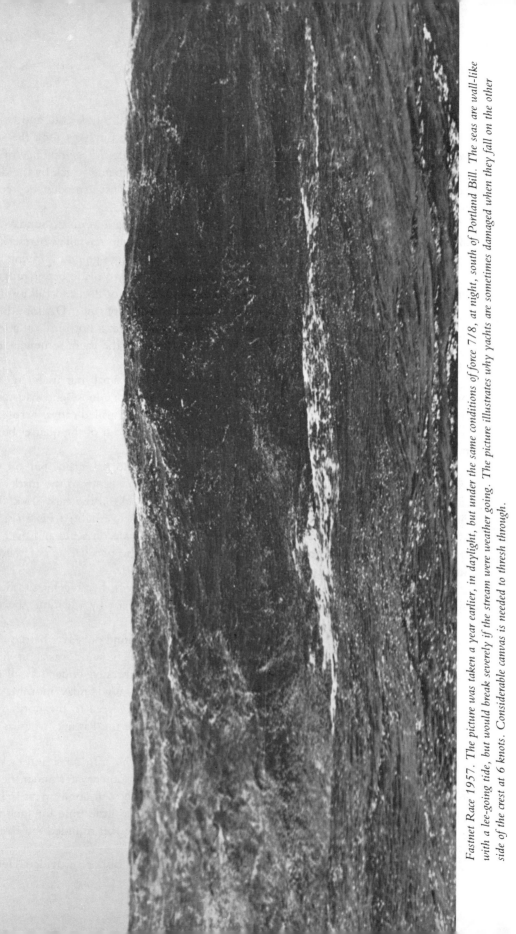

Fastnet Race 1957. The picture was taken a year earlier, in daylight, but under the same conditions of force 7/8, at night, south of Portland Bill. The seas are wall-like with a lee-going tide, but would break severely if the stream were weather going. The picture illustrates why yachts are sometimes damaged when they fall on the other side of the crest at 6 knots. Considerable canvas is needed to thresh through.

brightly. A lifeboat also passed on the same course. We tacked 5 miles out and I then handed over and turned in. Both quarter berths were flooded out, so I took the root berth in the fo'c'sle, which is supposed to be untenable in a gale. At each plunge of the boat I felt I was left in the air and put my hand up to prevent being struck by the deck above. Sleep was out of the question, though I managed to get a modest degree of much-needed rest.

The wind moderated in the early hours of the following morning (Sunday, 11 August), falling to force 7 and, after the change of watch, the staysail was reset and some of the turns in the mainsail were let out. It remained rough going in driving rain over forlorn grey seas the whole way across Lyme Bay, but we made better progress with more sail. To cut a long story short, the overreefing west of Portland Bill lost us a lot of time, for it caused us to miss our tide at Start Point. I put into Dartmouth for shelter during the foul tide. This gave us the opportunity to get a hot meal on a level table and, with the aid of two primuses, to dry the interior of the yacht somewhat and masses of wet clothes.

More important, it enabled us to carry out a number of small but useful repairs. Some of the screws in the mast track were beginning to work loose and had to be screwed up, repairs were made to the broken cabin door and I nailed canvas across the aft end of the coachroof and hatch to keep some of the water out of the quarter berths and chart table.

Once in shelter there is sometimes a reluctance to put to sea again, but on this occasion there was no delay, thanks partly to Alan, who kept us up to the mark. We left early on the tide and got past Start Point close to the rocks in the early slack. The wind had moderated and next morning (Monday, 12 August we sighted *Elseli IV*, the Swedish entry which rated at the top of the class. The race was on again and over the remaining 500 miles of the course, was bitterly contested, the two yachts rarely being out of sight of each other.

West of the Lizard the wind went light and *Elseli IV*, with her big masthead rig, gained on us steadily tack by tack. She had left us far astern by the time she had rounded Land's End.

Luckily for *Cohoe III*, the wind freshened again, and we found ourselves hard on the wind for much of the 180-mile stretch to the Fastnet Rock. The wind was about force 6 to force 7 (some say force 8), so it was a tough passage pressed under staysail and whole mainsail. Both yachts arrived off the Irish coast on Wednesday morning, 14 August, almost at the same time. Meanwhile the wind had veered to the north-west and headed us, so that our landfall was some 20 miles east of the Fastnet, whereas the bigger yachts had been able to fetch the rock on one tack.

We scored a few miles off *Elseli IV* by standing in on the Irish coast and tacking inside the Stag Rocks, thus getting an earlier fair tide and smoother water under the lee of the land. Off Baltimore we all looked longingly towards the soft contours of Ireland lying so alluringly close on the starboard hand, but there was little time for such idle thoughts, for *Elseli IV* was not far astern, and by the time we had rounded the Fastnet Rock at 1340 she was snapping at *Cohoe III*'s heels.

The run of 150 miles from the Fastnet Rock to the Bishop provided closely contested racing between the two yachts. *Elseli IV* had lost her spinnaker boom, but running under mainsail and boomed-out genoa she gained on us and steered a straighter course. *Cohoe III* carried all sail with her biggest spinnaker bellying out against the light blue sky and the dark blue white-crested seas. At 1610 we gybed and with an increasing wind she became almost unmanageable at times in the big quartering seas, forcing her to tack downwind too much to the westward. But for two hours she logged over 8 knots, far in excess of the maximum theoretical speed of a short-ended boat of only 26 ft waterline. For a while *Elseli IV* must have done better, for her log showed a steady 8 or 9 knots, and when surfing the needle was up to 11 knots. Gustav Plym describes the surfing in his book *Yacht and Sea* as 'really fantastic and something that none of us had ever experienced before in such a relatively large boat . . . it was fascinating and to tell the truth – slightly terrifying'.

At nightfall *Cohoe III* lowered her spinnaker and set her genoa. It was blowing hard at a good force 7, and higher in the gusts and the seas were building up. The night was rough and at 0515 next morning all hands were called to gybe on to an easterly course for the Bishop Rock. This was a tricky job in the seas which Gustav Plym described as 'high breaking mountains of water', because of the weight of the wind in the full mainsail and the risk of breaking something or broaching as the boom went over, but the crew managed it smartly.

When dawn came there was no sight of *Elseli IV* as we raced eastward. The wind had if anything increased, for it was blowing a good force 7, possibly force 8, and was reported as being force 9. *Cohoe III's* speed had risen again to over 8 knots. This was fortunate, because the two adversaries met again just west of the Bishop. *Elseli IV* had lost time when two hanks of her mainsail broke and she reduced sail to avoid being pooped, but she had steered direct as compared with *Cohoe III's* involuntary tacking downwind. There was a tremendous sea in the overfalls west of the Bishop, for here there was the full fetch of the Atlantic into a weather-going tide. The seas were positively tumultuous. *Elseli IV* was being driven through it all out. She made a spectacular sight and at times was almost lost to sight in the waves. Gustav Plym, her owner, told me afterwards that she broached-to twice, but neither yacht suffered any harm.

The seas were lower south of the Scillies and the gale gradually moderated throughout the day on the 80-mile run to Plymouth, though a gale warning was repeated at noon. Off Land's End the wind had softened sufficiently to enable *Cohoe III* to reset her spinnaker. *Elseli IV* could not respond owing to the loss of her spinnaker boom, and thus *Cohoe III* crossed the line with a lead of half an hour to win Class III by four hours on corrected time.

It had been a great battle between the British and the Swedish yacht which proved to be the only two small-class yachts left in the race. *Elseli IV's* owner (Gustav Plym) was elected 'Yachtsman of the Year' in Sweden and I received a similar honour in Great Britain.

Captain John Illingworth's *Myth of Malham* won Class II, but the first place overall in the race was won by the Class I American yawl *Carina*. It was a fine achievement on the part of her owner, Dick Nye, and his crew, for when emerging early in the race from the Solent into the Channel at the Bridge buoy, *Carina* fell off a sea and suffered considerable structural damage forward and her staysail tack parted. As *Carina* crossed the finish line at the end of the race her owner remarked: 'All right boys, we're over now; let her sink.' I admired his spirit.

Conclusions

Mr J A N Tanner, at that time the meteorological correspondent of *Yachting World*, says that the Fastnet Race of 1957 was sailed in variable conditions brought about by two depressions. On Friday, the day before the start, a deepening depression (982 millibars compared with 976 millibars in the Santander gale of 1947 and the Channel gale of 1956) was moving eastwards. At the start of the race on Saturday it was moving up the Western Approaches, with another trough about 300 miles astern, the winds in

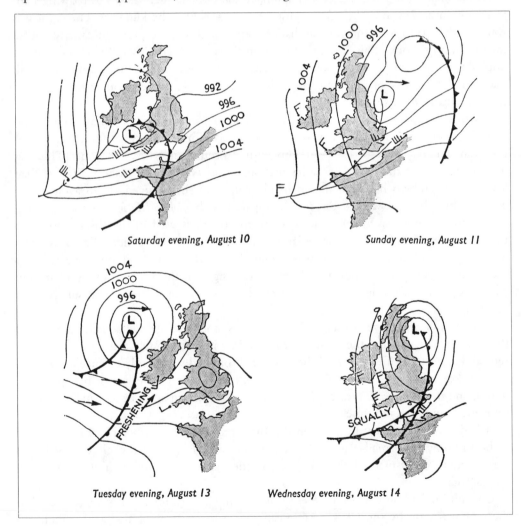

Fig 12.2. *Synoptic charts covering the Fastnet gales, 1957.*

this phase being even stronger than those which had gone before. The boisterous conditions continued for thirty-six hours and were followed by light and variable winds during Monday and the early part of Tuesday. The next depression (992 millibars) came in from west of Ireland and the wind set in from the north-west on Wednesday 'sufficiently boisterous to say that the race which started in a gale, also ended in one, and with nothing much different in between'.

Out of the forty-one starters in all classes only twelve completed the course. *Galloper* lost a man overboard. He had come on deck without a lifeline to empty a gash bucket and a sudden lurch sent him over the side. The one lifebelt thrown was so light that it was blown downwind too fast for him to get it, but he was able to secure a hold on the second buoy which was a heavy, sodden one of horseshoe type. It was only by superb seamanship under such wild conditions that he was rescued.

Inschallah had her deck-house ports stove in by a sea some 2 miles off Portland Bill and after exhibiting distress signals was escorted by the lifeboat into Weymouth Harbour. Maybe this was the yacht we had passed off Portland. *Maze* had the swaged end pulled out of a lower shroud and was dismasted. *Evenlode* incurred damage to her rudder, five of *Drumbeat's* winches broke under the strain and *Santander* developed a serious leak, found later to be due to a fractured keel bolt. Broken stays and torn sails were common and accounted for some of the retirements, but no doubt the principal cause was the old seafaring malady of seasickness. Lessons learnt were:

1 *Man overboard.* The accident in *Galloper* illustrates that personal lifelines are needed even for temporary visits to the deck, when driving to windward against gales. A sea must have struck *Galloper* on the windward side and I suspect that her crewman was catapulted overboard just as Geoff and I were in the Santander Race. The lee side of the cockpit is particularly dangerous and crew members coming up from below to empty the gash bucket or in a rush to avoid being sick in the cabin often forget their lifelines, which are laborious to don. The experience also suggests that two kinds of lifebuoy are required, light ones for use in ordinary conditions and a heavy one for use in high winds. Alternatively, a lifebuoy should be attached to a small drogue.

2 *Damage.* Long periods of beating into gales find out any weaknesses in hull, rig, sails and gear. Moderate winds prevail in most ocean races and hence some designers tend to keep everything above deck as light as possible. This lightness pays in ordinary

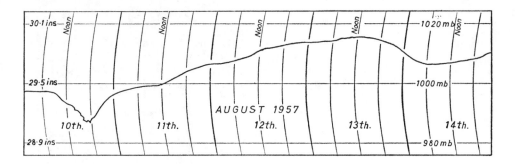

Fig 12.2. Barograph trace, Fastnet Race, 1957.

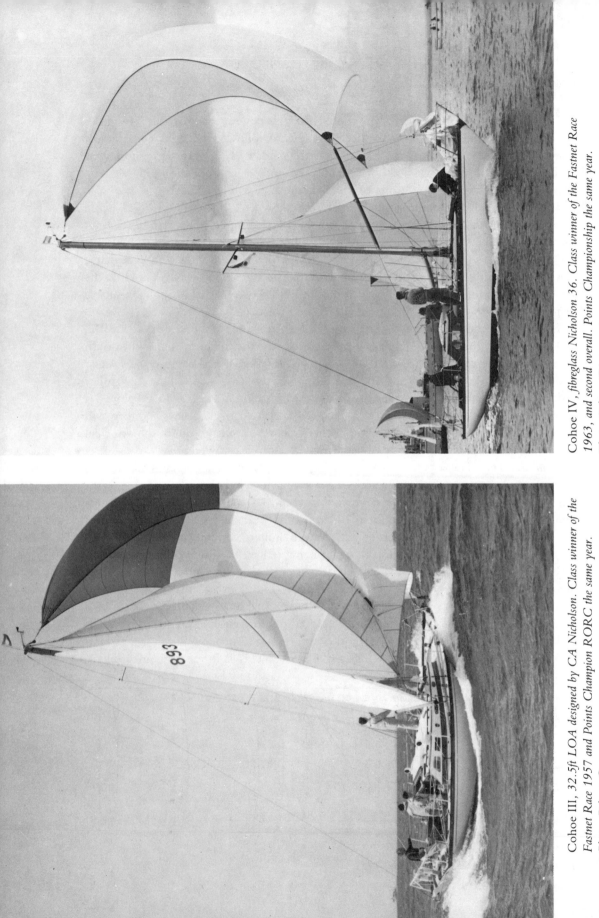

Cohoe IV, fibreglass Nicholson 36. Class winner of the Fastnet Race 1963, and second overall. Points Championship the same year.
Photo: Beken, Cowes

Cohoe III, 32.5ft LOA designed by CA Nicholson. Class winner of the Fastnet Race 1957 and Points Champion RORC the same year.
Photo: Beken, Cowes

racing, but a considerable factor of safety is required for punching to windward through gales of this kind.

3 *Pounding and falling off a wave.* In the old International metre classes the hull form was given to pounding and they did not appear to suffer much as a result. Ocean racers with their shorter ends suffer less from this complaint, but when they do so the shock is, or appears to be, much greater. John Illingworth in his book *Offshore* remarks that 'one wonders that every fastening in the boat is not loosened', but he adds 'the worst you are likely to do is to disturb the caulking in the forward part of the boat'. John Illingworth should know, because nobody has longer experience or drives a boat harder, but his ocean racers are designed and built for the job of hard racing.

Furthermore, there is a difference between pounding and what is called 'falling off a wave'. That is, breaking through the crest of a big sea at 6 to 7 knots so that the yacht literally falls into the trough on her stem. It was this that caused structural damage to *Carina*. A similar mishap occurred when *Bloodhound* fell off a wave off Berry Head. The impact damaged her stem scarf and caused a leak. She hove-to until dawn, when the extent of the damage could be identified, and she was able to carry on in the race, pumping only every four hours. *Elseli IV* broke three frames off Portland and it was thought that she had broken seven planks, but the damage was later found to be limited to surface splits at the nail-heads. Other yachts suffered damage, and it was this 'falling off a wave' that caused me occasional anxiety in *Cohoe III*. Much depends on the helmsman, who must luff to the crests and bear away at the top to let the boat down lightly in the trough. With a good man at the wheel the watch below will be able to relax, but with an inexperienced helmsman the impacts with waves will be terrific, enough to crack the heads in the fo'c'sle. The boat should be kept weaving in the seas, dancing all the time in the zest of it.

It used to be said that a yacht could take more punishment than her crew, but nowadays it is possible that, with picked crew regularly relieved, and able to drive a yacht all out against gales, damage can result. It is no longer certain that the ship is stronger than her crew. The forward sections of an ocean racer and such matters as the scarf, frames and keel bolts certainly merit special consideration on the part of the designer.

4 *Sheltering.* No cruising man, if he is in his right senses, will continue to beat to windward for long in force 8 if shelter is at hand. Ocean racers have to plug on, but I think it sometimes pays to shelter, as *Cohoe III* did, rather than beat round a headland against a strong foul spring tide. If one takes the strength of the foul stream as lasting four hours, all one has to do is to calculate the distance it will set the yacht back according to the tidal atlas and compare it with distance likely to be made good tacking at a wide angle to the wind on account of rough seas, surface drift and leeway. The answer depends upon the size, design and stability of the individual boat, but if little is to be gained by keeping at sea it may pay in a long race to give a respite to the crew, provided no time is lost in getting under way again when the stream weakens.

13 SWELL EAST OF USHANT

Adlard Coles

Swell is defined as being a wave outside its own original area of generation. It may confuse the pattern of the seas during a gale, but otherwise it is not associated with heavy weather sailing. On the contrary, the wind is often light when swell is at its worst.

In the ocean or in open sea there is no danger in swell. The yacht goes up and up and she then goes down and down. The boom jumps madly unless secured by a foreguy; the sails flog and flog. Wear and tear of sails and gear in a big swell during a prolonged calm may be greater than in a gale. Nothing is more sick-inducing than swell on a hot, windless day. Everybody is glad when it is over. However, swell is induced by bad weather somewhere. It occurs often as a fore-runner to a gale or is left over after a gale has passed on, so it has its place in this book.

In coastal waters swell is not entirely without its hazards. There are three. High swell finds its way into otherwise secure anchorages; it rises in height almost to point of breaking in shoal and rocky waters; it affects visibility and makes identification of a landfall difficult.

Ground swell is more pronounced on certain coasts, such as those facing the Atlantic. I have never seen a really high swell in the English Channel on the South Coast eastward of Start Point, but it is often encountered on the French side. West of the Casquets and Guernsey it is common. Off Ushant there is often a heavy swell, and in the Bay of Biscay ugly swells occur on the Atlantic coasts of Belle Ile, Ile d'Yeu and the other islands. On the north-west side of Belle Ile there is an anchorage which we sometimes use at Ster Wenn, in a narrow fiord branching off southwards in the rocky bay of Port du Vieux Château. Here one would expect a yacht to be safe, snugly sheltered by high land from east through south to north-west, at any rate in normal weather. Nevertheless, it is said that even on a fine day, swell may rise unexpectedly from some distant Atlantic storm and, by refraction and diffraction, enter the inlet. The swell may rise to such dimensions that is is said no anchor will hold.

Where I have found swell most pronounced is in Brittany between Ushant and Ile Vierge off the harbour of l'Abervrach. I thought this might be just luck until I met a very experienced French yachtsman at Roscoff who confirmed that it was a recognized local phenomenon. He added that another locality noted for swell was in the Bay of Lannion, which lies farther eastwards between Morlaix and the Sept Iles. So I think we can accept it as a fact that swell is more commonly found in some areas than others.

I will give a brief account of the swell which can be found east of Ushant.

In September 1957, my wife and I were cruising *Cohoe III* and the yacht was lying in the sheltered anchorage of l'Aberildut, east of Ushant, near the northern entrance of the Four Channel. It had been blowing fresh when we had arrived at midday. In the afternoon (6 September) the wind freshened to about force 6 and during the night the wind howled in the rigging, so that we were glad to be in a safe anchorage. At 1100 the following morning (7 September) the gale was over. It was sunny and the wind light to moderate. There was a forecast of gales and severe gales in area Sole to the westward and in the Fastnet area miles to the north-west, but none for Biscay or the Plymouth area in the south of which we lay.

I did not then know l'Aberildut well, and in the morning I rowed in the dinghy to the entrance to have a look at it at low water.

There is a narrow bottle entrance between a promontory on the southern shore and a high rock (Le Crapaud) on the northern side. Here there is a stony bar which nearly dries out at low water. Westward of the bar is an approach about a mile long and varying from 1 to 3 cables wide, between reefs of rocks to the north and south. There is Le Lieu beacon tower on the north side and an iron beacon on the outer rocks on the south side. I found that the bar was completely sheltered, but from the land I could see that a spectactular swell was sweeping in from the Atlantic. Huge waves were breaking higher than the off-lying rocks.

I had secured *Cohoe III* to a mooring, but at noon the rightful owner in the shape of a *gabarre* (ballast boat) returned unexpectedly, so we had to vacate the mooring and temporarily tie up alongside her. She was a vessel of about 50 tons, manned by a crew of three.

Hearing that we were about to get under way, they tried to dissuade us because of the swell. 'La Houle,' they clamoured indicating dramatic rolls with their hands.

'Difficile, mais elle n'est pas dangereuse,' I replied in my best schoolboy French.

'Mais les rochers,' they argued. When my wife explained that we were sailing for l'Abervrach to collect letters they became more emphatic about 'Les rochers dangereux'. 'Demain,' they said. 'Rester demain.'

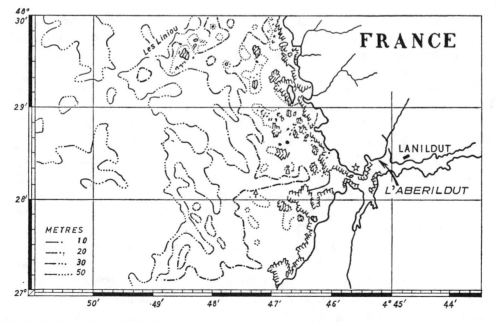

Fig 13.1. Approaches to l'Aberildut.

121

I have a great respect for Breton fishermen, from whom the crew of the ballast boat were probably drawn. They are as tough as nails. I often ask their opinions about the weather, but on this occasion we had a light or moderate fair wind and a fair tide and less than 20 miles to sail.

After lunch we cast off at 1400 (7 September) as planned. It detracts from this yarn, but I must confess we used the engine against the foul tide, which was a treat, as it did not always work.

We were soon at the entrance. It was narrow but quite easy and protected. Outside in the approach channel the rocks to the south at first gave protection, but swell was breaking heavily on the north side. There were fishermen on the shore and they appeared to be very agitated. *Cohoe III* must have made a picture on that sunlit summer afternoon in the narrow channel between the flashing white breakers.

The tricky part lay at the outer end of the approach channel. The swell was thundering on the rocks to port and the spray was being thrown in clouds into the sky. It was worse to starboard, for the tremendous seas were bursting high over the rocks marked by Le Lieu beacon tower.

There were spectacular seas on both sides, but my wife spotted the narrow channel left between the rough water on either hand. It was covered with great masses of foam the whole way across. I have never seen anything like it.

The yacht was safe enough because the water was not breaking in the fairway, though she was plunging in ever-increasing swell. It was not alarming, but I wanted to get clear of the proximity of rocks as quickly as possible. I set the mainsail and genoa. That speeded *Cohoe III* up. The wind was south of west and she could just lay the course against the westerly swell, so that under sail and power she was soon in the open.

Once outside we realized the full size of the swell into which we were heading. We stopped the engine and sailed on over the hills and down the valleys. All was not well, however. As we drew out we found the swell was so steep that in places it was on the point of breaking.

To starboard it was actually doing so, and away less than a mile and a half to the north-west lay the grim Les Liniou group of rocks. Here the picture was absolutely spectacular, as the swell reared up high into the air against them and the rocks in the whole area on the shoreward side. I see a note in my diary: 'At Les Liniou the spray made a fine mist the whole way from the rocks to the shore two miles away, with rainbow effects from the sun.' It was most remarkable.

Progress was slowed by the swell and we soon found we were being set down to leeward rapidly and unwillingly towards the 'rainbow effects'. The streams run very hard on this coast from the Four Channel.

I started the engine quickly again and it fired at the first press on the button. With this help the yacht cleared the dangers by an adequate margin. The swell still demanded attention, however. Hereabouts, though the water is deep the bottom is rocky and irregular. As a result the huge swell coming in straight from the Atlantic became exceptionally steep, seeming to be on the verge of toppling, though it did not in fact break.

Once well clear of Les Liniou we set a NW course to get out into deep water and give wide clearance to the rocks near the Four lighthouse.

This brought the wind nearer the beam and sheets were freed accordingly. The strange thing was that the mainsail then became completely useless, despite the boom being secured by the foreguy. The yacht was rolling so violently in the swell that, as she heeled to leeward to each wave this caused a draught of air against the lee side of the mainsail. The sail did no good at all and had to be lowered. The wind was then only about force 2.

We sailed on under genoa alone, which, being lighter, did not spill so much, but we had to use the engine again.

With a fair tide we made good progress to the outer Porsal buoy, where we altered course to the eastward for the entrance of the l'Abervrach channel, only 5 miles distant. Two miles to starboard lay the Roches de Porsal, a particularly grim series of rocks extending nearly 2 miles from the shore. The swell was breaking in magnificent formation over the rocks, and as the waves struck the lighthouse the spume rose three-quarters of its 52 ft height.

With the swell now brought aft, conditions for a while were much more comfortable, but we had a few thoughtful moments in the approach to l'Abervrach. The approach is flanked by the reef of rocks known as Le Libenter on the north side and the extensive rocks on the south extending some 2 miles offshore and marked by the spidery black and white buoy, La Petite Fourche. The channel has leading marks and is well buoyed. It is said to be safe under all conditions, and l'Abervrach is a regular port of call for most yachts bound south for the Bay of Biscay.

Our difficulty was that the swell was running so high that it interrupted our view of the proposed landfall. It is true that there was a bird's-eye view on the crests, but it was only a short one before it became blocked out by the wave as it rolled on and left the yacht in the hollow. The binoculars were useless, because before they could be steadied on anything the view had gone. The tall Ile Vierge lighthouse (249 ft) rose high on the port bow, but ahead lay nothing except big breakers with no gap that could be distinguished between them.

The yacht ran on, with the breakers heavy on the outlying rocks to starboard, and the white plumes of the seas growing higher ahead as we drew nearer. I suppose it was not until we were half a mile off that we sighted La Petite Fourche and soon afterwards the Le Trepied buoy glinting red among the blue seas and white breakers.

Once either buoy is identified, the rest becomes easy. We were soon in the channel, but the ebb tide had started, so that though the swell was lower it became remarkably steep and threw itself furiously in a confusion of flashing white breakers on the Libenter. To the west of Ile de la Croix, which marks the first turning-mark in the channel, there is a big plateau of submerged rocks and over this the seas were breaking wildly. In the channel itself it became rough, and the smaller seas were breaking the whole way across between La Croix and the Petit Pot de Beurre beacon tower and to the northward right across the Malouine Channel. However, for *Cohoe III*, in the centre of the main channel, all problems were over and she sped on soon to find sheltered water and bring up in the anchorage for a pleasant warm and sunny evening.

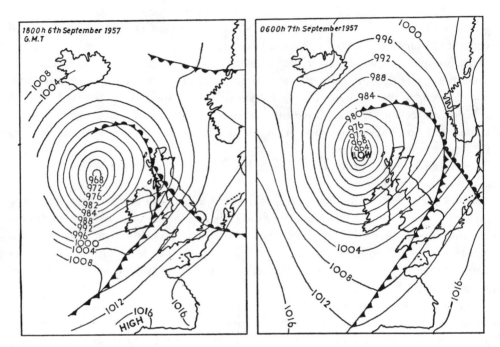

Fig 13.2. Synoptic charts, 964 millibar depression.

Conclusions

This story may sound as tall as the swell. The day was sunny, the wind light and no water broke over the yacht. If we felt any anxiety it was perhaps only because we had guilty consciences after ignoring the local warnings. Nevertheless, the passage did give us a close-up view of high swell where it meets ledges of rock and shallowing water. The scene was so impressive and at times so beautiful that on no account would we have missed it.

From the synoptic charts it will be seen that an intense depression (964 millibars, which is lower than any other mentioned in this book) had moved from NW of Ireland and was centred NE of Scotland by 0600 GMT on 7 September. The cold front would have passed over l'Aberildut an hour or two later and the wind soon moderated, for it will be seen that the isobars near Ushant became widely spaced. But the swell from the exceptionally severe storm to the NW would remain and only subside gradually.

The experience illustrated two useful points:

1 In a high swell with a light wind a *sailing* yacht can become almost unmanageable and lose headway. This can be dangerous in rocky areas if the tidal streams are strong and auxiliary power is not available.

2 High swell, even in fine weather, makes pilotage difficult owing to the interruption of vision.

14 OTHER FASTNET GALES

Adlard Coles

A remarkable kind of gale occurred in the Fastnet Race of 1961 (see chart, page 110). It caught most of the ocean-racing fleet between Mount's Bay in Cornwall and the Longships off Land's End. The gale was very localized and the leaders among the big yachts were sufficiently far ahead to avoid more than force 6 or force 7 as they sped into the Irish Sea, but the main body of the fleet were properly caught out in a potentially dangerous position between Land's End and the Scilly Isles.

The cause of the trouble was a small but intense secondary depression which moved quickly in from the Bay of Biscay. Gale warnings were received in advance. The barometer was falling smartly, the wind had backed to the SE and the sky had become heavy with rain when, in *Cohoe III,* we were off Mount's Bay in company with several other yachts of all three classes on Monday afternoon, 7 August.

It was not until the late evening, when we were about 5 miles north of the Wolf Rock lighthouse, that the gale really caught us. In the previous year *Cohoe III* 's sail plan had been altered by the addition of a masthead rig and a stainless steel bowsprit which enabled a huge spinnaker to be carried. We hung on to this big sail, because I had a stout-hearted crew, who optimistically hoped that we might slip ahead of the worst of the gale into the Irish Sea, for which the weather forecast predicted only force 6 winds. The leaders among the big class, probably 20 miles ahead, succeeded in doing this, but we had no such luck.

The wind rose quickly. *Cohoe III*, only $32\frac{1}{2}$ ft. overall, pressed by over 1,000 sq ft of spinnaker, became almost unmanageable. No sooner than I had given the order to lower it, at 1935 BST (1835 GMT), and as the watch below were tumbling out on deck, *Cohoe III* broached-to. In an instant the spinnaker split right across and the halyard had to be let fly to relieve the yacht of the strain. The pressure (probably nearly 2 tons), was so tremendous that the 4 in x $\frac{1}{2}$ in galvanized steel bolts which took the sheet blocks bent and drew half an inch through the toe-rail and the planking below, leaving half-inch fractures in the upper plank.

The spinnaker was quickly recovered, but, while the mainsail was being roller reefed, the leach line got snarled up in the folds and it had to be lowered. It was cleared by Tim Laycock, one of our crew, poised dangerously on the end of the transom. In these few minutes the wind had increased so violently that the new terylene storm jib had to be set, and under this small sail alone the yacht ran as fast as she had under spinnaker.

Everything was blotted out by rain. The gap between the Longships and the Seven Stones lightvessel is 12 miles wide. We set a compass course and raced in the gathering gloom seeing nothing at all, though we must have passed near the Seven Stones lightvessel (where the tanker *Torrey Canyon* came to grief in 1967) before emerging into the open sea in the Irish Sea approaches. The barometer touched its lowest at about midnight and the wind veered to SSW, followed by a veer to NNW two hours later.

I shall always remember that night. It was intensely black. The rain was absolutely torrential and the visibility nil. The wind was exceptionally strong and gusting well above gale force. The odd thing was that the seas never ran high. This was because the depression, although intense, was small in area and moved so quickly that the seas never had a chance to fully build up. The violent rain also flattened them.

I also remember wondering what would happen if any yacht ahead of us had hove-to. The visibility was so bad that we might have run her down. Equally, had *Cohoe III* laid-to she might have been run down by an overtaking yacht. We held on course, as in any case, we wanted to get clear of the land as fast as we could. I also wondered what would happen if one of the tankers bound for Bristol crossed our path. I wished we had *Cohoe I*'s false bow to take part of the impact. This gale was very comfortable for me navigating below, as the cabin was dry and with a following sea the motion was tolerable, but it was hell on deck for the crew, striving to keep a look-out in the blinding, searing rain. There was one brilliant flash of lightning, but it lit nothing but the rain-swept smoking sea.

When daylight came the wind moderated and the rest of the race was uneventful.

Conclusions

This gale was caused by a vigorous secondary of 996 millibars forming off the Bay of Biscay and moving rapidly across to the Scillies, intensifying as it went. The centre crossed immediately over the ocean-racing fleet at a pressure of 992 millibars, giving rise to sustained winds of force 9, possibly force 10 for a short while, with gusts approaching hurricane force.

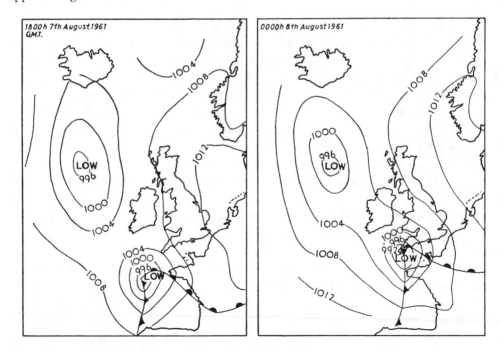

Fig 14.1. Synoptic charts covering the Fastnet Race, 1961.

The gale caused some anxiety to the Committee of the Royal Ocean Racing Club, because the yachts were caught in narrow waters between Land's End and the Scilly Islands. When the centre of the depression went through there was a risk of the shift of wind putting some of the yachts, which were under bare poles, on a lee shore. However, there were few mishaps and the tail of the fleet sensibly took temporary shelter or retired, rather than go through the gap when visibility was nil and with the prospect of a lee shore.

I have tabulated the wind direction and force for each hour (that is, the mean velocities over the past hour as recorded by anemograph) of the gale at the shore stations at Scilly and the Lizard, which are as follows:

Monday, 7 August 1961

Hourly values as recorded by an anemograph

	SCILLY			LIZARD		
GMT	Direction	Knots	Force	Direction	Knots	Force
1800	SE	17	5	E by S	21	5
1900	ESE	18	5	E by S	23	6
2000	SE by E	23	6	ESE	25	6
2100	SE by E	23	6	ESE	26	6
2200	SE by E	21	5	SE by E	33	7
2300	SSW	21	5	SE by S	32	7
2400	SW by S	21	5	S by E	26	6

Direction changed suddenly 2330 and 2345
Highest gust 2359, 52 knots   Highest gust SE, 49 knots

Tuesday, 8 August 1961

	SCILLY			LIZARD		
GMT	Direction	Knots	Force	Direction	Knots	Force
0100	N by W	35	8	S	24	6
0200	NNW	34	8	SW by S	15	4
0300	NW by N	32	7	WNW	40	8
0400	NW	29	7	NW by W	41	9
0500	NW	26	6	NW by W	35	8
0600	NW by W	27	6	WNW	28	7
0700	NW by W	24	6	WNW	30	7
0800	WNW	21	5	W by N	28	7

Direction changed suddenly 0155
Highest gust 0100, NNW 53 knots   Highest gust 0235, NW by N 63 knots

These records from the coastal stations are most interesting, but what we need to arrive at is the strength of the wind at sea, in the gap between Land's End and the

Scillies. Making a deduction of 25 per cent for height and multiplying by the appropriate factors, I arrive at the wind at sea at night to have been as shown in the right-hand column of the following table.

Time 8 August	Wind at coastal station		Wind at coastal station less 25% for height		Multiplication factor for wind at sea	Estimated wind at sea	
Scilly	Knots	Force	Knots	Force		Knots	Force
0100	35	8	26	6	1.8	47	9
0200	34	8	26	6	1.8	47	9
0300	32	7	24	6	1.8	43	9
0400	29	7	22	6	1.8	40	8
0500	26	6	19	5	1.6	30	7
0600	27	6	20	5	1.6	32	7
Lizard							
0300	40	8	30	7	1.6	48	10
0400	41	9	31	7	1.6	49	10
0500	35	8	27	6	1.8	48	10
0600	28	7	21	5	1.6	34	8

This gives an average of 44 knots (force 9) for three hours to 0300 off the Scillies. The full force of the gale was felt at the Lizard about two hours later and averaged 48 knots (force 10) for three hours to 0500, but this could be an overestimate owing to the great height of the anemometer at the Lizard, which is 240 ft above sea-level. Force 9 is probably nearer the mark.

Unfortunately I have only one record at sea and that is from the Seven Stones lightvessel in the middle of the gap. It is at 0600 on 8 August, after the worst was over, but the hourly value was force 8, which is two grades higher on the Beaufort scale than at the shore station at Scilly, only 10 miles to the westward, but almost agrees precisely with the Watts method of computation. This once again emphasizes that the wind at sea is higher than at shore stations, especially at night, though the gusts at sea are not necessarily higher than over the land. It also explains why in the Dinard Race of 1948 the ocean racers reported force 7 when only force 5 was registered at the shore station at Guernsey Airport only about 10 miles to leeward.

The features of the gale were the gusts and squalls when the wind veered, which are shown on pages 126-7 and also in the anemograph (page 129). For example, following the wind shifts at the Scillies near midnight on 7 August there were gusts up to 52 knots, but these were apparently not enough to put the average of the past hour above 21 knots, although wind increased in the following hour to a mean of force 8. At the Lizard there was a gust of 63 knots (nearly hurricane force) and several of near that velocity with the veer of the wind. These gusts were at more than double the mean

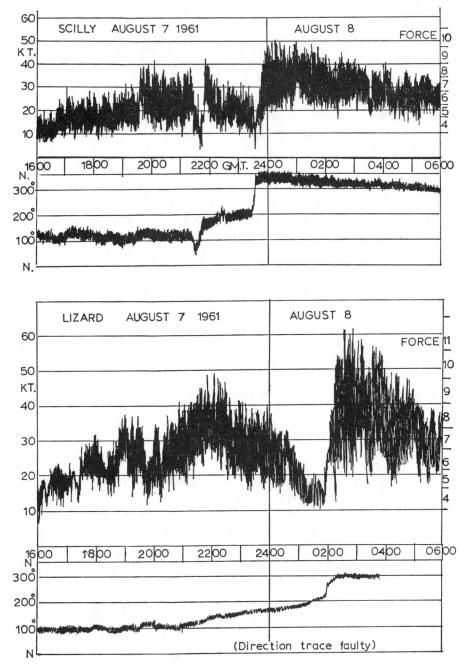

Fig 14.2. Anemogram traces, Scilly and Lizard, 1961.

velocity for the previous hour, but they may have been less at sea, where the wind is steadier, though the mean force is higher than over the land.

I mention that at no time did the seas run particularly high and on this occasion I can give the measured height. The figures from the Seven Stones lightvessel show that at 0600 on 8 August, after the gale had been blowing from NW for some hours, the period was 8 to 9 seconds and the mean height of the waves was 11 ft.

The following points are of interest:

1 Any boat which set more sail in the weaker winds preceding the veers would have been badly caught and possibly knocked down on her beam ends in the violent squalls which rapidly followed.

2 The anemograph shows that it is impossible to estimate the mean wind force on the Beaufort scale, which might be logged as a near hurricane, force 11, 10 or 9 according to the temperament of the skipper.

It shows why gales are so often described, even by experienced sailing men, in relation to the strength of the gusts rather than the mean force on the Beaufort scale. Without an anemometer it is sheer guesswork and later in this book this opinion is endorsed by experienced yachtsmen on the other side of the Atlantic.

3 *Quick reduction of visibility*. The speed at which the wind can rise in a small intense depression is remarkable. Visibility can be reduced equally quickly in the belt of rain accompanying it.

4 *Sail area*. Although *Cohoe III* sailed at her maximum speed under storm jib alone she would have been better with a corner of the mainsail left set. It would have enabled her to have pointed closer to the wind when it veered. This would have been of importance if, when the centre of the depression had passed through, the shift of wind had put her on a lee shore, but she was in open water by that time.

1963 – Inside Portland Race

All records up to that year were eclipsed in the Fastnet Race of 1963, when for the first time entries were well in excess of a hundred, and the Admiral's Cup was recovered from America, against hot international competition.

At the start of the race there was half a gale of wind from the south-west, which was quite enough with from forty to fifty yachts in each class milling around, battling to get into the narrow fair eddy inshore off Cowes Green. It was not surprising that in the mêlée there were two collisions, one of which was serious. The sailing conditions were similar to the early part of the Fastnet Race of 1957, but with less wind.

I was racing a new Class III yacht, the 11-tonner *Cohoe IV*. She was one of the successful class of fibreglass Nicholson 36-footers, a boat of about the same displacement and waterline as *Cohoe III*, but drawn out to provide a more easily driven hull. With greater overall length and beam and with higher freeboard she was a much roomier boat, and her longer immersed waterline when heeled gave her a higher theoretical maximum speed, but there was not any very marked difference in racing performance. The principal difference was that she carried a total crew of six instead of five, which made ocean racing lighter work, especially as her sail area was smaller. I had bought *Cohoe IV* for entry in the Bermuda Race 1964, but the CCA rule was altered, as a result of which she was too small, my mistake always being to build too near the minimum size eligible for that race. Accordingly, I sold *Cohoe IV* at the end of the season and kept *Cohoe III*, which my wife preferred for cruising, since she is more comfortable for two.

In the Fastnet Race of 1963 *Cohoe IV* started under small genoa and two rolls in her mainsail, the wind being about force 6. In the rough and tumble of breaking weather-going seas in the west Solent, and off Anvil Point and St Alban's Head, she worked out a lead in the same way as *Cohoe III* had done in 1957. In fact, she proved herself to be rather better, as off Portland Bill she was lying sixth in the whole fleet on elapsed time, boat for boat without handicap.

However, *Cohoe IV* probably saved at least an hour by taking the inside passage in the gap between Portland Bill and the Race. When bound westward in moderate weather this affords a short cut with a fair tide between one hour before and two hours after HW Dover. What is not so generally known is that this passage can also be used (with care and under suitable conditions) against a foul tide. There is a southerly eddy along the SE side of Portland as far as the Bill, which is where the danger lies as the stream sets strongly into the Race.

Besides Ross (who knew Portland and the Inner Channel well) I had a strong crew in Alan Mansley, Dr Rex Binning, David Colquhoun and Keith Hunt, so no one was perturbed at the idea of the short cut at night through the unlit channel at the wrong state of the tide.

When *Cohoe IV* approached Portland breakwater at 2236 we were lucky, as the wind moderated to little over force 5, permitting the mainsail to be unreefed and the big genoa to be set in smooth water under the lee of the peninsula. The lights on the breakwater were close when we tacked, and behind them lay the bright lights of Castletown with the high dark land mass of Portland itself fading into the sky to the southward. Four of us were on deck: the watch of two with Ross to hand the genoa sheets round the mast and act as local pilot, while I navigated.

At first progress was slow, as we had a weak foul tide against us, but off Grove Point we came into the fair eddy and passed through modest overfalls. The interesting part lay between Grove Point and the Bill, as we tacked between Portland and the Race. Except for occasional breakers it was not rough inshore, because the overfalls in the Race of Portland formed a kind of breakwater. On the offshore tacks the brilliant group flash of Portland lighthouse gave us something on which to take bearings. The inshore tacks were more difficult. The closer we could get to the cliffs the better, but here Portland light was obscured and it was difficult to judge distance at night. We used the echo sounder constantly. Soon we had over a knot of fair eddy, but the farther south we got, the nearer to the land we had to keep.

Progress as we approached the Bill was increasingly fast and each tack had to be shorter, for the Race lay close to the south. At the Bill we came close in to the rocks. I have never viewed the lighthouse at night from so near at hand. It was a lovely sight, with big windows, the upper one with a white light and the lower with a red. In the background were car lights and lights in a big building. To seaward all was dark.

It was only a glimpse, because we tacked close to the lighthouse just before midnight and only made a short board before coming round again. Here lies the crisis of the inside passage against the flood stream. At that state of the tide the whole weight of the easterly-going stream is diverted off the west side of Portland and accelerates southwards, pouring past the Bill at 5 knots or more headlong towards Portland Race, less than half a mile to the south-east.

Throwing *Cohoe IV* on to the port tack, we stood as close to the end of the Bill as we dared in the darkness; but very quickly the yacht was seized by the southerly set of the stream. We responded by easing the sheets and reaching at *Cohoe IV*'s maximum speed, but even with the strong free wind off the Bill and lee-bowing the tide we were set almost a cable towards the Race before we gradually broke through into the weaker stream. The seas were rougher SW of Portland Bill, but we had only force 5 SW and a left-over sea from the force 6 to deal with.

The paradox in using the Inner Passage against a foul tide is that a fresh or strong wind and a fast boat are needed to cope with the stream off the Bill (which on a big spring tide can attain 7 knots setting into the Race), but the stronger the wind the rougher the seas, which can be dangerous in gales. Everything turns on time and the tidal streams.

Portland Bill marked the end of the heavy weather part of the race so far as we were concerned, for over the rest of the course there were only moderate winds and almost

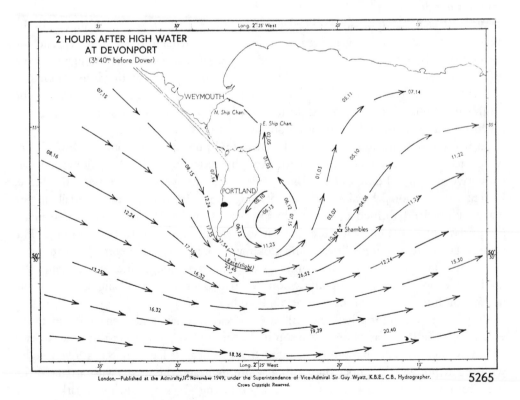

5265

Fig 14.3. Portland Tidal Stream - 3 h 40 m, Dover.

a calm off the Irish coast and near the Fastnet Rock. We managed to win in our class, with our sporting French rival *Pen-ar-Bed* second, but we lost the Fastnet Cup for first place overall in all classes by six minutes, being beaten by the Class II *Clarion of Wight*, then owned by Derek Boyer and Dennis Miller. This was the nearest we had ever come to winning the Fastnet Cup, but we lost no sleep over losing by so small a margin.

The winner of Class I was the American yawl *Figaro* owned by William Snaith, but the Admiral's Cup was regained by Great Britain.

Conclusions

From data supplied by the Meteorological Office, I find that the wind at the start of the race was force 6, and the maximum was nearly force 7 at 1600 with the highest gust 34 knots. The tidal stream through the Needles Channel and outside is about 3 knots, so that at times the *apparent* wind might have been as much as force 7. With wind against tide, conditions may fairly be described as rough.

It was in this part of the race, when all the yachts met the same conditions, that the results were so surprising. In view of the fact that they contributed towards an alteration of the Royal Ocean Racing Club's handicapping allowances for the first time in nearly forty years it is necessary to state them. The elapsed times at Portland, without adjustment of handicaps of the first ten yachts were:

1. *Bolero* (Class I) 7.05; 2. *Capricia* (Class I) 7.30; 3. *Dyna* (Class I) 8.45; 4. *Outlaw* (Class I) 9.35; 5. *Stormvogel* (Class I) 9.40; 6. *Cohoe IV* (Class III) 9.50; 7. *Clarion of Wight* (Class II) 10.05; 8. *Striana* (Class I) 10.00; 9. *Martlet* (Class II) 10.15; 10. *Figaro* (Class I), *Kay* (Class I) and *Belmore II* (Class III) 10.30.

It will be seen that out of 127 starters, two comparatively small yachts in Class II and two in Class III were among the leaders in the rough going, boat for boat without handicap allowance. For some years there had been growing feeling in the RORC, especially among the owners of the larger yachts, that the smaller boats came off best under the handicapping system. The results of the 1963 Fastnet Race may have brought matters to a head. The argument was that the time allowance favoured the small yachts in light and fluky conditions (as they do), but the large competitors no longer enjoyed the advantage in heavy weather hitherto conferred by their size and power.

In the words of the Committee of the Royal Ocean Racing Club: 'It now seems apparent that in a quarter of a century the small yacht has improved her efficiency more than the big one, so that the latter can only hope to win in exceptional circumstances.' Accordingly, the time allowances granted by large yachts to small were reduced. I will not comment on the appropriateness or otherwise of this decision, as we are not here concerned with racing rules. What is important was the official recognition of the increased ability of the modern smaller yacht.

It was a far cry from 1949, when Class III yachts were excluded as being too small and too slow for the Fastnet Race, to 1963, when time allowances were altered because they were considered too fast.

15 BERMUDA RACE GALES

Adlard Coles

The greatest ocean race on the other side of the Atlantic is the Bermuda Race, which attracts so wide an entry that it became divided into six divisions, in which the latest of design, sails and equipment is tried out. The Bermuda Race is 630 miles across the ocean, with the Gulf Stream as the principal hazard. The roughest Bermuda Race before the war took place in 1936, when there was a cyclonic low which produced a prolonged spell of heavy weather in the Gulf Stream.

It was not until 1960, twenty-four years later, that an equally severe gale occurred in a Bermuda Race. I record it here, in the light of what I have read by Alfred F Loomis and Bill Robinson in *Yachting,* because it was an important gale in which 135 competitors were caught out.

The 1960 Bermuda Race started off in a fog on Saturday, 18 June in a light south-south-westerly. Winds continued light and the first part of the race was slow going, the principal object being to enter the Gulf Stream at the right place, some 45 miles west of the rhumb line, in order to find the favourable meander which had been discovered by the Woods Hole experts.

It was not until 22 June that things changed. A cold front had been slowly moved southward owing to a 'high', which had drifted eastwards from the area of the Great Lakes to arrive over the Atlantic seaboard on 22 June. The front took a position extending more or less eastward near the thirty-fifth parallel. Meantime, an area of modest low pressure which was off Charleston on 20 June travelled NE slowly, deepening as it moved. Early on 22 June it lay SE of Cape Hatteras, and, deepening further as it went, it sharpened on a faster course eastward south of the cold front and over the racing fleet.

The strength of the ensuing gales was to vary considerably with the local frontal squalls, and their effect on the racing fleet was to depend upon the positions of the individual yachts relative to the centre of the low as it passed by. A considerable number of them went through the eye, experiencing abrupt shifts of wind from SE to SW, but others escaped the extreme.

Estimates of the wind strength varied considerably, in the same way as they did in the reports of the Channel Race. The velocities of the gusts would have varied locally, but broadly speaking it is clear that the lowest estimates come from the yachts which won places in their classes, among whom would be found some (but by no means all) of the most experienced owners. The highest estimates come from the yachts which hove-to, no less than twenty-three of whom reported gusts of near or above hurricane force. I noted Bill Robinson's comment that 'it is difficult for the average person, even an experienced seaman, to estimate wind accurately'. This is exactly what I have pointed out earlier; in the absence of a masthead anemometer, wind strength estimating must be sheer guesswork when it is above the forces commonly experienced. By taking the estimates of the class prize-winners and averaging them out I arrive at a mean speed

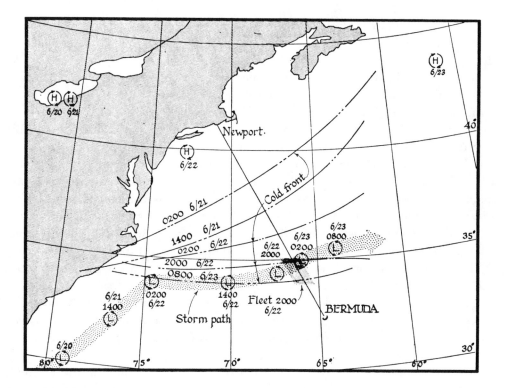

Fig 15.1. Development of weather, 1960 Bermuda Race.

of $42\frac{1}{2}$ knots, which is near the bottom of force 9 with gusts around 55–59 knots.

There were few retirements, but the Atlantic is like the Bay of Biscay in that there are no ports of refuge to retire to. More creditable is the fact that the great majority of the yachts continued to race and were driven hard throughout the gale, among them the British entry *Belmore,* skippered by Commander Erroll Bruce, placed second overall. Apart from allowance for the Gulf Stream, the best tactic appears to have been to hold on to the starboard tack, as the wind freed in time to bring the yachts that did so quickly to the finishing line. The class-winners were Clayton Ewing's *Dyna,* T J Watson Jr's *Palawan,* Henry B du Pont's *Cyane,* Fred Adam's *Katama* and Carlton Mitchell's *Finisterre.*

Conclusions

The lessons learnt from the Bermuda Race of 1960 were in general much the same as those of the Channel Race of 1956. They will not be repeated here other than as endorsements, together with some additional information derived from the reports.

1 *Man overboard.* The class A *Djinn* was knocked down flat in the water, presumably in the same way as *Tilly Twin* in the Channel Race. Six men were washed out of the cockpit into the sea. They got back aboard, aided by the lifelines which they were wearing. *Royono* likewise lost and recovered a man by his safety belt when he was thrown overboard between the lifelines when the yacht was caught aback at the outset

135

of the gale. A more dangerous accident occurred in *Scylla*. A member of her crew, having completed his watch, detached himself in the usual way from his personal lifeline and started to go below. At this moment the yacht lurched and overboard he went. It was night, but *Scylla* was equipped with an electronic flare waterlight, so brilliant that it is said it can be seen from the air at 50 miles range. This was thrown immediately and *Scylla*'s sails were lowered, but before the engine could be started the leads had to be changed over to a new battery. In all, it took about half an hour before the yacht returned to the scene guided by the brilliant flash of the light. The man was found, for in the meantime he had somehow managed to get out of his oilskins and strip to his shorts. His was a miraculous escape, all the more so because sharks are not unknown in these waters. It says something for *Scylla*'s crew that immediately after the rescue she continued in the race to finish sixth in her class. Except that it occurred at night, the incident in *Scylla* was on all fours with the accident in *Galloper* in the Fastnet Race. In each yacht a man was momentarily without a lifeline attached.

These near misses from disaster serve as a reminder that in a gale, when all seems well, there is always the chance of the exceptional squall or the freak wave (or both occurring at the same time) which may knock a boat flat on her beam ends and sweep her crew overboard. There were 135 yachts at sea and this accident happened in *Djinn*, one of the largest. It emphasizes the need to wear safety harness, however hampering and uncomfortable this may be.

2 *Masts and rudders*. It appears that only two yachts were dismasted in the Bermuda Race and one disabled. This is a small proportion when it is rememberd that 135 yachts were involved in the gale. I gather that one or both made St David's under jury rig, aided by auxiliary power.

What was more unusual was that two yachts lost their rudders. *Highland Light* refused to answer her helm. A long emergency tiller was shipped and handled by two men in relays, and in the closing miles, a spinnaker pole was lashed to the taffrail and manned by eight or ten men to assist in steering her. In the meantime her course had been maintained principally by trimming sails.

Cotton Blossom IV was caught aback, got in irons and with sternway drifted down on her rudder which broke. To be caught aback in a sudden shift of wind when a cold front goes through is not uncommon, but it has probably not occurred to many yachtsmen that this can throw enough strain on the rudder to break it. The experience also adds strength to the argument that lying to a sea-anchor head on in a full gale involves real risk of damage to the rudder.

The encouraging feature of these accidents, was that by good seamanship both these rudderless yachts were able, by careful sail trimming, to make port without assistance.

3 *Damage to transom*. *Stormy Weather* damaged her transom. It apparently occurred when she was caught aback and a great strain was thrown by the backstay on her bumkin, which in turn played havoc with her transom. Repairs were effected by wrapping cable around the stern overhang and securing the backstay to it.

4 *Rain*. Once again it is noted that violent rain has the effect of smoothing the seas, if only temporarily. It was reported from *Barlovento* that 'the biggest problem was vision with the rain lashing in horizontally'.

5 *Heaving-to.* Twenty-six of the competitors lay to during the gale. It seems that the majority who did so hove-to, whereas in the Channel gale the majority lay under bare poles, usually trailing warps.

The Bermuda Race 1972

Twelve years elapsed after the Bermuda Race of 1960 before there was another severe gale in this great blue-water event, for which again I thank Bill Robinson and *Yachting* for information and the accompanying weather diagram.

As in 1960 the 1972 Bermuda Race started in good weather with a forecast of a cold front advancing along the rhumb line with a northwester behind it. It was not until the last two days that there was a sudden change. This was due to the development of a slow moving but large tropical depression to the SW of Bermuda, bringing with it a 40 knot fresh gale giving SE winds over the Bermuda area. This 60° shift of wind greatly benefitted the boats which had navigated east instead of west of the rhumb line. The presence of hurricane (rare in June) *Agnes* over Georgia and few reports from the boats at sea, added anxiety among those awaiting their arrival on the usually sunlit shores of Bermuda.

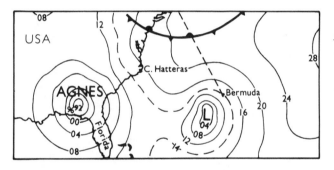

Fig 15.2. Weather map 0600, June 20, 1972, shows storm centre L to SW of Bermuda which caused SE winds on the rhumb line. Hurricane Agnes was then over Georgia.

The strength of the wind may in fact have been a little less than in 1960, but the race proved a greater test of navigation and seamanship owing to the truly shocking visibility in which the first half of the fleet ended it.

The lack of sun sights, the erratic RDF bearings taken in rough seas (confirming similar experiences in other gales) and the uncertainty about local currents (setting northwards) made the approach to Bermuda hazardous, coming in during a gale on far-extending reefs on a lee shore with confused seas over shallow water. Still more hazardous was the actual finish with land completely obliterated by driving rain and with the marker frigate at sea engaged in rescue operations. It was under such conditions that the Bermuda Race 1972 was won by Ron Amey's *Noryema IV*, skippered by Ted Hicks. This British success marked the first occasion that the Bermuda Race has ever been won by a foreign yacht.

Conclusions

Few lessons were to be drawn from the 1972 race other than those already learnt in 1960. There were three dismastings and the usual crop of minor troubles of stays breaking, sails tearing and the breaking of track slides and jib hanks. Less common were reports of wire rigging unlaying and in *La Forza del Destino* the steering quadrant came adrift and her emergency steering also broke. There were several reports of damage owing to falling off waves. *Windward Passage* brought down a lower shroud with the impact, *Crusade* broke her port intermediate shroud and cracked a bulkhead and *Nephente* fell off three waves in sucession and the third dismasted her.

To sum up, the 1972 Bermuda Race proved a big boat race and, although the smaller yachts had their share of the bad weather later, they missed the appalling visibility at the finish. Mishaps were relatively few and the race was a credit both to the ability of modern yachts and to their skippers and crews. Class winners were: R H Grant's *Robon* in A, Jesse Phillip's *Charisma* in B, Ron Amey's *Noryema IV* in C and overall, Stewart Green's *Dove* in D, Rodney Hill's *Maverick* in E and Alexander R Fowler's *Aesop* in F.

16 BISCAY GALES

Adlard Coles

When first I drafted this book I left out a description of a near gale in the Bay of Biscay, as I felt there was already a surfeit of gales, but owing to a parallel gale which has occurred since, in which there was a tragic loss of life, I think both gales should be recorded, because of the lessons to be drawn from them.

In the Cowes to La Coruña Race of 1960 there were twenty-four entries, divided into the usual three classes. In the south of the Bay of Biscay a near gale was experienced, but it was not a severe one and the forecast was of strong winds rather than of gales (see charts, pages 38 and 143).

I was racing *Cohoe III* and on Wednesday, 10 August in the south-west of the Bay of Biscay, there was a strong headwind from the south-west. The strength was logged at about 25 to 30 knots (force 6 to force 7). The sea was confused and the boat, driving fast under staysail and mainsail with two rolls in the reefing gear, occasionally fell off a wave with such a shock that it seemed best to heave-to for a while. The yacht lay very comfortably, but in the evening we took another two rolls in the mainsail and let draw.

We had a wild ride that night, with dense rain. I see that I logged two squalls at 'over force 8'. They were accompanied by hail and rain and occasionally thunder and lightning. *Cohoe III* hove-to during these, and I remember being surprised how quietly she lay despite having so much sail set.

After the worst squalls were over, we reefed the mainsail right down to the lower spreader and set the storm jib. I was on deck at the time, rolling down the mainsail, and then took the helm. We let draw, and I remember how impressive and beautiful the scene was as the yacht sped her way through the gale. Black clouds were tearing across the sky, with a brilliant moon breaking through between them and silvering their edges.

Next morning, 11 August, conditions remained rough and the yacht sailed very fast, but at times not so close to the wind as she should have done. There were occasional severe squalls.

At noon I again logged force 7, but the wind had veered to the north-west. There was a high swell and a confused sea, with small irregular crests. *Cohoe III* was still moving fast, with spray flying over her and heads of seas occasionally breaking on board, but conditions in the cockpit were not at all bad. At 1400 we hove-to again, but two hours later, as the barometer was steady and the wind had moderated, we made sail. During the night we had to lie a-hull again, but the midnight forecast gave only a continuance of force 7 winds for the Bay of Biscay. At dawn the mountains on the north-west extremity of Spain were sighted. Of the twenty-four starters all but seven completed the course and came to anchor off the Real Club Nautico of Coruña. If there had been any hardships at sea, they were soon forgotten in the happy surroundings. *Cohoe III*, incidentally, was beaten in Class III of this race by *Meon Maid II* by three-quarters of an hour on corrected time, which is as it should be, since *Meon*

Maid II had sailed harder. Class I was won by the French *Striana* and Class II by *Martlet*.

It was not until I returned to England that I learnt that a yacht had suffered serious damage during the gale. This was *Tandala,* a new Class II ocean racer designed by Fred Parker for Mr B V Richardson. She was a well-found vessel of 40 ft overall length manned by a crew of six. Here is her story.

'At 0245 on 11 August, we were some 100 miles NE of La Coruña with the wind blowing force 6 from the SE when the goose-neck of the main boom broke, the boom itself having commenced to show signs of deterioration and splitting where it was glued, with the track having pulled out some little time before this.

I thereupon gave up the race, and fixed a jury rig for the main boom, by means of a lashing with the mainsail well reefed down, and the small No 1 staysail set.

At 1300 it was blowing a full gale force 8, and I decided to set a course for Santander, the wind having gone to the NW. At 1400 the wind was gusting force 9, and I decided to lie a-hull with no sail set, and ride out the gale, with two crew with safety belts on duty in the cockpit.

At approximately 2230 I and three of the crew were sleeping below, when a terrific crash occurred, a wave having smashed in the starboard [lee] side of the doghouse, and carried away the cockpit coaming, and also the dinghy.

There was some 2 ft of water above the floor boards in the interior of the yacht.

The storm jib was used for closing up the gap in the broken doghouse, and in order to lash it, four holes were made in it, which as time went on enlarged owing to strain. The yacht was then put stern on to the following waves, with a looped warp towed over the stern, and the water eventually baled out, and the bilges pumped; the wind by this time having moderated to force 7. Some little later, we hoisted a small amount of mainsail and No 1 staysail and eventually entered Santander harbour in the early hours of 13 August, where arrangements for immediate temporary repairs were put in hand.

The 7 ft dinghy was carried on the coach roof, with webbing straps holding it down. The force of the water sheared the webbing and carried away the dinghy, and also bent the metal pedestal of the Sestral Moore Compass, the latter being thrown into the cockpit.'

This is a brief seamanlike report. *Tandala* was badly damaged (I saw her later) and the yacht and her crew had a narrow escape. The occurrence was exactly parallel to that of *Vertue XXXV* when she was struck by a sea in the Atlantic Storm north-west of Bermuda, except that it was not blowing so hard in the Bay of Biscay and there were more men in *Tandala* to cope with the inrush of water. It is to be noted that in both yachts the damage was done on the lee side.

The second Biscay race to which I want to refer took place in 1964 from Santander to La Trinité, a distance of 240 miles of open water. I was not in this race myself, and this report is based on the analysis of the race by the RORC coupled with information on the weather conditions which Commander Erroll Bruce obtained from the Meteorological Office and published in *Motor Boat and Yachting,* together with information which I have obtained myself from the Meteorological Office and other sources.

Forty-six yachts started the race at Santander on Sunday, 16 August, in light airs. By Monday the wind had freshened to force 7 with heavy rain squalls and in the evening there were gale warnings for many sea areas to the northward. Several yachts gave up the race, and the remainder carried on. There was a very big sea running when the leading yachts were approaching the finish, and this, together with the rain squalls, made it difficult to identify the lights around Quiberon Bay in the early hours of Tuesday morning. However, most of the fleet made La Trinité without mishap during the day.

Three Class III yachts, *Lundy Lady*, *Zeelust* and *Vae Victis*, hove-to with plenty of sea room, rather than close the coast with restricted visibility, and these experienced the worst of the weather. They did not finish until Wednesday afternoon and were the last to arrive. No doubt there was some uneasiness about the yachts which had not so far reported, but in heavy weather it is sometimes two or three days after the conclusion of a race before all the yachts are accounted for, as there are many factors which may cause delay before all are safely reported. But the La Trinité race was to prove an exception, for one yacht and five lives were lost.

At about 1800 on Monday evening the small French Class III yacht *Aloa* lay a-hull, heading about north magnetic on the port tack with the wind probably westerly, about abeam. Her position was then estimated about 25 miles SW of the Rochebonne shoals, and she remained hulling for nearly twenty-four hours. It was not until about 1700 on Tuesday, 18 August (when the weather was its worst), that a big sea threw her on her beam ends, which stove in a part of the lee side of the coachroof and part of the cockpit coaming.

The boat was flooded and in one report it is stated that her motor was put out of action, which may mean either that she was under power when the sea hit her, which is unlikely, as the propeller would have been out of water much of the time or, more probably, that the sea soaked the ignition so that it could not be started later. The crew pumped the boat out and attempted to cover the gap in the coachroof by plugging it with sails and mattresses.

At about 1800 *Aloa* shipped a second breaking sea which once more flooded the yacht. She was pumped out again, and the skipper decided to run off to the eastward under bare pole. As warps were not mentioned, it may be assumed that they were not streamed.

Two hours later, at about 2000, *Aloa* was struck on the port quarter by another breaking sea and this time she was capsized. This is evidenced by the facts that the wind vane and anemometer at the masthead were bent and the dinghy (presumably rubber) was caught in the cross-trees.

The three men in the cockpit were thrown into the sea. Two were able to get back on board by means of their safety harness. The safety harness of the owner was not attached to the ship, some say because he was emerging at that moment from the cabin, where possibly he had been navigating, since there was a lee shore only about 35 miles distant. When the boat righted he was 50 yards astern and was wearing a Swedish-type inflatable lifejacket.

The remaining crew pumped the ship dry, made emergency repairs, set the mainsail and hove-to. Rockets were fired, but there was little hope of recovering the owner, who by then must have been out of sight in the seas far out of reach to windward, and he was lost. It was not until the following evening, at 1800 that *Aloa* was sighted by the French trawler *Giralda* about 10 miles SW of the Rochbonne and taken in tow.

Marie Galante II, one of the smallest of the Class III ocean racers, gave up the race on Monday evening and ran for the River Gironde about 80 miles to leeward. She reached the entrance safely on Tuesday evening, 18 August, running under bare pole with two warps streamed aft, but at about 1800, when 4 to 5 miles NNW of Cordouan lighthouse, she was struck by three heavy breaking seas. The first smashed the transom and stove in the cabin doors and flooded her. The second capsized her and broke her mast, and the third caused her to fill and sink. The owner was wearing a safety harness which was made fast to the rigging and he was either unable or unwilling to free himself (it is stated in one report that he could not swim), and went down with the yacht. In the meantime the crew had attempted to inflate the rubber dinghy. It was punctured, but they were able to get it into the water and cling to it. One of the crew had been very badly injured by the ship's ice box when *Marie Galante* was turned over by the second breaker, and he and one other died shortly afterwards. The remaining two were wearing submarine diving suits which enabled them to survive the cold water. They were washed up on the beach in the early hours of Wednesday morning.

A third Class III yacht also got into difficulties. This was *L'Esquirol II,* which was making for shelter behind the Ile de Ré. Some reports state that the Pertuis Breton (which is the channel on the north side of the Ile de Ré to La Rochelle), was entered by mistake, as the usual channel is the Pertuis d'Antioche on the south side of the island. It is more likely, however, that the skipper (who was familiar with the coast) chose the Breton channel to avoid the rough water at the entrance of the Pertuis d'Antioche or the rocks off the Point des Baleines. *Esquirol II* safely gained shelter under the lee of the Ile de Ré, where she spent the night. It was the following morning, when she was under way again, that a jib sheet parted, and whilst it was being repaired, or for other reasons, she drove ashore east of the Grouin du Coup lighthouse about 5 miles north of the Ile de Ré. Her crew were able to walk ashore when the tide fell and the yacht was not badly damaged.

Conclusions

The features common to these two races in the Bay of Biscay was that damage was suffered in strong winds of force 7, under gale force except locally, probably force 8, with squalls and gusts probably about 40 to 50 knots. Damage was not anticipated by the Committee in either race, and it was only after the events were over that reports were received either of damage or casualties.

As will be seen from the weather charts for La Coruña Race, a shallow depression was approaching the English Channel on the morning of 10 August 1960. It deepened a little and moved slowly in a south-easterly direction to be centred off Ushant at noon on 11 August, with a complex frontal system stretching across the Bay of Biscay and crossing the course of the yachts racing to the NE of La Coruña. The isobars are fairly widely spaced and do not suggest more than strong winds of mean force 7 except frontally.

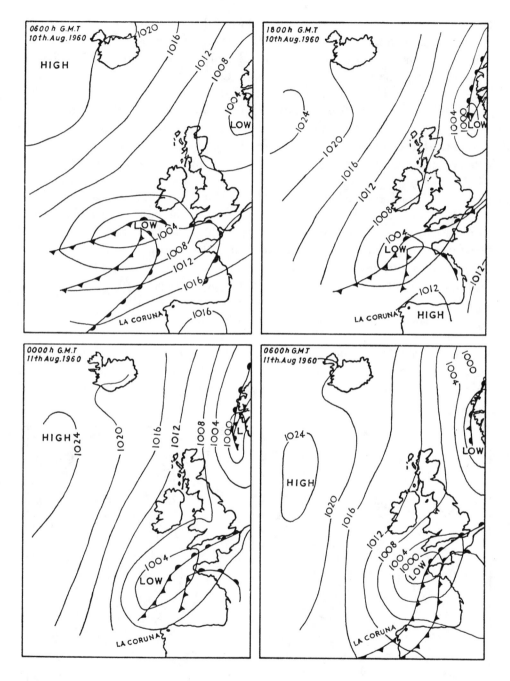

Fig 16.1. Synoptic charts covering La Coruña Race, 1960.

The cause of the bad weather in 1964 during the La Trinité Race was a deep depression which had moved in from the Atlantic and which on Monday evening of 17 August was centred near the Isle of Man with a minimum pressure of 979 millibars. This caused severe gales in the Irish Sea and the English Channel and strong winds up to force 7 hundreds of miles south in the Bay of Biscay. I understand that none of the French shore stations in the Bay of Biscay recorded more than force 7, but the wind is stronger at sea and off the coast on Tuesday evening it was probably gale force 8, and the seas in the Bay of Biscay with a westerly wind would have been very high.

The lessons to be learnt from these gales are mostly underlining of previous ones.

1 *Tandala*. Twenty-four yachts were in the l960 La Coruña race and, out of this number, only *Tandala* suffered damage, and she was an almost new and well-built boat. Damage such as this is fortunately rare. It depends upon the build-up and character of the particular freak sea and the angle at which it happens to strike the yacht. Unstable seas (and therefore potentially dangerous ones) are liable to occur with the veer of the wind when a cold front passes through, accompanied by violent squalls, possibly 50 knots or more, forming new trains of waves across the existing run of seas.

2 *Aloa*. The loss of her owner and the damage to this boat has been attributed to her being in the vicinity of the Rochebonne shoals, and I believe this follows on the official inquiry.

The Plateau de Rochebonne is situated in the Bay of Biscay, about 35 miles west of Ile de Ré, the island situated at the entrance to La Rochelle. The Rochebonne is a rocky shoal with heads, over which there is as little as 4-8 m, with deeper water

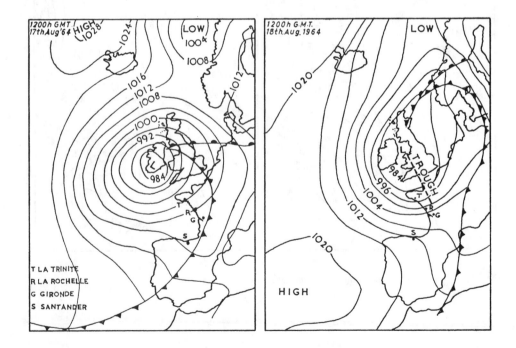

Fig 16.2. Synoptic charts covering La Trinité Race, 1964.

between them. The shoal extends in a NW to SE direction for about 5 miles and is about 2 miles wide. It is marked by four buoys, but these might not be seen during a gale, owing to the high seas restricting visibility until near approach. In the *Bay of Biscay Pilot* the Rochebonne is described as 'one of the most dangerous shoals off the west coast of France. The sea often breaks heavily on it and, in bad weather from the westward, is extremely dangerous.' It is stated that the WSW approaches to the bank are particularly dangerous in bad weather up to a distance of 4 to 5 miles from the bank itself. This is probably local knowledge, derived from fishermen, and is a warning, as although I know this part of the Bay of Biscay I would not have expected danger so far from the actual shoals.

The general opinion is that the accident to *Aloa* was caused by her being in the vicinity of the Rochebonne, but there remains a possibility that she was in deep water the whole time, as when she was picked up by the trawler she was reported to have been 10 miles SW of the shoals. Like *Tandala*, she may have been knocked down by freak seas when the cold front went through with a wind veer accompanied by violent squalls and gusts. A point which I have not seen mentioned is that the tidal stream had turned contrary to the wind, which would have caused a considerable increase in the height and steepness of the waves, even if the rate were as low as half a knot. There would have been a tumult of high, broken and dangerously irregular cross-seas in what, in westerly winds, is virtually the Atlantic ocean. The damage may thus have been caused by stress of weather, rather than shoaling water. This will never be known with certainty, but the tragedy underlines several lessons:

(a) During gales, areas where there is irregularity on the bottom or rocky ledges (even if deep down) cause seas to break and can be nearly as dangerous as lee shores.

(b) The experience of *Aloa* confirm those of *Tandala* that freak waves may attain dimensions dangerous to small yachts at force 8 (34-40 knots mean) and even lower, as will be shown in the next chapter.

(c) The damage is nearly always on the lee side of a yacht, and is caused by a breaking crest throwing her down on her side into the trough, striking the solid sea to leeward like falling on a pavement. It is usually the dog-house which suffers, and its construction is often a source of weakness.

(d) That despite damage, provided temporary repairs are made in time, the yacht may survive, as did *Vertue XXXV*, *Tandala* and *Aloa*.

(e) The accident to *Aloa* illustrates once more that the danger of losing a man overboard is greatest when the safety harness is temporarily disengaged, either when coming on deck or when changing helmsmen.

3 *Marie Galante II.* When the skipper of this yacht decided to retire, the strength of the wind was force 7, and he made for the nearest port, which was the Gironde. The yacht appears to have been well navigated, but her arrival off the Gironde coincided with the height of the gale, which off the coast may have been force 8, accompanied by violent gusts. The approach channel is dangerous even to large merchant vessels during onshore gales and it would be at its worst near low water, with wind against the ebb tide. *Marie Galante II* may have missed the deep channel in bad visibility during squalls

or been obliged to cut across the shoals in order to keep stern-on to the seas.

The seas in the channel might be enough to account for the loss, but it appears more probable that she was lost in the breakers on the Corduan shoals.

In either case the first two of the following points are clear:

(a) *Marie Galante II* suffered no damage when running before the gale streaming warps in deep water.

(b) It is safer to ride out a gale at sea rather than run for shelter to leeward. Even the deep approach to La Rochelle farther to the northward can be very rough when the wind is against the stream.

(c) The owner of *Marie Galante II* is thought to have been unable to unclip the hook of his safety line. His hands may have been numb from exposure, but hooks of safety harness sometimes get stiff from lack of use or from rust. The tragedy serves as a reminder that the hooks should be checked and greased so that they can be unclipped easily.

(d) The accident also serves as a reminder that life rafts should be capable of carrying the whole crew as prescribed by the RORC safety regulations.

(c) In the event of a roll over, with the yacht thrown by a sea beyond the horizontal, there is danger from anything heavy, such as a refrigerator or batteries, which can break loose and injure the crew unless they are really strongly secured.

17 MEDITERRANEAN MISTRAL

Adlard Coles

I have never sailed in the Mediterranean, but the following story by Mr Edward R Greeff which appeared in *Yachting,* and described an experience when his yacht *Puffin* was caught-out in a mistral, is instructive both in regard to local conditions and the wider implications of heavy weather sailing wherever it may be.

Puffin is a yawl with a masthead foretriangle rig and a total measured sail area of 778 sq ft. She is 40 ft in length overall, 29 ft 3 in on the water; the beam is 10 ft 7 in and she has a lead keel of 8,200 lb, relatively slack bilges and quite a raking rudder-post. Total weight 25,400 lb. The hull form was in line with Olin Stephen's current thinking in 1963 of an able ocean racer.

Edward Greeff had competed in the Bermuda Race of 1966, in which *Puffin* was third in class and twelfth in the fleet out of 176. He then crossed the Atlantic and sailed to Port Mahon on the Isle of Minorca, easternmost of the Balearic Islands. Here her crew consisted of her owner and his wife Betty, Braman and Marjorie Adams, together with David Smith and Kim Coit, who had been with the owner during the whole of the cruise.

Puffin left Port Mahon at noon on 15 August 1966, bound for Bonifacio on the island of Corsica, about 240 miles distant. The forecast from the airport said the weather would be fair, visibility good, with winds 10 to 12 knots, NNE, but would become variable when approaching the coasts of Corsica and Sardinia, with seas not exceeding 1-1½ ft. in height. I will now take up the story from Edward R Greeff.

'We had a fine breeze standing out of the harbour and found that we could nearly lay the course of 070°, and later the wind freed. We had the big No 1 on and were moving along at about 5½ to 6 knots. Later during the evening at 2000 the breeze died down and we started our old faithful Westerbeke, again making good our course and 5 knots.

At sundown I had noticed a rather heavy cover of clouds appearing in the west with some high cirrus preceding them. It looked like trouble, but the barometer still remained at 30.05 in (1018 mb) where it had been for weeks with only slight variations. In view of the weather report, I thought there was no point in alarming anybody by mentioning ominous clouds.

By 2050 the breeze returned and started to freshen. We also began to get a big swell from NNW. By midnight we had gone from the No 1 genoa to the No 3 and then to the working jib and soon decided to put a single reef in the main and lower the mizzen. We were still making good our course with our sheets eased a bit, wind N by E at about 30 knots. By 0200 we had a double reef in the main and ran into some rain. Sleep was rather difficult for me as I was worried about this change in the weather, but as there was no change in the glass I felt that it would probably not last very long. At 0400 I decided to take the main off and we proceeded on under working jib and mizzen, wind now N, 35-40 knots, which was a little forward of our beam. By this time we realized we were in for trouble as a French broadcast from Monaco confirmed the existence of a very severe storm moving east from the Golfe du Lion. All of this time the seas were building quite rapidly. I do not know how many people reading this account have cruised in the Med, but one of its peculiarities is the shortness and steepness of the seas.

By sunrise, Tuesday, 16 August, *Puffin* was again going too fast with the working jib and mizzen, so we shifted to the storm jib. I found that by keeping the speed down to 5 or $5\frac{1}{2}$ knots, *Puffin* did not labour and with two girls on board I naturally did not want to make it any more difficult or uncomfortable for them than was necessary. After dawn the skies cleared and the wind now west of north had increased to a good 40-45 knots and perhaps more in the puffs, and seemed to increase more as the morning wore on. I must say it was a magnificent sight to see the crest of these waves blow off in the sunshine. During this period I had been rereading the Hydrographic Office Sailing Directions about mistrals. They point out that a mistral can last for 25 days with winds as high as 60 or 70 knots. In the summertime, however, they do not occur very often and their duration on the average is not over 3 to 6 days. At a time like this, one always wonders what to do, whether to take everything off and run before it or heave-to. With the wind on our beam west of north we had Africa to leeward of us, and I therefore decided to continue. At 1200 we set the storm trysail and lowered the mizzen. We continued on this way, fortunately, with only one case of seasickness and

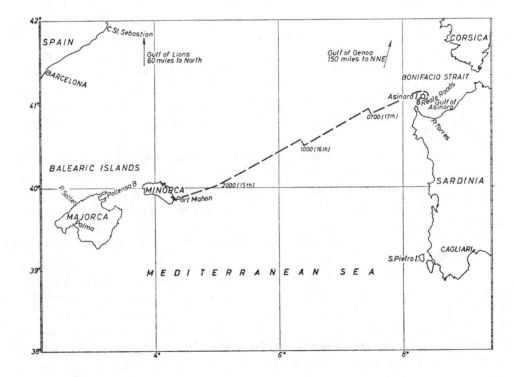

Fig 17.1. Western Mediterranean.

everybody else in good spirits, safety belts being worn on deck, *Puffin* dry below.

During the afternoon the wind increased still more and it seemed that it was blowing 50 and certainly higher in the gusts. By 1800 the storm jib was taken off and we kept on under storm trysail. The wind continued to back more to the north-west and by early evening was about west-north-west. Rather than continue on course for the night, I decided to trim the storm trysail flat and heave-to, but with a man at the tiller. *Puffin* rode very comfortably taking the seas about a point on the port bow and making perhaps a knot and a half to two knots through the water, but certainly not over the bottom. The seas, as mentioned, were very steep and occasionally a crest would break on deck partially filling the cockpit but it did not amount to more than a soaking. I was now pretty tired and got my first four hours of sleep. Dave Smith and Kim Coit were standing one watch and Braman and I the other, using the Swedish system. Tuesday night passed relatively comfortably except for a heavy cloud cover and some rain which followed tremendous lightning to the north. The Monaco radio at 0400 said that this very bad storm had stalled over the Gulf of Genoa and had done great damage along the French and Italian coasts.

When Braman and I came on deck for the 4 to 8 watch, the wind had moderated somewhat but there would be gusts which were quite severe. We both felt, as time passed, however, that these gusts were lessening and during the lulls the wind dropped to possibly 20 knots. At 0600 Wednesday morning, the seas seemed to be down as was the wind, and I felt that we could hold off and go on course again. It was almost impossible to take sights during Tuesday and, as there are no radio beacons on the west side of Sardinia, our position was only an estimate, which worked out to about 40 miles WSW of Punta Scorno.

The breeze by this time was on our port quarter and considerably down. With just the storm trysail our speed was about 3 knots, so I decided that we could set the storm jib with the hope of getting up to 5. Things shortly looked so encouraging that I thought we would set the mizzen rather than wake one of the boys to set a larger jib. About 0645 I went below to try to improve my estimated position, and while below had the feeling that the wind had increased slightly and so came up on deck to look around. It had increased but I still didn't think it was blowing more than 30 at the most in the puffs. Braman, who had the tiller, said that he felt she was all right.

About 0715 I was standing on the lee side of the main boom watching the seas and chatting when I heard a roar astern and turned to see a huge wave much larger than anything we had seen with a broken crest tumbling down the face. The width of it must have been about 75 yards and the height of the crest 8 to 10 feet. I yelled to Braman something to the effect that 'this is coming aboard, hold her off and steady'. I was not actually worried because everything was battened down, slides were in the companionway and all hatches shut.

As this sea broke over our stern I put my arms around the main boom. The sea went over both of us, and the next thing I knew there was a force of water around my waist so great that I was torn off the boom. Seconds later I was under water being pulled along by my safety line. *Puffin* had obviously broached, starboard side down, and was carried in the breaking sea on her beam ends, laying her masts in the water. I pulled myself up until I could get my hand on the rail, and by that time the boat had righted herself. Braman was still at the tiller, but we were laying in the trough.

I climbed aboard as Dave Smith came up to lend a hand and yelled to hold her off again. She paid off on course without difficulty though a bit sluggishly. Betty, who had been thrown out of her bunk smashing her nose, informed me there was about $1\frac{1}{2}$ feet of water over the floor boards. I could not believe my eyes when I saw that 8 feet of the trunk cabin on the starboard side was broken in. Water had apparently rushed through this hole with such force that it tore the plexiglass slides in the companionway out aft into the cockpit.

Realizing that we had to work fast as *Puffin* was in a very dangerous condition, we took the mizzen off immediately and lashed the gangplank (used for the stern-to moorings prevalent in the Mediterranean ports) along the side of the starboard trunk. I hoped that this might keep out another sea until we could get the water out of the boat. *Puffin* has two hand operated bilge pumps, one that can be worked on deck and one below. However, everything was such a mess in the cabin that gear had gone down into the bilge and the suctions of both pumps were clogged. Dave and I proceeded to bail with a bucket until we got enough water out of her so as to hold our hands over the pump suction to keep it clear without putting one's head under water. With Dave's long arms down in the bilge, I pumped and we soon were free of water.

The next job was to secure the gangplank properly to the deck house. It was 14 ft long by 18 in wide and $1\frac{1}{2}$ in thick. To do this we took the treads off, bored holes through the plank, put the treads on the inside of the trunk cabin in a vertical position, and then passed lines through the holes and around the vertical treads which held the plank very securely, Because the plank was so heavy, it could not conform to the sides of the trunk cabin, and we had to resort to stuffing towels in the open spaces. This was completed by about 0900.

As Betty was bleeding badly and we could not seem to stop the flow, I sent out a MAY DAY. I could not raise anybody and so kept calling periodically for about an hour and a half. Finally an Italian yacht answered and offered to relay our message. By this time things were under control, and I informed them that we were in no immediate danger and were proceeding to Bonifacio—we would like a vessel to stand by, however, and that we would need medical assistance on arrival.

Kim Coit who had been sleeping in the lower port bunk had been thrown out and landed in the starboard upper bunk and had received a bang on the head and shoulder, fortunately no permanent injuries. Dave Smith who had been in the port upper bunk was not hurt at all because the dacron bunk boards prevented him from being thrown out. Marjorie Adams had been in the starboard bunk in the forward cabin and had nothing more serious than some bruises. Betty, who had been in the port bunk in the forward cabin had been thrown out hitting her face on something – she doesn't know what.

In the meantime *Puffin* was moving along at about 4 to $4\frac{1}{2}$ knots on course, and I estimated our position to be about 20 miles WNW of Asinara Island which is the north-western tip of Sardinia.

After a quick breakfast I went on deck to take a careful look at the damage. Our

The 40 ft yawl Puffin, *designed by Sparkman & Stephens. After crossing the Atlantic and being involved in a tropical storm* Becky *she was knocked down and severley damaged by a freak wave in the Mediterranean. Photo: Morris Rosenfeld.*

survey showed that when *Puffin* broached, almost everything on the starboard side was torn off or damaged. This included the spinnaker pole which was resting in its chocks on top of the trunk cabin, the dorade type ventilators, one on the starboard side of the mainmast and the other on top of the lazaret hatch. The stanchions on the starboard side were bent and the life ring, Strobe lights and floats were missing. These could have been torn off by me as I went over the side. The mizzen rigging was all slack which was a puzzler. On checking we found that the mast had crushed the step which was a bronze plate and the mast was resting on the horn timber. The mizzen mast itself seemed to be intact as was the main mast. I think there are two things that contributed to the crushing of the mizzen step: the sea breaking into the mizzen sail created a tremendous compression load and also the force of the water on the top of the main boom contributed to the compression strain as the boom was hanging on a wire strap from the mizzen. Fortunately the mizzen did not break, and I think it was partly due to the fact that I had a mizzen backstay set. Incidentally one was set at all times at sea, whether I had a mizzen staysail on or not.

The dinghy which was a Dyer Dink sailing type was secured in chocks on top of the trunk cabin. These chocks were through bolted through the top of the trunk cabin and the line holding the dinghy went through the chocks over the dinghy, crossing it several times. The painter was secured around the main mast. It is interesting that the dinghy did not move in its chocks but both sides were crushed in so that the thwarts punctured and broke the fibre glass. This gives you an idea of the force that was exerted on the dinghy.

By 1100 our progress was quite slow as the wind had decreased further and Dave Smith said 'why don't you try the engine?' It never occurred to me to try it as I knew that the water had been over the battery – a foot and a half of water over the floorboards was enough to cover it. But miracles do happen and the engine started. It was probably lucky that the terminals on the battery were heavily greased but, more important than that, we had a diesel. With the engine running our speed increased again to over 5 knots and we picked up the coast and identified Punta Scorno, the northern tip of Asinara Island. Shortly after 1200 the Italian yacht called again asking for our latest position. At this time we were approximately 3 miles due west of Punta Scorno. The yacht directed us to Reale Bay on Asinara Island to meet a tug which would have a doctor aboard. We proceeded to Reale Bay and anchored at about 1430. The tug came alongside and put a doctor aboard who had no equipment whatsoever to make any examination. We were naturally all quite annoyed. We did, however, gather that he requested us to go to Porto Torres – 13 miles away – so that Betty could go to his office and to a hospital. We then asked the tug to give us a tow so as to save time as they could tow us at 9 knots versus our 5 plus. We arrived at Porto Torres at about 1730. Betty was under medical care shortly thereafter.

The principal thing that we all want to know is how did it happen—what to do at such times and how to avoid a similar situation. Normally if the going gets really tough, blowing very hard and dangerous seas, one is apt to take everything off and run before it and at times tow things to slow the vessel down. In our situation we were sailing because it was not blowing hard and we had good control steering. I might say that there were moments, however, as each crest passed under us that *Puffin* did not respond quickly. At all other times we had excellent control.

In my opinion there are two reasons for our broach. One is that the sea was so high and steep that it broke into the mizzen sail exerting great thrust which spun *Puffin* around. The other reason is that Braman Adams was hit so hard in the back by the sea coming aboard that he was thrown over the tiller, and to starboard. Although he never lost his grip on the tiller, it went down to leeward and we broached to port. Fortunately he had his safety line attached but perhaps he should not have had as much slack in it. If there had not been that slack he would not have been thrown so far.

I have no doubt at all that this could only happen in the Mediterranean where the seas are very short and steep. We had been through a severe gale in the Atlantic and at no time were there any seas that could have done this to us. I would hazard a guess that the seas in the Mediterranean are only half the length of those that we experienced in the Atlantic and twice again as steep. I think it was only because of this steepness that a sea would break as this one did with the wind decreasing and at most not over 25–30 knots.

One principal reason, however, for our trouble was that *Puffin's* trunk cabin sides did not have the vertical through bolts called for in Sparkman & Stephens specifications. These bolts go down through the sides of the trunk cabin, through the sill which sets on the deck, through the deck and through the header which is below the deck. The bolts called for were $\frac{1}{4}$ in silicon bronze and there should have been at least six in the 8 foot span of trunk cabin which was broken in. Obviously these bolts would have added considerable strength.'

Conclusions

The notable feature of this experience is that the sea which did the damage occurred after the storm was over at no more than 25 to 30 knots, indeed not more than 30 knots in the gusts, which suggests about force 6 on the Beaufort scale. This is confirmed by the fact that *Puffin* was only making 4-4$\frac{1}{2}$ knots under storm jib, trysail and mizzen, for had there been more than a strong wind one would expect at least 6 knots. It was a huge wave with an 8 to 10 ft crest tumbling down its face, much larger than any which occurred even when the mistral gusted over 50 knots with the crests of the seas blowing off in the sunshine, or during the severe weather experienced by *Puffin* in the Atlantic. In the previous chapter examples were given of freak waves in winds of only force 7 or force 8, but this experience shows the possibility of a rogue sea occuring even at force 6. I do not believe the Mediterranean can be blamed for

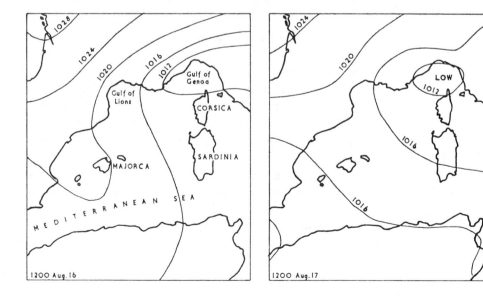

Fig 17.2. Synoptic chart, Mediterrranean.

this. It can occur anywhere, though happily very rarely. Once again, the damage was done on the lee side.

There was no warning of this storm either by weather report or by barometer. As can be seen from the synoptic charts, the depression developed very quickly. It is the same story as occurs in several other gales described in this book. Meteorological experts can (though rarely) be foxed by sudden and unpredictable changes in meteorological conditions, especially locally.

Among the observations Edward Greeff makes in his article are the following, to which I have added my own comments in italics where appropriate:

1 If *Puffin* had been a centreboard boat, I doubt very much that she could have come back with as much water below as we had. *Puffin* snapped back very quickly. I strongly recommend a keel boat to anyone planning an ocean cruise in a small boat such as *Puffin*. *This is interesting because a number of ocean cruising men have come to the same conclusion.*

2 When running off in a big sea never carry any sail aft. *I have recommended the yawl rig because Cohoe II ran so well under mizzen and jib. Puffin's experience is a logical objection to the use of a mizzen which I had not foreseen, but it applies only if seas become really dangerous.*

3 Safety belts should be of the harness type. The snap hooks should be very strong—not the bronze boat snap kind, but a drop-forged galvanised snap.

4 Sheets of $\frac{3}{8}$ in plywood should be carried as wide as the cabin is high and besides plenty of screws and bolts of all sizes—nails if you have a wooden ship. *This confirms advice already given, but note I have carried them around for twenty years without using them.*

5 I would run a $\frac{3}{8}$ in line fore and aft on deck, port and starboard sides, very taut inside the shrouds and made fast to padeyes or stanchions on the foredeck and near the transom, this to which to attach the snap of your safety line. This then gives one freedom to move fore and aft and to be hooked on all the time. *Note also that Mr. Greeff recommends that the helmsman's personal lifeline should be short and have no slack in it.*

6 The Bermuda Race requirements on pumps are very sound—two hand operated pumps. Don't rely on an electric one, because if you get into trouble you may need the juice to call for assistance or to start the engine.

7 The area between floors in a wooden vessel or where the bilge pump suction is located should be covered with a wire mesh to prevent material from clogging the suction hoses. *It is, of course, essential that pumps must not clog. This can be achieved by copper-mesh strum boxes at the foot of the pipe, or as suggested here or better still by using both methods.*

8 Strong wood lath is handy for battens holding down cloth besides many other uses. A full set of tools is needed for working in metal or wood, including three types of drills: electric, hand 'eggbeater type' and brace and bits.

9 Emergency running lights and binnacle light – battery type – should be carried.

10 Carry at least two buckets – one can be of plastic and one of canvas.

I would like to add the following notes:

(a) *Heaving-to.* During the worst of the storm *Puffin* hove-to, and she did this under trysail only, trimmed flat, but with a man at the tiller steering. Edward Greeff had adopted this system when *Puffin* was caught-out in the tropical storm *Becky*, five days out of Bermuda. He found that by this means the boat could be manoeuvred sufficiently to minimize the danger of bad breaking seas. *Puffin* had no trouble in the Atlantic storm because the seas were so much longer and therefore less steep than those in the Mediterranean.

(b) *Coachroof.* Edward Greeff's recommendation of the Sparkman & Stephens method of strengthening the coachroof sides is interesting, and it is adaptable where the sides are 1 in teak, as in *Puffin*. The alternative is steel, bronze or aluminium stiffening to the whole coachroof and coamings, carried down to the deck beams and frames, thus tying the superstructure to the hull. The form of strengthening is a matter for the individual designer, but what is clear is that coachroofs and doghouses can be a source of weakness in the event of meeting freak seas and receiving a knockdown.

18 TWICE ROLLED OVER

Adlard Coles

For a yacht to be completely rolled over through 360 degrees is a rare occurrence, but it does occasionally happen in the ocean. The two classic examples which come immediately to mind are Voss's 19-ft. *Sea Queen* in the North Pacific and the 45 ft ketch *Tzu Hang*, which were both pitchpoled and later rolled over in the South Pacific. There are several other examples.

The following is an account of a yacht which was completely capsized twice off the coast of North Carolina. I am indebted for the particulars to her owner and to the editor of *Yachting,* in which an article describing the incidents first appeared. The experiences are of particular interest, as the yacht involved was a normal ocean racer of moderate size and the incidents occurred while making a normal passage, subject only to the special hazards of the Gulf Stream.

Doubloon is a 39 ft centreboard yawl designed in 1957 by Aage Nielsen and then owned by Joe C Byars. She measures 27.5 ft on the waterline and has a beam of l0.8 ft, draught 4.5 ft and a CCA-rated sail area of 823 sq ft. Her displacement is 9 tons and the ballast keel about $2\frac{1}{2}$ tons.

The yacht departed from St Augustine, Fla, on Saturday morning, 2 May 1964, bound for Morehead City, NC, where she would join the Intra-Coastal Waterway and thence sail to Newport, RI, to arrive in time for the start of the Bermuda Race. The distance from St Augustine to Morehead City (some 50 miles south of the notorious Cape Hatteras) is about 360 miles, and as the course was on the axis of the Gulf Stream, the passage would, in the ordinary way, take about a couple of days. (See chart, page 161.)

Doubloon was skippered by her owner, Joe C Byars, who had as crew Gene Hinkel and two other young men, Mel Burnet and Roger Ryll, who were inexperienced at the start, but finished as the owner puts it, as 'graduates of some of the rougher courses the sea has to offer'. When she left St Augustine at 0600, there was no wind. The marine weather forecasts had not been kept up with, but from Daytona a commercial report forecast no bad weather and the only hint of trouble came from another commercial station which reported a small craft warning for the Charleston area. The barometer was about 1,013 millibars.

The engine was used until 0800, when the wind came in SE moderate, dying away about 1100. It then came in as a whole sail breeze from the east, but later hauled round to SE and freshened, necessitating close reefing. During the night *Doubloon* sailed at 8 knots under mizzen and jib. On Sunday morning (3 May) the wind moderated to 12-14 knots (force 4), and shifted to west. Later in the morning a swell was encountered which Joe Byars described as the largest he had ever experienced. It came from NNE and the height of the waves was estimated at about 15 ft. The wind was blowing their crests backwards in large sheets of spray and, although *Doubloon* had a good wind, it was almost impossible to keep her sails filled, as the huge swells were large enough partially to blanket them. A preventer was necessary to keep the main boom outboard and progress through the water was poor. This is an interesting corroboration of my own experiences off Ushant, described in Chapter 13. It confirms the little-known fact that a very big swell can make it almost impossible to sail in light or even in moderate winds.

Byars at first assumed that the swells came from an expired gale to the northward, but by 1500 the barometer had fallen to about 999 millibars, and when the wind shifted to the north-east (the same direction as the swell came from), he decided to alter course for Charleston, which was about 105 miles to the west-north-west. It was just on 1700 when the skipper noticed what looked like a line squall or a front to the north. Preparations were made for a blow. All sails were lowered and the storm jib hanked on in readiness. All hatches and ports were secured.

A few minutes later the wind backed to north and freshened to 35-45 knots (force 8 to 9) and *Doubloon* was run off under bare poles. A north-easter had started in real earnest and soon rose to 50-70 knots (force 10, gusting above hurricane force). The tops were blowing off the seas, filling the air with spray.

By sunset Byars estimated the height of the seas to be between 15 and 18 ft and by the morning 18-25 ft. The seas, larger than he had ever seen before, were vicious, very steep and white capped. Now and again *Doubloon* took a sea over the stern, half filling the cockpit. This is normal in yachts during gales, but what followed during the night was near disaster. The storm jib was set to steady the yacht on her helm, and her skipper thought that it helped, but the boat began to surf, so it was lowered again. Later and during the night followed 'five full smashes from breaking seas'. Let us examine them.

The first smash. The sea came from dead astern shortly after 1800 and engulfed the cockpit. Byars was wrenched from the tiller and thrown hard against the aft end of the deckhouse. Burnet went overboard to port, but his safety-belt line held and he was quickly back aboard. Neither was hurt. The cockpit was full to the top and as usual, the cockpit drains proved to be too small. Nearly all yachts are the same in this respect. A large stewing pot was brought into operation. The cockpit was emptied and the bilges were pumped out. There were only about 10 gallons of water in them.

It was then decided to put *Doubloon* on a south-west heading, bringing the wind and seas on her starboard quarter. She rode well and this seemed to be the best chance to work out of both the storm and the Gulf Stream.

Second smash. This course was held for nearly three hours until about 2100 taking no seas aboard. Then *Doubloon* was struck by the second full breaking wave. Once more the cockpit was completely engulfed and the yacht went right down on her beam ends. Byars and Burnet were swept out of the cockpit and overboard, but their safety-belt lines held and they floated back aboard as she righted. The stewing pot and the pump were brought into operation again. A porthole which had been smashed was repaired by nailing the lid of a bunk bin over it with large nails, and later a piece of plywood was also nailed over it from the outside.

The skipper then decided to try to get *Doubloon* closer to the wind, and put the helm hard down. The best she would lay was about 60 degrees to 70 degrees from the wind, about 300 degrees on the compass, which is rather better than I have usually found when lying a-hull, and was perhaps due to the windage aft on her mizzen. The centreboard was then let down to try to make her hold up closer. The yacht seemed to handle better lying a-hull heeling about 20 degrees and taking over no seas for four hours.

First roll over. About 0100 the following morning (Monday, 4 May) there was a tremendous crash. The boat was slammed down on her side to port. It seemed that she might have paused for an instant, but instead of coming back she went right over. Her crew were all thrown to the port side, then on to the cabin ceiling. Before they could think about it, they were back upright. The complete 360 degree roll over was estimated to take only three to five seconds.

Joe Byars was in the cabin at that time, but he knew what had happened and his immediate thoughts were of the survival of the crew. He felt certain that *Doubloon* was lost and hurried up the companionway. He can remember the dreadful feeling of seeing the tiller moving back and forth with no one at the helm. He yelled for Gene Hinkel, who had been steering, but got no answer. A quick survey showed the main and mizzenmasts and booms were gone. Byars rushed to the stern and tried to remove the U-shaped life-ring, but the stern pulpit was bent over it, so that it could not be detached. A light buoy was bobbing to leeward which must have been the other life-ring, and as it was not drifting away it was hoped that Gene would make for it if possible. Byars shouted for Gene again and again, but there was no response.

Down below it was reported that the bilge water was not rising. With the aid of a portable pump the water in the bilges was kept under control. Gene had to be given up for lost, because there is no chance of survival for a man overboard from a disabled yacht at night.

The first thing to do was to cut away the masts and rigging. They were alongside, mostly to windward, but they were not hitting the boat, and it was decided to make no attempt to clear the rigging until daylight. Byars went below, but before he did so he noticed that the styro-foam life-raft, which had two other life-rafts lashed to it, was gone. The starboard padeye had pulled out, releasing the tie-down strap.

Roger Ryll and Mel Burnet asked where Gene was, and Joe Byars had to reply that he was gone. They were stunned at the loss, but there was nothing that could be done.

Fifteen minutes later their spirits were immeasurably raised when who should appear in the companionway but Gene himself. They were astounded and overjoyed.

'I heard you, Joe', said Gene, 'but you couldn't hear me. I had her helm at 300 degrees and I got the life-rafts.' He had been overboard on the windward side, where he had found the main raft, with two inflatable rafts strapped inside, wedged in the tangled mass of rigging. His safety line was on, still secured to *Doubloon,* but he could not get himself free. The screaming wind and the pounding seas together with the darkness of the night had put him out of communication, so that he could not be heard, but he managed to get himself back aboard *Doubloon.*

Gene immediately got to work. The batteries had fallen out of their boxes in the roll over, and he got them back in place so that at last light could be restored in the cabin. He wanted to go on deck to cut the rigging away, but the skipper wouldn't let him.

All turned in in their bunks to rest and Gene picked up a book and started reading. As the skipper puts it, 'This was the same man whom minutes before we had given up for lost.'

Fourth smash. Shortly after dawn there was another tremendous crash, *Doubloon* went over to her beam ends, but came quickly back. The rigging had been washed completely over the yacht to the lee side. The mainmast was in three pieces with a sizeable piece held vertically against the lee side protruding 6 or 7 ft above the water, but it was not pounding the boat.

Second roll over. Between 0800 and 1000 *Doubloon* received the hardest blow of all. She went over on her side and continued going. The cabin darkened for an instant, and then she came upright again. All hands were below at the time. As the yacht rolled over Byars fetched up hard twice, once against his head and once against his body. He was bleeding about the face and felt he had been hurt, but it was not until later that he found he had broken a rib. Everybody suffered cuts and bruises in the roll over and later it was found that Roger Ryll had suffered three partially crushed vertebrae.

It is surprising that despite this complete roll over *Doubloon* had taken in very little water, which says a lot in favour of her construction and hatches. The water never rose above 3 or 4 in below the cabin sole, which could easily be dealt with by the crew.

It seems that *Doubloon* could stand anything more that would come her way, and the spirits of her crew rose a little. The next thing to do (it was now daylight) was to clear the rigging. The wire cutters had been in the cockpit and had gone to the bottom when the yacht rolled over. Nevertheless, Gene managed to cut the halyards and remove all clevis pins to rigging except the forestay, which was attached to a good piece of the mast which acted as a sea anchor. He then rigged another sea anchor out of a No 2 genoa, a sail bag and an anchor. *Doubloon* was riding 60 degrees from the wind. Yet another sea anchor was put out comprised of a working jib, with the head attached to the tack to create more drag. Two mattresses were lashed on what remained of the stern pulpit, in order to create windage aft to help hold her head up. The effect was to keep the bows at about 50 degrees to 60 degrees from the seas and wind.

All that remained to do was to check over other damage. Both ventilators had been torn off, but the holes were plugged. Both spinnaker poles had torn loose, and no spars were left other than the 2-ft. high stumps of the main and mizzen. A starboard winch was partially uprooted and every stanchion on deck was bent, as were both pulpits. The tiller was broken off at the head, while the binnacle was gone.

The barometer had risen to 1,004 millibars, but wind and sea remained at full force for some hours. However, *Doubloon* was making appreciably better weather of it and took no more knockdown blows from seas. Her skipper attributes this to the sea anchors holding her up to the wind. Her weary crew turned in to rest in the cabin, which was now a shambles.

That was the end of the adventure, but it was still blowing hard on Tuesday morning. It was not until around noon on Wednesday that her crew got in the sea anchors, and a jury mast was rigged out of the only spar left. This was a $6\frac{1}{2}$ ft. aluminium spinnaker guy strut for keeping the aft guy off the shrouds. It was rigged inside the stump of the mainmast and put the head perhaps 10 ft above the deck. To this a mizzen spinnaker was set, which was sufficient to give *Doubloon* a speed of 4 to 5 knots on a westerly course.

To cut a long story short, two ships passed *Doubloon,* one only a mile away, but neither saw her flares. It was about 1600 on Wednesday that the freighter *Alcoa Voyager* stood by for rescue operations. Joe Byars was pressed to abandon *Doubloon,* but he refused, as she was not leaking and was making progress towards the coast under jury rig and had plenty of supplies. The *Voyager* then made a wide circle and came back close to leeward and took the yacht in tow until relieved by the coast-guard cutter *Cape Morgan,* who gave her a lively ride at 12 knots to Charleston Harbour.

Doubloon had been towed for over 160 miles east of Charleston, her position when picked up by the *Voyager* on 6 May being 32° 41′ N 76° 36′ W. Her skipper notes that the Gulf Stream had carried the yacht to the north-east in spite of gale winds against it. The sea anchors must have accounted substantially for a part of this.

Doubloon was skippered with tremendous determination, for despite the ordeal, Joe Byars refused to abandon her when asked to do so by the *Voyager* and despite somersaults and dismasting he managed to get the yacht to Newport RI, and have her repaired in time to take part in the Bermuda Race in June and to win second place in Class E.

Conclusions

I have obtained particulars of this storm from the US National Weather Centre. It was an extra-tropical cyclone which formed as a wave on the polar front over northern Florida and adjacent Atlantic waters. Whole-gale winds were reported from several ships in the area of cyclogenesis over the warm Gulf Stream waters. SS *Santa Rita,* 34.1° N 75.7° W at 1800 on 3 May logged Bar 1009, wind E 50 knots, and at 1200 on 4 May SS *Platano* 33.0° N 75.5° W reported Bar 1009, wind ENE 50 knots.

Doubloon must have been nearer the centre of the low when her barometer reading was 999 millibars and may have experienced higher winds of 55 to 60 knots. A fair guess would be force 10-11, gusting at hurricane force.

The storm seems to bear a marked similarity to that experienced by the British yachts north of Bermuda. Both were extra-tropical cyclones with predominantly easterly or NE winds. In neither storm did the barometer fall exceptionally low, only to 993 millibars in 1950 and 999 millibars in 1964.

The principal difference between the experiences of the yachts was due to the Gulf Stream. Of the British yachts, only *Vertue XXXV* and *Samuel Pepys* appear to have been in the Stream (near its southern boundary) during the earlier storm, whereas *Doubloon* was on the axis of the Gulf Stream, where the current was running near its maximum speed directly contrary to the wind, so she must have experienced phenomenal seas.

Joe Byars estimated the height of the waves at 18-25 ft, which is remarkably accurate, for the guesses are confirmed by SS *Platano's* report of 16ft and SS *Santa Rita's* of 24 ft height.

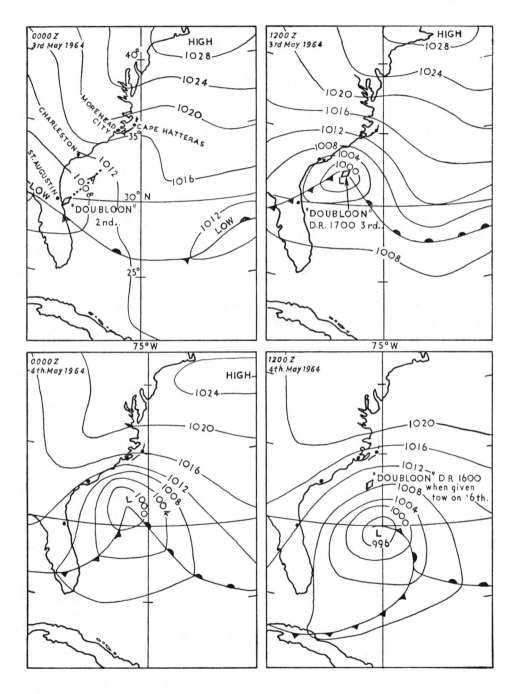

Fig 18.1. Synoptic chart covering the Gulf Stream storm, 1964.

Comments are as follows:

1 *Gale tactics*. *Doubloon*'s skipper does not like running dead before a major gale, but he did not try towing warps, which are certainly a help. Streaming warps might not have saved *Doubloon* from her roll over in the exceptional seas of the Gulf Stream, but, as her skipper says, they might have worked well enough running in more normal seas in a gale. Despite his experiences, he thinks that once the wind had risen to force 10, *Doubloon* would have been in greater danger running off either with or without warps. He is firmly of the opinion that he would rather take his chances in a gale by lying as near as possible head-to-wind. For a boat of *Doubloon*'s size, a storm mizzen is recommended with 9 ft on the hoist by 4 ft on the foot, made out of 12-oz Dacron with wire mast-hoops and rollers which can be shackled around the mizzenmast. A sloop could hoist a storm jib with the main halyard, with the luff hanked to the backstay and with the clew forward sheeted in flat. Byars believes that this would hold the bow up to about 30 degrees to 50 degrees from the wind without a sea anchor, and that the boat should make the proper leeway, giving with the seas, although a small sea anchor might be necessary to slow her and to reduce leeway or hold the bow closer to seas.

The rudder would have to be adjusted to suit the requirements of the individual yacht and conditions, lashing it amidships or allowing it to swing through a small angle. He considers this should work for most yachts without endangering the rudder by backing down on it.

It cannot be denied that the bow is the stronger end of a yacht and less vulnerable to seas than the stern with a cockpit which is an invitation to the head of any 'freak' sea to flood it and sweep the crew overboard, as has happened time after time in severe gales and storms. The crew can usually be recovered by their safety belts, but it is preferable that they should not go overboard at all. It is instructive to note that on the second occasion that *Doubloon* was rolled over she was without masts. This contradicts the theory that a hull without masts will float like a cork through anything.

2 *Centreboard*. From this experience it appears that the centreboard helped to hold *Doubloon*'s head closer to the wind, but her owner thinks it probably contributed to the roll-overs by tripping the yacht. The centreboard was $\frac{3}{4}$ in bronze and it was found to have been bent about 30 degrees, probably when it hit the water as the yacht came back to even keel after the roll over.

The beamy centreboard yawl is a proven type of good seaboat, tested by many gales, among them the 1957 Fastnet Race, which was won by the American yawl *Carina*. It is possible, however, that in 'survival' storms there comes a point of heel where stability is lost compared with the conventional deep-keeled yacht. *Doubloon*'s draught is 4.5 ft and her keel is $2\frac{1}{2}$ tons. Once thrown so far over on her side that most of the stability afforded by her 10.8 ft beam was lost, the righting moment of her ballast keel would be far less than that of a narrower boat with say $3\frac{1}{2}$ tons of lead at the bottom of a 6 ft keel. It is purely a matter of conjecture, as 'survival' storms are rare, and some authorities believe that reasonably shallow draft is no handicap. On the other hand, a centreboard or a deep fin keel tends to provide a pivot on which a boat can be rolled over.

3 *Doghouse windows.* The year before the storm *Doubloon*'s four fixed windows had been replaced by smaller ports. The seas struck with tremendous force, giving somewhat the impression of an explosion on deck. It seems doubtful whether she would have survived if she had still been fitted with big windows, such as are found in the doghouse of the average yacht. In my own boats since 1950 I have carried shaped plywood panels, with holes drilled in them, ready to be quickly nailed over a broken coachroof window or port. I have never had to use them, but they are light and stow easily under the quarter-berth mattresses and might prove invaluable in an emergency.

4 *The watch below.* *Doubloon*'s owner recommends staying below to get rest after making everything on deck secure, thus eliminating the chance of injury on deck, or being washed overboard.

5 *Other recommendations.*

(a) Cockpit hatches (lockers) should be raised, strong, and have a good means of fastening to be watertight. Cockpit-hatch leaks will allow a significant amount of water in. This adds to the burden of keeping buoyant. The usual gutters and small drains are near useless under such conditions. Two additional cockpit drains were installed after the gale.

(b) The bilge pump was altered, so that instead of emptying into the cockpit it discharges on deck. The emergency pump was changed to a large-capacity Edson Diaphram pump, which is nearly 100 per cent clogproof.

(c) Everything on deck must be well secured. Ventilators should be removed and the pipes stuffed with rag if they are not fitted with deck plates to screw down to make them watertight.

19 SEPTEMBER HURRICANE

Adlard Coles

Tropical storms are known in the northern Indian Ocean simply as cyclones, in Australia as 'Willy Willies', in the west and north Pacific as typhoons and in America and the Caribbean as hurricanes. It is the latter which yachtsmen have most often to reckon with, because they so often move up the eastern American seaboard, leaving a trail of havoc among yachts in harbour or those unlucky enough to be caught out at sea in one.

Hurricanes are not entirely unknown even in European waters. For example, hurricane *Carrie* in 1957, after meandering thousands of miles, ended her travels in southern Ireland, though over a month too late to add to the entertainment of the Fastnet Race of that year. In 1958 *Helene* developed east of the West Indies and proceeded off the New England coast to Newfoundland. She then went ocean cruising towards Greenland before changing her mind and paying a courtesy call at Cardiff. Happily by the time hurricanes reach our waters, most of the sting has been taken out of their tails and there is more buzz than bite.

Hurricanes are carefully tracked by the American meteorological experts and repeated warnings are given of their anticipated course, but they are temperamental, wayward things that from a met. man's point of view are perhaps not quite house-trained. Their courses are sometimes unpredictable. Here is the story of one such which appeared in the American magazine *Yachting*.

The yacht concerned in it was *Force Seven*, owned by Warren Brown of Bermuda. She is an ocean racer, designed by William Tripp and raced in Class E of the Bermuda Race under the Cruising Club of America or Class II under the RORC. She measures 40 ft in overall length and 27 ft 6 in on the waterline. Her beam is 11 ft 8 in and draught 5 ft 10 in. She is typical of the modern boat that races on both sides of the Atlantic. I know *Force Seven* and I exchanged greetings with her owner and crew when she and *Cohoe IV* were off the Lizard in the Fastnet Race of 1963, and the two yachts lay alongside later at Plymouth.

Force Seven took her departure from Bermuda on 30 August 1964, bound for Newport, RI. The owner was skipper/navigator and with him were Herbert Williams as mate and four undergraduates as crew. There were also two youngsters on board, the mate's daughter and a schoolboy. Both skipper and mate had many year's experience of ocean sailing and racing, and the yacht was in first-rate condition and well fitted out.

The September hurricane belt lay between Bermuda and the United States and a careful eye had to be kept on the progress of storms coming up from the south, though with the new weather satellites in orbit most of them can be tracked more accurately than in the past.

Departure had been delayed a day because hurricane *Cleo* was at that time lying between Bermuda and the mainland, but on Sunday morning, 30 August, it was

The 40ft masthead sloop Force Seven, *designed by William Tripp. She survived hurricane* Cleo *running under bare pole, taking the seas on her quarter. A nearby tanker reported the gusts at 83 knots. Photo: Beken, Cowes.*

reported that the winds had dropped to about force 9, and that *Cleo* had moved over North Carolina and had started to break up. So with no indication of any further disturbances *Force Seven* left Bermuda in beautiful weather.

It was decided to enter the Gulf Stream approximately 30 miles to the west to allow for it to set the yacht back to a rhumb-line course. There was enough fuel on board for the diesel engine to give a cruising range of almost 400 miles, because during September are sometimes calms which may last two or three days. Winds proved to be light and variable and the engine had to be used at times to push ahead as fast as possible. (See pages 170-1)

The first indication of an approaching storm came on Wednesday morning, 2 September, at about 0800, when the barometer started to fall very rapidly. *Force Seven*

165

was then under full main and genoa, averaging about $7\frac{1}{4}$ knots with the wind on the starboard beam from the east. As the barometer fell the wind increased and by 1030 the jib had been lowered and stowed and the boat was making roughly the same speed under mainsail only. Here I will pick up Warren Brown's account in his own words of what was to follow.

'Soon after this, the sky started to darken and contrails high in the sky were replaced by low cloud and rain squalls. The wind increased rapidly from 30 to 45 knots, and we dropped our mainsail at noon, just before a vicious squall hit, with gusts of well over 50 knots. The seas by this time had built up considerably, and with the barometer dropping alarmingly, I realized that we were in for what I thought at the time would be a severe depression of relatively short duration. We were on the southern edge of the Gulf Stream by our dead reckoning position and water thermometer readings, which we had taken since early that morning, which showed that the water temperature had increased six degrees.

Shortly after 1300, when we had been under bare poles for one hour, I picked up a radio report from the United States that hurricane *Cleo* had reversed herself off the coast and had picked up speed and intensity. She was once again a full-fledged hurricane. Her estimated position was, according to my dead reckoning, almost exactly the same as ours! I now realized why the barometer had fallen so rapidly, and knew that we were in for a jolly good pasting.

Not wishing to alarm the crew, I did not report this fact to them, especially since we had youngsters aboard. However, we set to work immediately to rig the ship for severe weather conditions. All ventilators were removed and stowed below, and the appropriate ventilator plates were fitted to replace them and screwed into position. The deck was cleared of everything movable and the main boom was lashed amidships by handybillies, making it completely immobile. All jibs were stowed below and the mainsail itself was securely lashed so that no matter how hard it blew, it could not shake itself loose. All hatches were secured.

By four o'clock in the afternoon the seas were masthead height, the wind was still increasing, and visibility had diminished considerably. Tops of seas were already breaking over us. One factor which worried us at this point was the dinghy lashed to the top of the cabin. We decided that, secured in this way, the dinghy definitely placed extra strain in the cabin roof, so we moved the lashing to the base of the stanchions. By five o'clock we were in the full force of the storm, and averaging six to seven knots downwind. One minute we were on top of a huge sea, and the next we were down in a hollow out of the wind. The only visibility we had was on top of each wave. In the late afternoon we sighted a huge tanker, also running away from the storm, keeping the wind on her port quarter.

In the meantime we tried to keep the seas on our starboard quarter. The wind by this time had swung from the east into the north so that we were running approximately southwest, and I realized that the storm must be passing very close to the south.

Since I had experienced gales off the coast of the States as well as in the North Atlantic, I knew *Force Seven* was encountering a very different set of sea conditions.

The ultimate storm. This, together with the photos on pages 175, 185, 187 and 188 were taken by Captain de Lange in North Atlantic gales of about force 10/11, between 35° and 45°N. Seas such as these are experienced by freighters on the West Indies run about once in four years, and hence might not be encountered by a yacht even in a lifetime of voyaging. A yacht might ride the huge seas in the foreground but look at the 'freak' wave coming up astern.

The reason for this was that the wind was blowing against the current of the Stream, and the seas were very much shorter between the crests, and very much deeper than they would have been in the open ocean. Every sea was extremely steep and every wave was breaking at the top.

How hard was it blowing? Quite frankly, I do not know. When conditions increased above 65, it is impossible to tell from the deck of a small boat whether it was blowing 70, 80 or 90. Visibility was so restricted and the weather so dark that everything was a confusion of sea, wind, foam, and driving rain. We learned afterwards, however, from the Coast Guard in New London, that the tanker we had spotted reported winds of 83 knots at the time.

By 1700, steering had become extremely difficult, and of concern for the deck watch. We criss-crossed the cockpit completely with rope, giving hand-holds for every movement in this area as a safety measure additional to safety belts. Watches were put at two hours, one man watching the sea while the other steered. The problem facing us was keeping the boat going fast enough in order to keep out of the way of the huge breaking seas by sliding down their sides and keeping them on the quarter. If we failed to react fast enough, the sea would break completely over us and would fill the cockpit. If, on the other hand, we kept the sea directly on the stern, we were in the dangerous position of surfing on a downward slope for a quarter of a mile at a speed of about 15 knots. This was not a safe thing to do, as we might have had the full weight of one of these seas break on top of us.

The question of being scared never came into the picture of our predicament; it was more a question of apprehension. We knew that if we broached at any time we could very easily lose the top of our cabin, the most vulnerable part of the boat.

At 1745 one exceptionally bad wave hit us on the stern and put us on our beam ends, filling the cockpit from leeward as we went over on our side. Luckily we righted in time to keep out of the way of the next sea.

We learned one very valuable lesson. In order to keep manoeuvrability we should have had larger cockpit drains. The two that we had could not drain the water fast enough and the boat became sluggish as the cockpit filled. After this experience, I would never build a boat with a large cockpit in which the forward end leads directly into the cabin. If the hatch were to be broken by a heavy sea, any flooding of the cockpit would drain directly belowships.

It was so rough by 1800 that the constantly filling cockpit and our being knocked down by breaking waves became a matter of routine. The brackets on the stern in which the life-rings rested became useless, as the rings were plucked out by the sea. One was washed away and the other was grabbed by an alert crew member as it went over the side. It was obvious that our life-ring brackets were not designed for this kind of weather.

The radio insulation blocks on our backstay started to break up under the strain but, fortunately, the backstay did not slacken enough to allow the mast to whip around. Our greatest danger was the possibility of broaching and being rolled over sideways or being turned head to wind. It would have been virtually impossible for any sailing boat to heave-to under the conditions we faced that evening, without being torn apart completely. The waves were so steep and such a short distance apart that the seas would have broken over our decks and would have swept our small boat clean.

A second great hazard was our dinghy. We were tempted to throw it over-board, but did not do so as we felt it might be picked up and a report given out that we had been lost at sea. Given the same circumstances again I would not hesitate to do so, as the dinghy constituted a danger, not only to the boat, but to the crew, in case it was swept aft. I would also strongly recommend two life rafts in place of a dinghy for anyone making a sea voyage at this time of the year.

If we had dragged warps, we would not have been able to control the speed of the boat to get out of the way of the huge breakers and it would not have been long before we had the full weight of one, which would have caused major damage.

From 1800 until about 2200 that night it was a constant battle, with one man watching the sea and the other steering. The boat was continually being flooded by breaking tops of the waves, or being filled from leeward by knock-downs. Every small opening in the superstructure was leaking and despite the small hatch opening, it was a continuing race to stop gallons of water from flooding below. By 1100 the following morning [3 September] the wind had abated to about 45 knots, but we did not put the sail back on again until midday as the seas had still not improved. To have made any headway directly into them would have been impossible.

During the course of the storm we had travelled under bare poles approximately 120 miles in a south-westerly direction, but had actually covered only about 70 to 80 miles over the bottom, as we were running against the flow of the stream.

It was not until Saturday, 5 September, at about 1600, we found ourselves just off Block Island, at which time the Coast Guard picked up our weak radio. They forwarded the news to Bermuda and sent out a plane which flew over us a few hours before our arrival in Newport.

Upon our return to Bermuda we plotted our position against the track of the storm with the local CG units and with one of the CG ships stationed in the Colony. The plottings confirmed that *Cleo* was indeed a fickle lady. Not only had she reversed herself and turned into a full fledged hurricane again, but had actually gone south to pick up speed. At one point she had moved over 100 miles in just six hours. Our estimated position when she crossed our bow at approximately 1800, 2 September, at the height of the blow, was approximately 45 miles from the centre, and the fact that she was moving so rapidly probably kept us from a worse beating than if she had stopped in her tracks.'

Conclusions

This is one of the most modest and instructive accounts which I have read of the experience of an averaged-sized yacht in a hurricane. Let us attempt to assess the force of the wind when hurricane *Cleo* passed close to the yacht.

First of all, it will have been noted that Warren Brown states frankly that he does not know the strength of the wind. I have long put forward the view that one cannot estimate wind forces above those with which one is familiar. For the average cruising or ocean racing man, this is about force 7 to force 8. As Warren Brown says, how can one possibly tell from the deck of a small boat whether it is blowing 70, 80 or 90?

I have obtained full particulars of *Cleo* from the US National Weather Center and the track of the hurricane is reproduced here by their permission. Superimposed on this, are the approximate positions of *Force Seven*.

The early stages of development in hurricane *Cleo* moved off the West African Coast south of Dakar on 15 August, with a minimum pressure of 1006 millibars. The disturbance intensified rapidly with a central pressure of 992 millibars and with winds of hurricane force by 21 August. It swept through the West Indies and the bulletins described it as 'a small but extremely dangerous hurricane'. The highest winds were estimated to be 120 knots near the centre, with hurricane winds extending 40 miles from the centre in all directions and gale winds extending 150 miles north and 100 miles south. The coast of Florida was reached on 27 August, and sustained winds of over 80 knots were recorded at Miami on the western edge of its eye as it proceeded along the coast. On 28 August *Cleo* moved inland near Savannah, Georgia, and the wind diminished. By Sunday, 30 August, no strong winds were left, but the depression moved out to sea between Norfolk and Hatteras early on 1 September. By the afternoon tropical storm force had been regained and on 2 September *Cleo* had intensified and regained hurricane status over the warm water of the Gulf Stream. The highest winds in the centre, just outside the eye, were 80 knots and gales extended 200 miles from the centre in all quadrants. On 3 September, *Cleo* then curved north-east

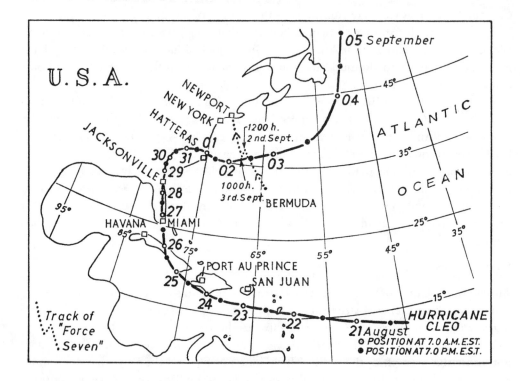

Fig 19.1. Track of hurricane Cleo, 1964.

and northwards and the weather conditions remained much the same. The bulletin on 4 September at 1100 still gave 80 knots near the centre and gales extending 250 miles in all directions. A diminution of the wind force was forecast at the centre, but gale force winds were expected to extend 500 miles to the south-east of the centre.

The words 'highest winds' used in the American bulletins appear to mean sustained winds and not merely the gusts at the highest velocity. The official reports therefore corroborate everything that Mr Warren Brown has stated. It is probable that when *Force Seven* was nearest the eye of the hurricane she experienced winds which, according to the accompanying graph, were of a mean strength of about 68 knots (hurricane force) with gusts of 83 knots as reported by the tanker when near by.

I have seen the list of the observations from ships passing in the vicinity of *Cleo*, but unfortunately there is a gap between 29 August and 4 September. After *Cleo* had passed *Force Seven* and proceeded to the north-east, there is a report on 4 September from SS *American Challenger* in position 45.5° N and 50.0° W of the westerly winds of 80 knots. The master of the *Queen Mary* reported it as the worst blow he had experienced since 1939, and that the *Queen Mary*'s stabilizers had rolled out of the water.

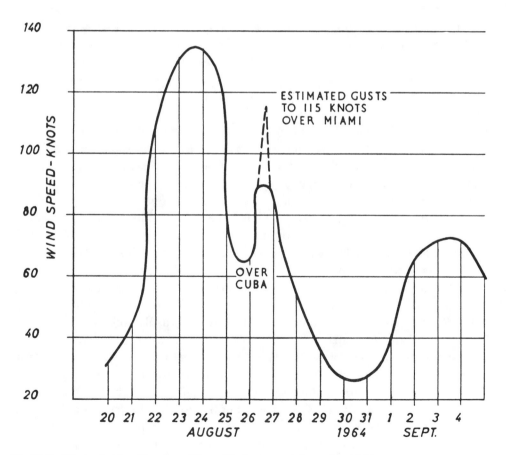

Fig 19.2. Wind velocities of hurricane Cleo, 20 August to 4 September 1964.

The height of the waves was observed to be about the height of *Force Seven*'s mast, which would be about 50 ft if my formula of three-fifths the observed height of a wave is accepted as a fair estimate of the real height, that gives a height of 30 ft. However, recent oceanographic research and methods of measurement suggest that the height of the highest waves might have been considerably greater.

The principal lessons to be learnt from the experiences of *Force Seven* are:

1 *Storm tactics*. Warren Brown had given considerable thought to what he would do if he was caught out in a yacht in such an exceptional storm as this. He adopted what appear to me to be new tactics for survival in a hurricane. He considered the seas in the Gulf Stream were too dangerous for heaving-to or lying a-hull. He did not experiment with the general accepted practice of dragging warps. He had two very long heavy warps in the cockpit ready to stream astern in a bight, but he did not use them, because had he done so, he believes he would not have sufficient control of the boat to avoid the worst of the seas. He says that if he had trailed warps it would not have been long before the full weight of a breaking sea would have caused major damage.

He found that when running with the seas directly on the stern the yacht surfed on the downward slope of a big wave for a quarter of a mile at a speed of about 15 knots, risking being pitch-poled. Warren Brown is a master of understatement when he writes: 'This was not a safe thing to do . . . as we might have the full weight of one of these seas break on top of us.' Accordingly, he altered course to bring the seas on the quarter, though I get the impression that he kept *Force Seven* maintaining about 5 to 7 knots under bare pole. He states that he found speed essential for safety. If the helmsman failed to react fast enough, 'the seas would break completely over the yacht and fill the cockpit'.

In Chapter 8 in this book I commented that *Cohoe* got pooped in a moderate gale because she was not running fast enough to give the helmsman sufficient control to hold her stern to the bigger seas. But this is the first time I have read of a yacht running through a hurricane under bare poles at such a speed without streaming warps. Some may consider this rank heresy, of course, but I would point out that, apart from knockdowns, *Force Seven* came through the hurricane unscathed in conditions in which many yachts might have gone to the bottom. More evidence in favour of this tactic in exceptional gales is afforded by *Joshua*, when caught out in a storm in the South Pacific, to which I refer in the next chapter.

2 *Preparation for gales*. As soon as the radio warning was received by *Force Seven* everything on the yacht was properly secured and she was prepared to cope with the approaching storm. In this connection, two points are of particular interest.

The first is that the dinghy lashed to the top of the cabin was regarded as a potential danger, because of the strain it would have imposed on the cabin structure if it had been struck by a heavy sea, and because if the dinghy were torn away it might injure the crew. Yachts have carried dinghies on their cabin tops without experiencing trouble in countless gales of force 8 or force 9, but the possible danger of a dinghy, if a yacht is caught out in a survival storm, may not have been apparent. This is a point stressed both by Joe C Byars and Warren Brown. Happily, most yachts now carry rubber boats which make less resistance than a rigid dinghy and do not endanger the crew.

The second point is that Warren Brown gives a useful tip as a safety precaution in a storm. By criss-crossing the cockpit completely with ropes, he provided handholds for his crew to grip when *Force Seven* was knocked down. It is not uncommon for part of the crew to be swept out of the cockpit if a yacht is knocked down in a heavy gale, and several went overboard even in the gale of the Bermuda Race of 1960. In most cases, crews are saved by their safety harnesses, but obviously it is preferable not to go overboard at all, and the criss-cross of ropes seems a practical idea to help prevent accidents of this sort.

3 *Fear.* A major storm or a hurricane is something of an ordeal in a small yacht. However, I think Warren Brown hits the nail on the head when he says: 'The question of being scared never came into the picture. It was more a question of apprehension.' The word 'apprehension' is the right one, because there is always an element of the unknown in a real gale.

4 *Cabin entrance.* Warren Brown states that he would never in any circumstances build a small boat with a cabin entrance (with doors or wash boards) entered directly from the cockpit. *Force Seven* has a very small companionway entered from above and this probably saved her, as the seas which broke into the cockpit time after time might have stove in doors from the cockpit to the companionway as arranged in most yachts.

5 *Working on deck.* The tremendous violence of a hurricane will be appreciated when it is stated that it took twenty-five minutes to move the Beaufort liferaft from its position forward of the mast to the cockpit, where it might be wanted. A man had to inch his way forward with two safety lines. Lines had also to be run forward so that the raft would not be washed away. Difficulty in working on deck occurs at force 10 (as was experienced by *Tilly Twin* in the Channel Storm of 1956), and at force 12 any movement on deck becomes dangerous.

6 *Other points.* *Force Seven* confirms experiences in many lesser gales. The bad visibility when seas are very high, the way water finds its way below like penetrating oil, and the inadequacy of drains of the ordinary 'self-emptying' cock-pit. It is also interesting to note that *Force Seven*'s liferings were washed away, like those of *Doubloon* when she was capsized. Normal brackets for lifebuoys appear to be useless in this kind of weather, though it is difficult to think of a substitute, as if lifebuoys are lashed down they cannot be cast off quickly. A final point is that radio insulation blocks on the backstay can be a source of weakness. I have never had these blocks in my own boats as I always suspected them. Warren Brown has confirmed that these doubts are not without foundation.

20 SURVIVAL STORMS 1938–1985

Adlard Coles & Peter Bruce

The difference between a gale and what has become known as a 'survival' storm is that in the former, with winds of force 8, or perhaps 9 (say 30 to 45 knots mean velocity), the skipper and crew retain control and can take the measures which they think best, whereas in a survival gale of force 10 or over, perhaps gusting at hurricane strength, wind and sea become the masters. For skipper and crew it is then a battle to keep the yacht afloat. There is no navigation, except rough DR, because the course is dictated by the need to take the breaking crests of the seas at the best angle.

In this chapter I shall deal broadly with a few further experiences of yachts involved in survival storms and hurricanes and the lessons to be drawn from them.

Where yachts more commonly get caught out in storms and hurricanes is on the western side of the Atlantic. It is for this reason that I have turned to America and the American magazine *Yachting* for information about survival storms, as the principal danger area lies on the route between the New England yachting centres and Bermuda, Florida and the Caribbean. The principal hazards are the tropical storms, with closed isobars and mean wind force between 34 and 63 knots, and the hurricanes with wind force of 64 knots or more, with gusts possibly reaching 170 knots. The maximum velocity is not exactly known, as most anemometers disintegrate at about 125 knots. Hurricanes as a rule occur between June and November and principally in September, but there can be out-of-season ones at pretty well any time of the year. According to Captain Edwin T Harding, US Navy (the author of *Heavy Weather Guide* and a meteorological specialist in the subject), waves of 35 to 40 ft are not uncommon in an average hurricane and, in giant storms, build up to 45 to 50 ft. Waves even higher have been reported, but happily they are very rare.

Bermuda yachtsmen tell me that during the winter there are severe storms in the Atlantic which, while not termed hurricanes, are equally formidable. They may last for a duration of three days with winds reaching the vicinity of 85 knots, well above hurricane force. Yachts do not always survive such severe storms. For example, the 70 ft schooner *Margot* or *HSH* (Home Sweet Home) left Bermuda at 1700 on a January evening bound south. The weather forecast was good for fifty miles south of Bermuda, but by 2000 the same night the wind reached 85 knots locally and for the next week the minimum winds recorded at Bermuda were 50 knots. The schooner has never been heard of since. It is thought that she went down on the first evening while running under bare poles, and that her extra large cockpit and unsafe companionway resulted in her being flooded by a following sea. It is not known whether she was towing warps or whether her loss could have been averted had she done so.

Another example of a winter storm was that experienced by the 68 ft schooner *Curlew,* which left Mystic, Connecticut, in a fresh north-wester on Sunday, 11 November 1962, bound to the Caribbean for charter service. She was skippered by Captain David Skellon, an Englishman, and the mate was Ed Lowe, a Connecticut

The schooner Curlew in distress in the Atlantic storm, north of Bermuda, with winds gusting up to 85 knots. Photo: USS Compass Island.

sailor. The two of them were the only deep-water sailors aboard and they took the helm in turns during the whole of the nights in the bad weather which followed. By Wednesday morning the wind was northerly, about Force 10, and the yacht was running under bare poles. A number of troubles had already developed, the most serious being the failure of the braking screw that kept the propeller shaft from turning and a bad leak in the propeller-shaft packing. The bilge pump operated by the main engine was only just capable of keeping ahead of the leak.

The storm steadily increased during Wednesday and throughout the night. *Curlew* had entered the Gulf Stream, where the seas became more dangerous. In the second watch the following morning she suffered her first real broach-to, and was knocked flat on her beam ends for almost three minutes before she slowly righted. After straightening the yacht out before the storm, the crew streamed a 3 in warp astern in a long loop, with drags lashed to it.

On Thursday the seas were higher than ever and the wind was estimated as gusting 75-80. At 0700 a mountainous sea broke over the full length of the ship and stove in the main cabin skylight. As a result of MAY DAY calls to Bermuda, *Curlew* was spotted by a search plane and at 1400 the USS *Compass Island* hove into sight.

The yacht then continued to run under bare poles on her course for Bermuda, with *Compass Island* standing by and giving course instructions by radio telephone. This says a lot for David Skellon's tough efficiency, operating the telephone and navigating below with water almost up to his waist, swilling over the charts and chart table. That night *Curlew,* under a lee created by *Compass Island,* succeeded in getting within a quarter mile of the flashing buoy off St George's Harbour. Shelter was at last at hand. But the wind must have shifted and it was so violent that no further progress could be made against it, even with the help of her powerful engine. Like *Dancing Ledge* off Cherbourg it was impossible to gain harbour and *Curlew* had to run off. By then the yacht's condition was critical and, as the weather forecast predicted a continuance of the storm for another 24 hours, there was no alternative left but to run off and abandon her.

Curlew manoeuvred alongside, under the lee of *Compass Island,* but in doing so broke her bowsprit and carried away her foremast and shrouds against the ship's side. Nevertheless, all the crew were rescued by *Compass Island* without injury by means of cargo nets—a creditable performance at night with wind little below hurricane force.

Now comes the strange ending to this story. Three days later it was reported that *Curlew* had been sighted. She was located and towed back safely into St George's Harbour. By then there was some 5 ft of water above the cabin sole and everything below had been smashed, but after survey it was found that the hull was undamaged. All her seams and fastenings were as good as new. She is Everdur fastened, mahogany planked over white oak, with teak decks.

Curlew's was a remarkable survival of a storm stated to be the largest low-pressure in the area for forty years. The 56 ft schooner *Windfall,* which left Mystic at the same time as *Curlew* on the same course for Bermuda, was never heard of again, as she broke up. All four of her crew were lost, and when last seen by a freighter they were hanging on to wreckage, but the weather was so bad that the freighter was not able to do anything to help them. Nine other ships were in distress at the same time as *Curlew,* and altogether the sea claimed over 144 seamen.

What is confirmed by the experiences in this storm is that once wind and sea have risen to or near hurricane force there is no knowing what will happen. *Windfall* was sunk, but *Curlew* survived, despite being partially waterlogged. Her tactic of running off, streaming warps, may have saved her, but despite this she broached several times. It is probable that *Windfall* may also have streamed warps, because it was the recognized method of coping with following seas.

Captain Skellon's log of *Curlew's* ordeal appeared in *Yachting*, February 1963. There

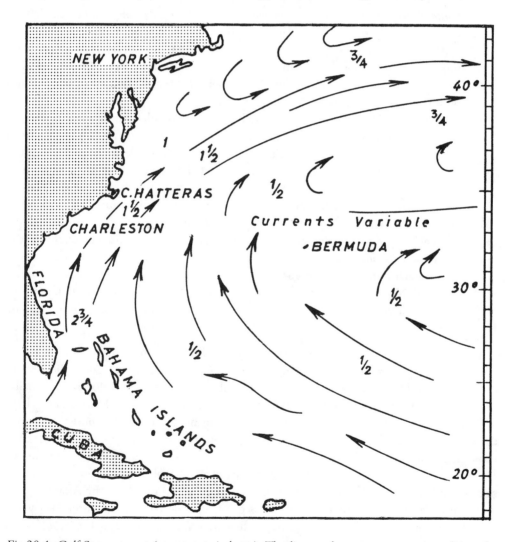

Fig 20.1. Gulf Stream current (approx. rate in knots). The diagram shows average summer conditions, but the stream varies in direction and rate in knots and is often much stronger than shown. It causes exceptional seas when the wind is contrary to the stream and its meanders provide a time-honoured speculation for navigators in the Bermuda Race.

are several comments I should like to make:

1 *The value of mechanical aids.* It was the engine-operated bilge pump in *Curlew* which enabled the leak to be kept under control, and the radio telephone which summoned assistance.

2 *The weakness of many steering wheels.* Five spokes of *Curlew's* wheel were broken when a man was swept against it by breaking seas. Damage to wheels is by no means rare in gales.

3 *Broaching-to.* In a storm, however violent, broaching-to is not necessarily disastrous.

4 *Partial waterlogging.* A yacht may survive despite being partially waterlogged. This is evidenced both by *Tzu Hang* (to which I refer later in this chapter) and *Curlew* in circumstances where one would imagine survival to have been impossible. After she had been abandoned, *Curlew* must have drifted out into the full force of the storm and yet left to herself she remained afloat lying a-hull with water several feet above her cabin sole.

5 *Coming alongside.* The rescuing ship has great difficulty, sometimes finding it impossible, in hurricane conditions, to lie alongside without damaging a yacht in order to take off her crew. A ship has to lie close alongside to ensure saving the men, and in heavy seas there is grave risk of the yacht's mast breaking against the ship's side. The mast may then provide an additional hazard, as the broken part and rigging will flail around, and may prove a lethal weapon if it strikes the crew or rescuers as they are climbing the scrambling nets. I am told that an alternative procedure which might be used today for rescue (such as the survivors of *Windfall* clinging to wreckage) would be to try and make a lee and drift inflatable liferafts to anyone overboard, as ships' lifeboats are useless under these conditions. The difficulty would be that the ship would drift rapidly to leeward down on the yacht, so great skill in handling the ship would be required.

Curlew may not have experienced the worst conditions of the Gulf Stream, as in her position the wind appears to have been blowing across rather than against the stream, though I am told that there is nothing certain about its course or velocity. Normally, the maximum strength of the Gulf Stream is found in the Straits of Florida and where it flows northward towards Cape Hatteras. Here, in northerly gales against the stream, the seas in the axis of the stream are more dangerous to a yacht in an ordinary gale than those in a storm or possibly even a hurricane in the open Atlantic. This accounts for *Doubloon's* experience of being twice turned over and possibly for the loss of *Revenoc*, one of the most deeply felt yachting disasters on the American side of the Atlantic.

Revenoc was a Sparkman & Stephens designed, outside-ballasted, centreboard yawl. Her dimensions were 42 ft 7 in LOA, 29 ft 7 in LWL, 11 ft 10 in beam, 4 ft 6 in draught, with a sail area of 883 sq ft. She was built to the highest specification and particularly well equipped for cruising and ocean racing.

On 1 January 1958 the yawl sailed from Key West bound for Miami. In her sailed her owner and skipper, Harvey Conover, his son Lawrence Conover, their two wives, and William Fluegelman. The crew were highly experienced. Harvey Conover was a veteran deep-water sailor who had sailed since boyhood and was a former commodore of the Cruising Club of America. His son, aged 26, had been brought up with boats and was a first-class seaman. William Fluegelman had sailed a great deal with the Conovers and was a former Coast Guardsman. Mrs Harvey Conover was an able and experienced hand and Mrs Lawrence Conover had also sailed extensively in the two *Revenoc's*.

On 2 January a NNE gale gusting 65 knots struck the area without warning. The Weather Bureau summary as reported in *Yachting* read: 'A big high pressure area over south-eastern U.S. was pushing back a broad, not especially severe, cold front south-easterly across Florida, the Florida Straits and the Bahamas. Meanwhile a small, intense low pressure centre [not reported until Thursday, when *Revenoc* was expected at Miami] suddenly developed on Wednesday over western Cuba, moving north-east across the path of the front. As they approached each other, the clockwise wind pattern of the front and the counter-clockwise wind around the low centre, both blowing from a generally NNE direction, combined to set up a sudden gale with terrific gusts in the Florida Straits before daylight Thursday.'

The seas under these conditions on the axis of the Gulf Stream would have been fantastic. It was in this storm that *Revenoc* was lost with all hands. No trace of her was ever found except for her swamped dinghy, which drifted ashore near Jupiter Inlet on 6 January, having been carried northward by the current.

The loss may have been caused by many things. Most probably, as *Yachting* suggests, it could have been by being run down by a ship at night, because shipping is heavy where she was caught out and, as I have remarked, a yacht's lights may be lost in the seas and spray of a gale. It could also have been accounted for by a mast over the side damaging the hull beyond repair before it could be cleared away, by the yacht being driven on to outlying coral reefs, or by a rogue sea stoving in the superstructure or decks or rolling her over like *Doubloon*.

The answer will never be known, but the tactics adopted in *Revenoc* can be guessed at. In an article by Carlton Mitchell in *Yachting* of June 1956 an extract was quoted from a letter by Harvey Conover, after he had been caught out before in a Gulf Stream storm in *Revenoc*. In this gale (gusting 56 to 65) he had run at 2 or 3 knots under bare poles towing warps so satisfactorily that he thought the yawl would take almost anything by this means. So, provided he had enough sea room, it is most probable that *Revenoc* was running before the storm trailing warps (and perhaps sails) on her last voyage. This is conjecture and the only lesson to be learnt from the tragedy is that no yacht, however sound, and no crew, however experienced, are immune from the dangers of the sea.

Returning now to the subject of hurricanes, let us consider Jean Gau's 29 ft 6 in ketch *Atom*, caught out in the path of hurricane *Carrie* in September 1957, about 360 miles south of Montauk Point, Long Island. She survived lying a-hull streaming one warp and with oil bags secured to the weather rigging screws. There are four points which may have helped her in weathering the storm. After crossing the Atlantic, her bottom was so foul that she lay almost dormant in the water. Her draught was only 4 ft

6 in and she had 2 tons of iron on her keel and 2 tons of inside ballast. She was thus of a type that would be difficult to capsize and there would be less tendency to be tripped by a long shallow keel than a deep narrow one. It is probable that she had a low rig, though I have no particulars of this. *Atom* was in the track of *Carrie*, but about 160 miles to the northward of Bermuda *Carrie* changed her mind, and sped off eastward across the Atlantic and finally made a precise landfall on the Fastnet Rock off south-western Ireland. As I have no particulars of *Atom's* course to compare with the track of the hurricane, I do not know whether she was anywhere near the centre, but even if she avoided the worst she would have been involved in gales so severe that they might come into survival storm category.

Atom's evidence is thus in favour of lying a-hull, but on 26 February 1966 the ketch was caught again between Durban and Cape Town. She lay a-hull with wind and sea on her port beam, but on this occasion she was completely rolled over through 360 degrees. She lost all her spars and sails. Jean Gau was asleep below at the time and fortunately suffered no incapacitating injury. He spent the next fourteen hours pumping the bilges (filled to the cabin sole), cutting away spars threatening to damage the hull, clearing rigging and finally coping with the horrid job of drying out the engine. Then, under power, he managed to get to Mossel Bay, some 75 miles distant.

The *Atom* had a reliable anemometer, which recorded the wind velocity of 60 knots. Whether this was in gusts or at the mean speed on the Beaufort scale (which at 60 knots would be force 11) I do not know.

Jean Gau's experiences show that one can voyage time after time across the oceans without harm, but it takes only one freak wave of particular size and shape, catching the boat on the wrong foot, to do real and sometimes disastrous damage.

One of the most remarkable stories of yacht survival in a hurricane was that of *Pendragon,* which was involved in hurricane *Carol* in 1954.

Pendragon was lying in the somewhat insecure Gosport harbour at the Isles of Shoals, situated in the Atlantic to the north-east of Boston. She is a cutter measuring 41 ft LOA, 30 ft LWL, 10 ft beam and 6 ft 3 in draught, built by Nevins in 1935. Her crew consisted of William H Mathers and his wife Myra and two friends, Mr and Mrs Smoot.

Meanwhile, hurricane *Carol* had stalled over North Carolina, but on the Tuesday morning, 31 August, came the startling news that she was well on her way again, moving up the New England Coast. The warning arrived too late to make a better harbour. In *Pendragon* all preparations were carried out for *Carol's* arrival, but at the height of the storm a kedge began to drag and *Pendragons* cable had to be cut to clear her. The wind was too strong to enable them to regain shelter under the lee of the breakwater upwind under engine, so in order to avoid going on the rocks *Pendragon* proceeded to sea in the dangerous quadrant of the hurricane.

Once in the open, the seas were found to be mountainous and the full weight of the hurricane winds was experienced. What we are concerned with here, however, is the method adopted so that the yacht could survive. The extraordinary thing is that she was handled under engine. This is the only occasion that I have heard of where an engine was used in an auxiliary yacht in open water during a gale, let alone in a hurricane. It seems to me that the average auxiliary engine would be useless even at force 7, though I have never tried the experiment at sea. So I wrote to Mr Mathers and from his answer it appears that it was done in this way.

The engine was a four-cylinder Gray with a rating of 25 hp. It was fitted with a 2:1 reduction gear giving a propeller speed of about 800 rpm in the hurricane. The propeller was 18 in diameter with a relatively flat pitch. In the hurricane, *Pendragon* was steered in the troughs beam to the seas. The crests were about 300/400 ft apart. In the troughs (where the yacht was under the partial lee of the waves), the engine gave enough power to give a speed of about $2\frac{1}{2}$ knots, which gave her skipper sufficient control to luff to the breaking crests of the seas, sufficiently to prevent her from being rolled over.

Pendragon's course is shown on the accompanying diagram. First she ran before the storm and her rather fine bow seemed to bury and 'her stern would lift, which made the rudder almost useless. At the bottom of each wave she would turn one way or another at her own discretion and roll badly. In one of the troughs she rolled so far that she took in green water over the cockpit coaming.'

Owing to the dangers and the outlying rocks off Duck Island, *Pendragon* could not be run off to east or north-east away from the centre of the hurricane. For this reason, after clearing Appledore Island, William Mathers rounded on to the port tack heading SW, although this would bring her nearer the centre of the storm. At about 1400 the sky cleared considerably in the west and the wind definitely let up. Half an hour later the wind veered from SE to S and SW, and *Pendragon* was able to lay to the E and SE. A tiny patch of blue – a beautiful deep blue – appeared and disappeared before returning again and gave moments of warmth and better visibility. At 1545 the lighthouse on one of the islands in the Isles of Shoals was seen. The yacht was run off

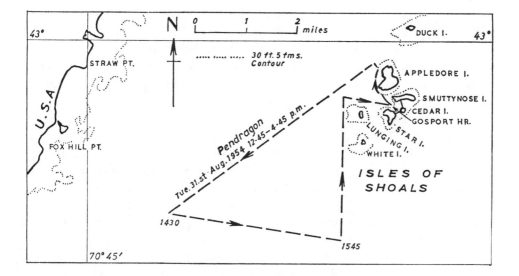

Fig 20.2. Course of Pendragon *during hurricane* Carol.

to the northward and an hour later she was safely back in harbour. As Myra Mathers puts it 'it did seem somewhat incongruous to crash around for four hours unable to see anything and end up exactly where we started'.

The only accident occurred when a sea hit *Pendragon* on the port quarter, just as the helmsmen were being changed and the yacht was momentarily off course. Myra Mathers had handed the tiller to her husband and was sliding past him when she was catapulted head first into the sea and drifted 25 ft to leeward. Her head struck a stanchion as she went overboard.

By a stroke of good fortune, the sea had knocked the yacht to a standstill and the hurricane winds quickly drove her to leeward to Myra Mathers, who was then picked up. Much the same thing happened when *Tzu Hang* was pitch-poled in the Pacific, to which I refer later, and Beryl Smeeton was thrown 30 yards to leeward. As the yacht had been dismasted, she was partially waterlogged and lay motionless. Beryl Smeeton, although injured, swam to the floating wreckage of the mizzenmast and pulled herself along to the side of *Tzu Hang*.

In both yachts the difficulty was to pull the survivor, weighed down by sodden clothes, out of the water into the safety of the cockpit. Myra Mathers was temporarily entangled round a stanchion, but once this was realized, she was quickly hauled on board *Pendragon*. The rescue of Beryl Smeeton in *Tzu Hang* was even harder, because she was unable to help, with only one arm owing to the injury to her shoulder. It took the combined strength of the two men to get her on board.

Hurricane *Carol,* in which *Pendragon* was involved, was intense, causing tremendous damage and insurance claims. During its unexpected dart up the New England coast, it blew down a 630 ft television tower on the roof of the radio station at Lynn, Massachusetts, so no further weather forecasts were received from that source. It also toppled over a crane on the breakwater at the Isles of Shoals just before *Pendragon* started on her hurricane cruise. It is remarkable that, although *Pendragon* was rolled over until her spreaders were in the water and the cockpit filled, she survived with no more damage than might occur in an ordinary gale. She fared better at sea in the hurricane than most of the yachts which had taken shelter in harbour, where houses broke up and floated down on the anchored vessels and boards, planks and other land objects took to the air, striking rigging and endangering crews. Great numbers of yachts broke their moorings or dragged their anchors and went ashore.

Points to note are:

1   It was only William Mathers's quick decision to put to sea that saved *Pendragon* from the fate that befell many yachts in harbour. Big ships sometimes leave port when hurricanes are anticipated, as they are safer in deep water far from land, but I have not heard before of a yacht doing so. However, I am not familiar with hurricane areas, so I write with no authority on this matter.

2   Attention is drawn to the difficulty of recovering a man or woman over-board even when alongside the yacht. The added weight of sodden clothes and the violent motion of a yacht in a storm combine to make the task unexpectedly difficult. This is particularly significant when a yacht's crew consist of two only, such as the owner and his wife. If one goes over the side, even if a hold is

retained, the weight could be too much for the one remaining on deck. In such an event a rope ladder or steps could make all the difference in an emergency.

3   A yacht provided with enough power and with a big slow-revolving propeller, may be able to survive a hurricane, as *Pendragon* did. In her case, the seas seem to have been so immense and so long that enough speed could be obtained in the troughs to luff to the crests, but I doubt whether this would be possible in short and very confused seas. Furthermore, I do not think *Pendragon's* tactics could be adopted by a yacht equipped with a high-revving propeller, as the strain would be too great on the shaft with the propeller alternately racing in the air and at full load in the water.

Most of the gales and storms which I have described occurred in waters frequented by yachts and thus they afford practical examples of what can happen when engaged in ordinary cruising on the American side of the Atlantic, but some of the worst storms and the highest seas are found in the high southern latitudes, where yachts rarely sail except for the occasional world voyagers or Cape Horners.

The classic example of a storm of supreme violence in the South Pacific Ocean was afforded by William Albert Robinson when he was caught out in one in 1952 about 40° 45′ 50. Robinson had circumnavigated the globe and is one of the best-known and most experienced deep-sea cruising men of this generation. He had had experiences of other storms and recorded hurricanes, but the storm which he describes as 'The Ultimate Storm' in his book *To the Great Southern Sea* was by far the greatest that he had experienced in a lifetime of deep-sea voyaging.

His yacht was named *Varua*. She was a brigantine of 70 ft overall length, designed by the late Starling Burgess in consultation with her owner for deep-sea voyaging, to be capable of weathering exceptional gales and storms and able to run before them cleanly with little risk of broaching-to. As yachts go she was a big ship, and she owes her survival to size, design and her owner's experience. I do not think any ordinary yacht, such as yours or mine, could have lived through the ordeal which she encountered.

During the storm *Varua* lay-to under forestaysail and lower staysail until the seas reached such a height and steepness that her sails were alternately blanketed in the troughs and blasted by gusts on the crests. They were then lowered and *Varua* lay a-hull. Instead of lashing the helm down (as I do, rightly or wrongly) the wheel was lashed amidships and, finding her natural drift, the brigantine fell off several points and headed slowly downwind with the seas on her quarter. Oil was used and Robinson states that the slick was more effective than when the boat was hove-to, when most of it was blown to leeward.

The gale backed slowly from north-east to north, and towards midnight *Varua* began to get out of control. 'The seas were so huge and concave at this point that the whole upper third seemed to collapse and roar vertically down on us. Our oil had little or no effect now, as the surface water was all being blown to leeward.'

Robinson unlashed the wheel and ran her off downwind dead before the storm, gathering speed under bare poles to 6 or 7 knots. As he considered this dangerous, he let go five 75 ft lengths of 2 in warps plus 200 m of smaller lines. This reduced her speed to 3 or 4 knots and she steered under perfect control, and the oil slick seemed more effective at this lower speed. Nevertheless, at times she ran down a sea and buried her bowsprit in the trough before rising again. Robinson says that if *Varua* had not been trailing drags she might have run right down. As he puts it, 'When a fifty-ton, seventy-foot vessel surfboards shudderingly down the face of a great sea on its breaking crest, you have experienced something.'

The detailed description of this storm should be read in Robinson's book, but briefly his points are:

1   Conventional methods of riding out the average gale are totally inadequate in exceptional storms in the ocean. He absolutely condemns sea anchors for deep-keeled hulls in extreme storms. He contends that 'the greatest effort of wind and wave-crests is exerted on the forward part of the ship, which has the least grip on solid water. Thus as the vessel makes sternway, as it is bound to do while riding to a sea anchor, the bow falls off, pivoting on the after-part of the hull, which has deeper grip on the water.' This explains the reason for what many cruising men have found in lesser gales. He goes on to question the value of a riding sail in an ocean storm, as, even if it stood up to the blasts of wind on the crests, it would be becalmed and ineffective in the troughs. He also draws attention to the risk of the sternway causing the rudder to break.

2   Robinson states that no ship would have been capable of holding her bow up into such seas as *Varua* experienced without sustaining major damage, and he proved his theory that to survive in such a storm, a yacht must be allowed to take her natural position, running dead before the seas under bare poles, 'moving just fast enough to retain good steering control, using drags as a brake to prevent going too fast'. For *Varua* this meant 3 to 4 knots. As he puts it, 'The ship was alive and responsive . . . we had flexibility, choice of action when the wind shifted, freedom to swerve to meet a great sea coming in out of line with the others. And when a monster of a sea did come along and break over us we met it end on, offering the least possible resistance and gave with it.'

3   *Varua* used oil with some effect when she lay-to at the start of the gale, but she was forereaching too fast and much of the oil was left astern. When she was lying a-hull the seas were so concave that the surface water and the oil on it was blown to leeward. Oil seems to have helped when running streaming warps, but a considerable amount must have been used, for besides two oil bags on each side, oil was pumped out through the forward heads.

Tzu Hang in her attempts to round Cape Horn from the Pacific to the Atlantic was involved in survival storms in much the same latitudes as *Varua*. On the first occasion she was manned by a crew of three, consisting of her owner, Brigadier Miles Smeeton, his wife Beryl, and John Guzwell of *Trekka* fame. In this attempt to round Cape Horn she was pitch-poled, stern over bow, when running streaming 110 m of 3 in hawser. During the second attempt, when she was sailed by Miles Smeeton and his wife alone, another storm was encountered. This time *Tzu Hang* lay a-hull, but she suffered a complete roll over. On both occasions the yacht was dismasted, severely damaged and partially waterlogged. The seas which did the damage must have been 'freak' waves,

Close up picture of a following sea. It is difficult to understand how any yacht, or even a small ship, can survive in such seas. Photo: de Lange.

formed by a combination of wave trains with unlimited fetch in the wastes of the Pacific. 'Sometimes', wrote Miles Smeeton, in *Once is Enough*, 'a wave would seem to break down all its front, a rolling cascading mass of white foam, pouring down the whole surface of the wave like an avalanche down a mountainside.'

In weather conditions such as these, few yachts would live without sustaining damage, whatever their type and whatever the method of defence adopted. The astonishing thing was that she survived at all. *Tzu Hang* had a tough crew, and temporary repairs were effected on each occasion, but had the seas which did the damage been quickly followed by others equally formidable she surely must have gone to the bottom. Possibly the dismastings enabled her to ride the seas better and thus saved her from this catastrophe, but *Doubloon* was rolled over again after she had been dismasted so one cannot be sure of this.

On the other hand, there are many examples of yachts which have sailed in the Roaring Forties and round Cape Horn without incident or near disaster. This was the 'impossible route' chosen by the Argentine, Vito Dumas, for the great single-handed voyage he described in his book *Alone through the Roaring Forties*. Dumas's yacht *Lehg II* was a 31 ft 2 in Norwegian type from the board of Manual M Campos, and was a modernized version of the old Rio de la Plata whaleboats, somewhat akin to a Colin Archer double-ender. She was designed for the purpose of ocean voyaging, with a long keel, and to be easy to steer in all weathers. The ballast keel was $3\frac{1}{2}$ tons of iron and the design provided a high degree of reserve buoyancy. No inside ballast was carried. The success of the design was proved by the apparent ease of steering single-handed without the modern aid of self-steering. There must have been plenty of buoyancy in her pointed stern to have survived the seas which Dumas experienced. In his voyage it was not a matter of a gale here and there but of almost continuous dirty weather, with winds on occasions estimated to be gusting up to 70 knots. His tactics in heavy weather were original. 'As regards a sea anchor,' he writes, 'I have one point of view which settled the question for me; I would never give such an object sea room. I am convinced that a boat can stand up to any sea, comfortably enough, under sail. She has freedom of movement and can lift to the sea. Should the wind force exceed 50 knots I would say, contrary to the opinion that following seas play havoc by breaking on deck, that one of my favourite pleasures was to run through squalls on a mattress of foam. My speed on these surf-riding occasions exceeded 15 knots: I then presented the stern to another wave and began this exciting pastime anew.'

Some readers may regard the speed of 15 knots as exaggerated, but the exact speed is immaterial and what is clear is that *Lehg II* experienced the right length of ocean sea which enabled her to surf for appreciable periods at far above her theoretical maximum speed. Dumas does not give any detailed description or advice on how this surfing was accomplished, or of how he managed without self-steering. Surfing in ocean seas can be dangerous on account of the risk of being thrown down into the trough and pitch-poled. However this may be, Dumas ran before gales at about 5 knots and succeeded in sailing round the world in the most dangerous waters that can be found and arriving at the end of the voyage with his boat in perfect condition.

There are more recent examples of yachts which have voyaged in the dangerous

In storms of extreme violence, seas can become absolutely chaotic. Note the perpendicular wave rising against the sky at the left centre. This remarkable picture looks as if it comes from another world. Photo: de Lange.

Bernard Moitessier describes the seas which Joshua encountered in the South Pacific as 'breaking without interruption from 650 to nearly 1,000 ft.' He ran under bare pole taking them 15° to 20° on the quarter to avoid being pitchpoled. The yacht was of steel construction, steered from within a steel cupola. Photo: de Lange.

waters of the South Pacific and rounded the Horn without resort to streaming warps or any of the conventional methods of weathering gales and storms. When Sir Francis Chichester rounded the Horn in *Gypsy Moth IV* in March 1967, after making a remarkably accurate landfall without a fix of sun or stars for three days and little sleep for a week, he was running under storm jib. It was evidently blowing very hard, with the violent gusts and high seas for which the Horn is notorious. *Gypsy Moth*'s cockpit was filled on five occasions and once it took fifteen minutes for the water to drain away, which provides further evidence of the inadequacy of the drains in self-emptying cockpits.

Gypsy Moth IV ran under storm jib and there is no mention of her streaming warps. Her speed seems to have been between 5 knots and later 7 knots, so this affords another example of a yacht maintaining considerable speed when running in heavy seas before a gale.

Much valuable evidence on the subject of running before storms comes from Monsieur Bernard Moitessier, who, in 1967, was awarded the Blue Water medal of the Cruising Club of America and the Wren Medal for Seamanship of the Royal Cruising Club for his outstanding voyage from Moorea to Alicante via Cape Horn. He has written a book entitled, *Cap Horn a la Voile*, on his experiences but, in the meantime, I am indebted to him and the *Royal Cruising Club Journal* and *Cruising Club News* (published by the CCA) for the following information.

Joshua, in which the voyage was made, is a 39.6 ft double-ended Bermudian ketch of 12.1 ft beam and 5.25 ft draught, designed by Jean Knocker. She is of steel construction and has a fixed keel. The sail area of 960 sq ft is considerable for a yacht undertaking such long-distance voyages. A feature of her design is the 'pilot's post', which is a metal cupola from which she is steered. *Joshua* left Moorea (the island west of Tahiti) on 23 November 1965 and rounded Cape Horn on 11 January 1966, forty-nine days later. As it chanced, the wind was moderate from the NW when Cape Horn was rounded. No difficulty was experienced and as with *Tzu Hang* it was in the South Pacific that *Joshua* encountered a 'survival' storm in which she nearly foundered. This storm lasted six days and was caused by two low-pressure systems. Moitessier had no anemometer, but he estimates that the wind was at hurricane force in the gusts, which suggests that the mean strength would be about Force 10 or perhaps force 11 on Beaufort scale. In the South Pacific, given six days for the seas to build up, this would be a 'survival' storm. The seas were reported to have been absolutely gigantic and their length was estimated to be about 500 to 560 ft, breaking without interruption for from 650 ft to nearly 1,000 ft, leaving acres of white water behind them. They were described as being 'absolutely unbelievable'.

Joshua at first ran before this storm towing five long hawsers, varying in length from 100 to 300 ft, with iron ballast attached and supplemented by a heavy net used for loading ships. These afforded so much drag that the yacht failed to respond quickly enough to the wheel. They also failed to prevent her surfing on the crests of the gigantic waves. On one wave, which Moitessier says was not especially large, but just about the right size for surfing, *Joshua* took off like an arrow, with the warps behind her as if 'dragging a fishing line', and buried herself in the wave at an angle of 30 degrees, so that the forward end of the boat was buried up to the ventilator abaft the mast.

Another wave taken in the same way might have caused *Joshua* to pitch-pole just as *Tzu Hang* had done and been dismasted, but happily the next really dangerously breaking sea caught the yacht at an angle and Moitessier thinks it is this that saved her from surfing and pitch-poling. It was then that he remembered the Dumas technique of running in gales at about 5 knots and putting the helm down on the arrival of each wave sufficiently to luff a little so as to take the wave at about 15 degrees to 20 degrees on the quarter. By this means a yacht will not be thrown forward surfing and in danger of pitch-poling because she is at an angle to the sea, nor will she be rolled over as she is not abeam to the sea, but there must remain some risk of a broach-to.

Accordingly Moitessier cut the warps and released *Joshua* from the drag. From that moment she was safe, following the Dumas technique of running fast and taking the seas 15 degrees to 20 degrees on the quarter. Moitessier says that *Joshua* could not possibly have survived except by this means. This ties up very closely with the experiences of Warren Brown, who ran in *Force Seven* before a hurricane at speed taking the seas on the quarter. There is much to be learnt from Moitessier's theory of running to which I refer below:

1   In extreme storms in the South Pacific, the ordinary methods of surviving storms are not enough. A sea anchor would be useless. Heaving-to would be out of the question. Streaming warps may not prevent surfing at perhaps 15 knots with the risk of pitch-poling.

2   *Joshua* was of steel construction and in her owner's opinion would not have survived without it. She was constantly struck by seas sweeping over the whole vessel up to the mast. These would have carried away any timber deckhouse and disaster would have followed. *Joshua* was steered from a steel pilot cupola. It was not safe on deck, even with lifelines, as the yacht sometimes disappeared completely below the sea. Moitessier also recommends that any yacht planning world voyaging should have a flush deck if she is of wood construction.

3   It is recommended that storm sails should be small and not too heavy and cumbersome. I think this applies to any sails such as a storm trysail. The old-fashioned ones used to be very heavy for the sake of strength, but cumbersome heavy-weather sails are difficult to handle and to set. There is no need for very heavy sails, especially now that terylene is used, because the area is so small that canvas of even moderate weight is very strong in relation to the area.

4   Moitessier states and then repeats that 'no one can claim he will not founder in these latitudes'. This confirms the opinion of other deep-sea cruising men that a time can arrive when no yacht can be certain of survival, whatever her size or rig may be.

5.   Moitessier does not say a great deal about the form of the seas which he found most dangerous. He describes such seas as 'crazy', which is, of course, the equivalent of 'freak' or 'rogue'. Such seas usually have heavy cascading crests, but Moitessier states that this is not necessarily the case, as the most dangerous seas in the South Pacific are very steep, not necessarily breaking, which pick up and throw the boat into the trough and can cause pitch-poling. He also refers to

waves much bigger than the others and coming from a different direction. For a yacht to be pitch-poled requires a sea of such immense size that it is not likely to be encountered except in very exceptional storms and with the unlimited fetch of the ocean.

Four Pacific Ocean accounts are added to Adlard Coles' collection of survival storms as further examples of events in extreme circumstances.

The first is a story where survival involved the use of oil and warps. On 2 May 1938 Group Captain Geoffrey Francis was caught out in a typhoon whilst taking his new 57 ft ketch *Ma-On-Shan* on passage south from Hong Kong. Happily he had an abundant supply of the two items, rope and oil, which he used to good effect. He had taken the oil in case he ran out of wind and the rope because he felt it was sure to come in useful. During the typhoon he deployed between 900 and 1200 ft of a huge length of hairy natural fibre rope from the stern, and used a marlin spike to puncture in turn the 25 four gallon cans of tractor vaporizing oil which had been stowed aft of the mainmast. Though typhoon-sized waves were breaking all round the vessel as she ran under bare poles at hull speed, significantly none broke immediately behind her.

The seas were kept as much as Geoffrey Francis dared on the quarter as he believed he was in the dangerous quadrant of the typhoon and the steady pull of the warp was most effective in preventing *Ma-On-Shan* from broaching. He remains convinced that it was the combination of the large quantity of oil used, and the warp streamed in very long lengths, that saved the boat. It was not easy to say how much rope was streamed, but if the lengths were between 450 and 600 ft, as seems possible, they would have probably been long enough to span a whole wavelength of a typhoon wave.

The *Ma-On-Shan* had been spotted by the P & O passenger liner *Rawalpindi* shortly before the arrival of the typhoon and, incidentally, when *Ma-On-Shan* eventually made Saigon, Geoffrey Francis was unable to cash any cheques. His friends and associates back at his home base in Singapore had been told by the *Rawalpindi's* officers that the *Ma-On-Shan* must have been lost in the typhoon, and a stop had been put on his bank account.

The second story comes from Alby Burgin, an Australian who sailed the 37 ft *Rival*, a Vashti design by Alan Buchanan, through cyclone *Emily* on a race from Brisbane to Gladstone, in 1972.

'The wind was recorded at Bustard Head lighthouse at 132 knots and within a radius of 15 miles from our position two large fishing trawlers, a steel yacht and a trimaran broke up and sank with the loss of 12 lives. The waves running in shallow water against the current were 30 ft high, very steep and coming to a point where the top three or four feet was being blown off. Horizontal water was everywhere, visibility was about 50 ft and the noise deafening.

We were under storm jib and I was the only person on deck at the time, the crew being down below behind securely fixed storm boards. If I'd held my head up, the wind and water would have cut my eyes out; as it was, the force of them was hurting even through my wet weather gear and undergarments. Waves were crashing

191

continously over the bow and rolling along *Rival*'s deck across the overflowing cockpit. The compass in front of the tiller was mainly under water.

The boat was managing the conditions fairly well until an exeptionally large wave, like an enormous green and white breaking veranda, crashed on to the deck, taking handrails, the liferaft and breaking the plate glass windows of the coach roof. The tremendous weight of water buckled my knees and rolled *Rival* through 360°. My lifeline had been fastened to a large teak cleat but when the yacht was upside-down I found myself loose in the turbulent sea. The first thing that came into my mind was to stay under water to allow the sea to wash me away from the yacht. If I surfaced too quickly, I had visions of bashing my head on the rolling yacht or hanging myself in the rigging. When I estimated I was clear, I found I could not make headway to the surface. I realised that I must be water-logged, having been exposed on deck for many hours. My next thought was to get rid of my wet weather gear which I managed to do whilst still under water – I was glad that I never wear sea boots which are almost impossible to get off under water. Only then was I able to get to the surface, to find the dismasted *Rival* 20 ft away. I swam to the yacht and my crew, being on deck by this time, helped me back on board. I had lacerations to both arms and face. When the cyclone had passed over we jury-rigged and sailed into Gladstone. Then I had time to wonder at the serious injuries I would have received from the deck gear had my lifeline not broken.'

Many years later, another Australian race will long be remembered for the loss of four yachtsmen who were drowned when their yachts sank. This was the JOG Tasman cup on 15 April 1983 with a 44 mile overnight course from Sydney harbour, south along the exposed coastline to a buoy off Port Hacking, and back. The forecast was for 20 – 30 knot south to south-west winds as well as south-easterly swell, giving a beat down to Port Hacking. It seems that the wind that materialised was much as forecast, but it created a secondary swell across the one from the south-east. In addition, there was a south-going current running offshore – locally known as the 'set' – and the whole combination created some very steep breaking waves.

The first yacht to get into trouble was the *Montego Bay*, a Hood 23 production boat with an experienced crew. After falling heavily off several waves she was found to be making water. The crew decided to abandon the race but, in spite of some spirited work with a bucket, the boat rather quickly sank. Just beforehand they had succeeded in firing two flares but had obtained no acknowledgement from their MAYDAY transmission and were too late to reach their lifejackets in the cockpit locker. There seems to have been some confusion in the two boats of the racing fleet that saw the flares as to whether there was a genuine emergency, and it was not until three hours later that a yacht returning to Sydney after abandoning the race happened to encounter two of *Montego Bay*'s crew in the water, and raised the alarm. Two more of the crew were picked up half an hour later, but a fifth man was never found.

Meanwhile another yacht, the Farr 727 *Waikikamukau* with a crew of four, had also sunk. She had been capsized by a particularly large breaking wave, and with no wash boards in place she had filled rapidly through her main companionway. The crew were attached to the boat with their safety harnesses, and one of them was unable to detach his safety harness before the stern of the yacht submerged and he was drowned. Another crewman had great difficulty in kicking off his seaboots which were hampering his ability to swim. None of the crew were wearing lifejackets, and the horseshoe lifebuoy stowed in the stern sheets of *Waikikamukau* had gone down with her. The three remaining crew quickly separated in the heavy seas and darkness.

In the end, there was only one survivor who had to swim for ten hours before being picked up by a passing fishing boat. He reported that both helicopters and rescue craft, who were still looking for the fifth man from the *Montego Bay*, had repeatedly passed close by him, but had failed to spot him or hear his cries.

But for the warmth of the water, no one would ever have known what had happened to the two yachts and their crews, and of the many hard lessons learnt from the tragedy of this race, the importance of lifejackets is clearly one of them.

A less harrowing event took place in the South Pacific in February 1985 when an experienced Australian couple, Alan and Kathy Webb, with their 16 year old daughter, Portia, were overtaken by a survival storm. The heavily-built 45 ft steel cutter *Supertramp*, in which her owners had great confidence, had been designed for world cruising with a moderate fin keel, a skeg hung rudder and immense strength.

They arrived in the Roaring Forties with Easter Island 450 miles to the west and Chile some 1400 miles to the east (40° 20' S, 101° 37' W) to be confronted by a plummeting barometer and 80 knot winds which continued for 36 hours. The self-steering soon broke and Alan Webb was left to steer for the duration as they ran under bare pole in winds they estimated at force 12. Huge breaking seas developed, aggravated and confused by squalls. The Webbs make a practice of lying a-hull with the wheel lashed to windward in gales but in this instance Kathy Webb says it would have been impossible to heave-to as they would certainly have been capsized. Alan comments:

'When running down waves of this extreme height, *Supertramp* developed too much speed, but by surfing down them at an angle, I found I could remain high on the crest and also take speed off the yacht. It was vital to keep her on top of the first tier, high upon the shoulder of the wave, by steering about 20° across its face rather than let slip into the trough. We still stood a chance of being rolled over, but a slow roll-over would do much less damage than being pitchpoled at speed.'

The Webbs have since wondered whether they might have hoisted their storm jib in the lulls to give steerage way, but conclude that their speed under bare pole seldom dropped below four knots, and for the rest of the time any sail would have been too much.

Several other interesting points came from the experience. Firstly, Alan Webb felt his beach surfing experience was helpful in negotiating the breaking seas. He was also helped by the fact that *Supertramp* was extremely controllable under bare pole, and could even be tacked when lying a-hull. Another good feature of the boat was her centre cockpit with 10 in coamings and large diameter drain holes. Finally they mention 'Polar Mitts', given to them by a Canadian icebreaker in the Magellan Straits, as the only effective gloves they have ever come across for wet freezing conditions.

21 THE 1979 FASTNET

Peter Bruce

The Fastnet storm of 14 – 15 August 1979 is still regarded as the standard by which other yachtsmen's storms are judged. This seems reasonable as, though short-lived, at between force 10 and 11 it was a very severe storm for August. In addition, three hundred yachts were involved, and, with 15 lives lost plus the crew of four from a trimaran following the race, it became a major international news item.

In order to gain a perspective on its severity, it is useful to have been given the actual wave heights by Sheldon Bacon (whose Wind Waves chapter – Chapter 31 – gives more details on wave measurement). He says that the closest scientific instrument was installed on the Seven Stones lightvessel, 130 nautical miles south-east of the Fastnet Rock. Although the light vessel was not under the worst of the storm, the readings show some hefty wave heights. The table (Fig 21.1) shows monthly maximum values of significant wave height taken every three hours recorded between 1962 and 1986. It can be seen from these values that a typical winter maximum lies in the range 5 – 9 m with a few extreme values up to 11 m; and that a typical summer maximum lies in the range 2 – 6 m, with a few values up to 9 m, mostly in the April and September months. Within this pattern, the measured maximum during the 1979 Fastnet at the Seven Stones lightvessel was 7.8 m; the height of an ordinary severe winter storm, but an unusually severe summer storm. As all the reports suggest that wave heights further north were rather higher than those experienced at the Seven Stones, there can be no doubt that the 1979 Fastnet storm was of rare intensity for the time of year.

This chapter contains descriptions of incidents aboard a number of different yachts, and a summary of the knowledge gained from experiences of the whole fleet. By way of introduction there follows my own account of the storm, written a few days after the event, and as seen from the 39 ft British Admiral's Cup Team yacht *Eclipse*. The boat in question was not far from the centre of events, and completed the course without any major mishaps; incidentally, she finished first in her class, first of the Admiral's Cup fleet and second overall, behind Ted Turner's 61 ft yacht *Tenacious*.

'At 0900 on the morning of the 13 August, *Eclipse* was 127 miles short of the Fastnet Rock in company with the similarly sized yachts *Casse Tete* and *Regardless*. The weather was dull and visibility poor; it was curious that the wind was unexpectedly from the north-east. During the forenoon the barometer dropped steadily (Fig 21.2), the wind swung falteringly to the south-west via east, and there was something of a brooding menace in the clouds. Preparations were made for bad weather and the 1355 shipping forecast was awaited with interest. This gave south-west, force 4 – 5, increasing force 7 and becoming west.

As the afternoon wore on, the wind freshened and some wild surfing at 15 knots caused us to lower the spinnaker. An hour after this a No 2 genoa and 2 slabs in the

Monthly Maximum H_S (m)

Year	1	2	3	4	5	6	7	8	9	10	11	12
						Month						
1962	(4.06)	7.77	7.17	8.95	6.11	3.48	3.86	4.44	6.47	6.03	7.66	7.55
1963	5.18	-	-	-	-	-	-	-	-	-	-	-
1968	7.07	4.53	6.18	4.99	4.45	3.08	2.79	4.98	6.81	5.47	5.27	8.24
1969	7.81	5.70	3.63	5.52	3.46	4.57*	3.69	3.18	3.40	5.08	8.03	6.98
1971	-	-	-	-	-	-	2.36	3.60	4.28	5.72	6.32	7.70
1972	7.97	7.59	9.36	7.56	6.85	4.16	3.40	4.01	2.61	5.37	6.89	7.00
1973	7.12	7.66	5.59	5.05*	5.00	3.43	3.80	4.32	5.35	3.83	5.06	6.39
1974	9.84	8.67	5.48	3.57	5.20	4.01	-	-	-	-	-	-
1975	-	-	-	6.30	3.33	2.70	3.58	3.72	6.90	4.74	6.00	6.03
1976	7.30	6.11	10.68	4.48	3.79	3.67	3.15	2.05	5.70	7.71*	6.15	9.07
1977	7.33	7.01	7.37	6.41	4.13	4.09	3.36	5.05	4.48	5.67	6.41	6.81
1978	7.44	(2.74)	-	-	-	3.09*	4.08	3.25	5.33	3.60	5.34	9.34
1979	6.71	5.38	6.50	4.99	5.33	3.32	3.34	[7.80]	(2.63)	5.88	5.75	10.57
1980	7.55	8.16	9.28	3.49	3.99	5.14	2.94	3.72	7.60	7.64	6.12	7.55
1981	6.11	5.57	6.53	6.39	5.56	(2.95)	-	(3.80)	(5.96)	(5.19)	4.93	8.25
1982	5.96	7.17	6.88	4.39	3.91	4.31	2.53	4.57	4.77	11.13	7.64	8.75
1983	9.19	7.49	6.27	4.54	6.13	4.23	3.53	2.71	9.01	7.93	6.55	8.67
1984	9.64	10.22	6.08	3.87	5.07*	-	-	-	-	-	-	-
1985	(6.37)	6.28	7.18	9.37	6.21	6.97	3.05	5.34	4.86	5.74	5.37	7.17
1986	8.36	5.87	7.93	5.10	(5.82)	3.70*	(2.72)	(6.29)	3.55*	(5.62)	(10.24)	8.79*

*: missing 10-20% data

( ): missing >20% data

Fig 21.1. Monthly maximum significant wave heights in metres at the Seven Stones lightvessel between 1962 and 1986. There is a box around the height recorded in the Fastnet storm of August 1979.

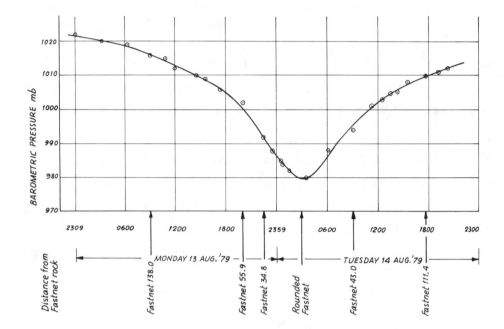

Fig 21.2. The plot of barometer readings taken in Eclipse *by Peter Bruce.*

mainsail were too much, so much so that we changed right down to the storm jib. The 1750 shipping forecast gave a gale warning with south-west force 5, locally 6, increasing force 8 and veering north-west later. By 2330 the barometer had dropped to 988 mb from its 2005 reading of 1002 mb. We climbed to windward of the rhumb line so that if the wind continued to increase we could run down to the Rock. The wind strengthened steadily and the seas built up rapidly. At about 0100, 12 miles off the Rock, a curling wave top caught *Eclipse* beam-on and she was thrown over to about 70 degrees. Evidently we had too much sail up, but though a messenger was rove to the third slab it would have been a daunting task to get it down in the normal way so we lowered the mainsail. There was then a discussion as to what should be done next. On the one hand it was pointed out that it was getting very rough and the seas between the Fastnet and Cape Clear Island might be dangerous enough to warrant giving up the race. On the other hand, from the race point of view, it was important to get round the Rock before the wind went round to the north-west. Eventually, the mainsail went up with the third slab, and we reached on to the Fastnet which has to be left to port. We took it a little wide, bearing in mind the shoal patches round it. Even so, when dead downwind of the Rock another curling top caught the boat beam-on and threw us sideways.

We tacked easily enough and passed to weather of the Fastnet at 0255 on Tuesday. The breaking seas, illuminated by the light, were a magnificent and awe-inspiring sight. The new course was quite comfortable and the wind was whistling less. There was even talk of hoisting the No 4 genoa, but then, at 0330, with the barometer at 980 mb, the wind returned with renewed strength from the west. We dropped the mainsail in haste and lashed it to the boom with a spare genoa sheet. Our speed with just the storm jib was about seven knots, such a speed giving useful control in the wild seas. It seemed that a wave with a breaking crest would mount up from one direction, then moments later another wave would quickly follow from a different slant.

Fortunately, *Eclipse* had a good supply of skilful helmsmen who coped well with these frequently breaking and confused seas. Also she had a tiller, rather than a wheel, and though heavy work, the short response time of a tiller may have been an advantage over a low-geared wheel in this particular situation.

We all got wet and some were seasick, but an effort was made to keep things tidy, and hot soup was brewed for those who could take it. To do this, the galley stove had to be reshipped, having leapt from its mountings at some point, despite a preventer wire intended to stop this happening. Water was finding its way down below, though operation of the electric pump with its wandering suction hose (fitted with this situation in mind) kept the amount in the bilges to a minimum.

After listening to the 0015 Tuesday forecast, which gave us south-west veering west force 7 – 9, locally 10 (something we had already well appreciated), both radios were left off for the night; but we did wonder anxiously about the smaller yachts in the race. Meanwhile, *Eclipse* continued for the rest of the night under storm jib. Things were uncomfortable but we did not appear to be in acute danger.

At daybreak the seas were spectacular. They had become very large, very steep and breaking awkwardly; but the boat was handling well. We could see the 46 ft Hong Kong Admirals's Cup team yacht *Vanguard* running under bare poles close to starboard, and later heard that she had been nearly rolled right over running back from the Rock with her No 4 genoa up. With this sail she had been surfing much faster than her young helmsman felt was comfortable or wise; then after a momentary lapse in his concentration, the boat turned broadside on in a trough to be rolled over by the breaking crest of the next wave. The entire watch on deck went over the side to the limit of their lifelines, and all winch handles and torches on deck were lost overboard. Fortunately, the deck crew were able to climb back on board suffering only bruising and shock, and there were more winch handles and torches stowed down below in this well equipped yacht. Following the capsize, the No 4 genoa was lowered and *Vanguard*'s speed dropped to a comfortable and much more manageable six knots, with an occasional surf to ten. As luck would have it, a wave dumped heavily on board as the sail was being taken down through the companionway, bringing a huge amount of water below that filled the boat to two feet above her floorboards. It took a strenuous half an hour to pump out.

From *Eclipse* we saw no other yachts, apart from *Vanguard*. This was probably due to the fact that we had tended to keep to the south of track having allowed, as it turned out, rather generously for leeway and surface drift. The many yachts which had not managed to make the Fastnet Rock would have been driven to the north-east in the night and were probably many miles distant.

The barometer had risen steadily since 0330 and though the 0625 shipping forecast was still giving force 10, we felt the wind would moderate before long. Sure enough, at midday, the gaps between the gusts became longer and their strength reduced. The

The American yacht Ariadne *dismasted and abandoned. Photo:* RNAS Culdrose.

Rescue of three of the crew of Trophy *by* HNLMS Overijssel. *Their sole support is the upturned upper ring of the liferaft, the lower ring with the topping-up pump and the canopy having broken away hours earlier. The survivors have tied themselves on to the ring with their lifelines. Photo: Peter Webster, Lymington.*

need for more sail met with little initial enthusiasm, but at 1220 we decided to put up the No 4 genoa. No difficulty was experienced with this much larger sail, and within a short time the watch on deck had also hoisted some mainsail. The 1355 forecast once again gave force 10 winds, but with the barometer now at 1003 mb this was ignored. On the news following the forecast we heard of the havoc in the fleet behind us for the first time, but by then, there was not much we could do except sail on.

The usual Admiral's Cup team's 'roll call' by the Dutch guardship *Overijssel*, obviously busy with rescue work, did not take place so we reported to Land's End radio that *Eclipse* was safe. Seemingly, this message got no further. As we rounded the Scillies under full sail the seas became much less violent and the wind went light during the night. The next morning a spanking breeze came up and we had bright sunshine. By now the reports of the disaster, to which we listened with dismay, were streaming through.'

This story, when compared with the harrowing experiences of many other yachts in the race, suggests that the crew of *Eclipse* was either clever, or lucky, or both. We were certainly lucky that no-one came to grief on the sharp corners of the galley stove when it parted company with its bearers. Moreover, it was only being restrained from tumbling further by a highly stretched rubber gas pipe. In the light of the thorough analyses of the race, it does seem, as was thought at the time, that having rounded the Fastnet Rock before the crew was exhausted, and run before the storm going in the right direction, concentration and enthusiasm within the top-notch Admiral's Cup crew remained high enough to keep the boat out of trouble by good steering. By good steering one means working the boat over the waves to keep her on her feet, in the same way as the helmsman of a dinghy does when planing, besides paying careful attention to what might be coming from up-wind. Swift helm movements were sometimes required to keep the stern on, or nearly on, to the big breaking seas, and it was helpful for a non-steering crewman to be sitting on the cockpit floor looking astern to call the whoppers that could broach the boat. It may be significant that our course brought the wind and the majority of waves about 20° off the stern.

Whilst it was a great psychological advantage to be still in the race, confidence was not so high as to prevent the liferaft being taken from its stowage and placed in the cockpit ready for immediate use. Besides, the boat might not have come well out of a total inversion as internal lead ballast had already started to come loose after a knockdown. This could have been a most unwelcome addition to our difficulties had the boat been rolled right over. It is probable that there were some waves about that night which would have capsized a yacht of *Eclipse*'s size, regardless of her number of exceptionally able helmsmen. For example, in the case of the Class I yacht *Jan Pott*, due to the confused nature of the sea, it was impossible to judge the angle of the approach of the wave which went on to roll and dismast her. In *Eclipse* we noticed the wind becoming lighter as we approached the Fastnet Rock at the very time others were experiencing their highest wind speeds, though the wind soon became stronger again as we headed south-east. Perhaps being closer to the centre of the low may have given us a brief intermission from its full strength. The yachts in the general area of the

Labadie bank had no such respite; indeed the evidence is that many of these yachts experienced a wind field with markedly more severe conditions. Alan Watts calls it a 'cyclonic pool' and has gone to some lengths to work out how a 'storm within a storm' can come about (Chapter 30). There were large boats in the cyclonic pool area which had already rounded the Fastnet Rock, and, as one would expect, they generally fared better than the smaller boats. Even so, the 46 ft Argentinian Admiral's Cup boat, *Red Rock*, was knocked down beyond the horizontal, and the 45 ft German Admiral's Cup team boat *Jan Pott* was rolled through 360 degrees.

The Fastnet inquiry report did not find sufficient evidence to indicate particular tactics to adopt in very severe conditions. It did state that 'there is a general inference that active rather than passive tactics were successful and those who were able to maintain some speed and directional control fared better'. Passive tactics generally amount to lowering sails, lashing the helm, closing the hatches, going below and hoping for the best. No yacht sank by adopting this tactic, and of the crews who felt that it was the most sensible thing to do, some found that there was as much of a need for a good look-out in bad weather as at any other time. *Sarie Marais*, a long-keeled wooden boat of traditional racing design lay a-hull for 12 hours retaining a two-man watch on deck. With tiller lashed to leeward rather than amidships, this being a matter of consequence, says her experienced skipper, she made 1.5 knots to leeward with few problems. The position of *Sarie Marais* was 37 miles to the north of Round Island, outside the cyclonic pool area, and the skipper believed the weather he experienced was not as vicious as further north.

Another yacht that lay a-hull successfully was Brian and Pam Saffery Cooper's 30 ft yacht *Green Dragon*, a newish and highly successful Doug Peterson design intended for top-class competition, but with more moderate features than some. At 2.15 am on 14 August, when 50 miles from the Fastnet Rock and therefore probably to the north of the cyclonic pool area, her crew sighted a red flare to leeward. They took down the mainsail and bore off downwind to render assistance. Under No 4 genoa alone the boat was travelling too fast and when she fell off a wave the foremost bulkhead and ring beam broke, allowing the hull to pant. The crew made a temporary repair using floorboards, sawn off jockey pole, a Spanish windlass and numerous drill bits. By this time they no longer felt able to render assistance and thereafter left the boat to lie a-hull with the helm lashed halfway to leeward, this being the angle at which she rode the waves best, besides protecting the damaged starboard bow. Though 'tossed about like a shuttlecock' during the next ten hours, the mast did not touch the water and the Saffery Coopers would use this method again to ride out such extreme conditions in a similar yacht.

Nevertheless, the accounts from within the cyclonic pool area suggest that there was a high chance of yachts lying a-hull being rolled right over, and that being rolled over could mean losing the mast, real chaos down below with consequent crew disorientation and dramatic reduction of morale. Thus, whilst lying a-hull was a much wiser choice than taking to a liferaft, this manoeuvre was not necessarily more successful than the active tactics used by some of the other yachts.

Such measures can be roughly divided into six categories: running before the waves with sufficient speed to steer out of trouble, much as did *Eclipse* under storm jib and *Vanguard* under bare poles; sailing into the wind under storm canvas; heaving-to under engine; lying to a sea anchor, or trailing drag devices; and, lastly, heaving-to under sail.

The Fastnet inquiry report shows only 26 yachts as having hove-to under sail, clearly a less popular tactic than either lying a-hull or running off. Thirteen yachts continued under jib alone, six under reefed mainsail or trysail alone, and seven under both jib and a reefed mainsail or trysail. Reading the comments in the inquiry report it could appear that the ones that managed to keep sailing with the helmsman steering to luff to the breaking waves may have done best. This applied particularly to Contessa 32s, which are stiff boats. Their stiffness must have been an advantage, though the Contessa 32s may not have reached the area of the strongest wind and worst seas. There is no information available on the whereabouts of the yachts that hove-to in the traditional sense, ie under mainsail/trysail with backed headsail; and without knowledge of their position the results are not clear. In the worst area it seems probable that there was too much wind either to heave-to or sail to windward under storm canvas. It may be significant that two yachts reported that, though they used a storm jib on its own, using a trysail on its own might well have been preferable.

The French 36 foot Sparkman & Stephens designed yacht *Lorelei*, of moderate to heavy displacment compared with many others, lay a-hull aided by her engine after picking up *Griffin*'s crew from their liferaft. The skipper, Alain Catherineau, had at first tried to get to the liferaft under triple-reefed mainsail, but he was travelling too fast and, after one unsuccessful rescue approach, he started his 12 hp engine and lowered his mainsail. He then found he could not make into the seas under power and had great difficulty in turning through the wind to effect the rescue, but he could make across the seas to the disintegrating and unstable raft. The engine, with its automatic variable pitch propeller, gave him the vital control needed to attach a line; and after being obliged to use it during the rescue the skipper must have felt confident about the way the boat handled under engine, as he continued to use it for the duration of the storm. It is noteworthy that *Lorelei*'s skipper, who was widely acclaimed for his most opportune rescue, modestly attributes his success to the good design of the boat for heavy weather.

Whilst engaged in another rescue, the crew of the Nicholson 55, *Dasher*, found that their boat would fore-reach under bare poles at 1.5 knots. This gives credence to the view that the addition of engine power can enable the yacht to manoeuvre quite effectively across the sea in high winds. Fore-reaching under bare poles seems to be a useful characteristic of a modern yacht, perhaps especially those that are stiff and have keels of generous lateral area.

Most yachts regarded themselves as still racing well into the storm, and did not use their engines. This was true of Graham Laslett, a veteran of the 1956 Channel Race, who was skippering *Bonaventure*, an Ohlsen 35. Stemming from his long-keel experience, he was an advocate of keeping up minimum sail, but on this occasion the boat seemed uncontrollable in the confused sea of the cyclonic pool region. At first *Bonaventure* was most comfortable running under bare poles. As the seas built up, increasing numbers of warps and sailbags were towed astern, eventually including a spare mainsail in its bag to give sinkage. Nevertheless severe knockdowns and near inversion still occurred, as the small rudder only worked at speeds above five knots when the boat was charging down the front faces of the waves. At other times the helmsman was helpless, particularly in the troughs, when he was unable to turn the boat to avoid being beam on to some of the waves, different angles being a feature of this sea. Significantly, the rudder was later modified on this design.

With hindsight, Captain Laslett feels an engine might have maintained steerage way at this critical point but wonders whether the towed ropes would have remained clear of the propeller. Certainly, in the case of the yacht *Autonomy*, a warp was washed forward by the following sea and jammed between the rudder and the counter, thereby making steering impossible.

Eighty-six yachts lay a-hull and 46 yachts ran before the wind towing warps or other impedimenta astern. Of particular interest was the OOD 34 *Windswept*, skippered by her owner George Tinley, who not only tried both of these actions, but a sea anchor as well.

After lowering his oversize storm jib, which alone had been too much sail for the boat, he at first tried lying a-hull. This seemed comfortable until the boat was caught by a breaking wave and rolled well beyond the horizontal, sweeping the two crew on watch overboard to the limit of their harnesses. Both men had difficulty in getting back on board when the boat righted herself, and they received no help from those down below, who had problems of their own, such as flying jam pots.

Not wishing for this episode to be repeated, but with continuation of the race in mind, George Tinley then tried to lie to an improvised sea anchor led over the port bow with a view to minimising leeway. The crew were called below and a five minute look-out system initiated. *Windswept* lay like this for a while in relative comfort with her port side to the wind and with the helm lashed to leeward. Then a sea knocked her round so that the starboard bow faced into the wind. As the helm was now the wrong way she became very uncomfortable, and George Tinley went on deck to put her through the wind under engine to lie as before. Susequently, a jib sheet fouled the propeller, denying the use of power. George Tinley went below again, leaving all the tapered washboards in position and the main hatch open a crack for ventilation. A little later, even from down below, the crew clearly heard a wave coming, and this rolled the boat right over with a shuddering crash. George Tinley, who had been standing in the middle of the saloon, broke his right arm. Moreover, the lower washboards were lost, the liferaft came out of its stowage and a large amount of water came down below. As lying to a sea anchor over the bow had not prevented another capsize, the crew brought the sea anchor round to the stern, whereupon, with careful steering, the crew felt for the first time reasonably under control and safe. Ten years later George Tinley remains convinced that his last remedy was the best of the three actions taken, and the one he would employ in the future, if faced with such extremes in a similar boat.

After being rolled over twice, the situation down below in *Windswept* was really very unpleasant. In particular the lavatory had not been pumped through fully, a

A crew member of Camargue *is encouraged to jump overboard to allow rescue by an aircrewman. This Wessex V of 771 Squadron rescued all eight crew of this yacht. Note the presence of the mast preventing the helicopter coming nearer to the yacht. Photo: RNAS Culdrose.*

An aircrewman is about to 'double lift' *this survivor off the dismasted* Grimalkin. *Photo: RNAS Culdrose.*

misdemeanour common enough even in harbour, and sewage was mingled with the sodden mass of clothing, bedding and loose equipment sloshing about at bunk level. The main pump handle, normally held by two plastic clips on the under-side of the cockpit locker lid, had somehow been lost in the capsize. When a broken off tiller extension was found to fit, the pump still would not work as the strum box was choked with tea bags. The secondary pump, incorporated with operation of the semi-submerged lavatory, did not seem a good alternative at the time. Driven by acute discomfort and fear of further capsizes, consideration was given aboard *Windswept* to abandoning the boat for the liferaft tugging on its painter beside the boat but, no doubt fortunately, the liferaft defied all attempts to make it inflate. Incidentally, some yachts' liferafts did inflate when not called upon to do so, thereby creating a tremendous problem, overcome in the case of the United States yacht *Aries* by crew members stabbing it into submission.

Of the active tactics employed, running before the storm without any drag device was employed by 57 boats. This tactic worked well for most of the yachts with a sufficient supply of alert and skilful helmsmen, capable of maintaining the concentration necessary to sail the boat out of trouble. Even members of the Australian Admiral's Cup team, who collectively managed the storm best of the international contenders, commented upon the intense effort they found necessary to cope with the conditions whilst on the helm.

Andrew Cassell, skipper of the light displacement J 30, *Juggernaught*, spent most of the storm on the tiller. After trying the boat at every angle to wind and wave he found favour in running with the seas on the quarter. He says:

'Running dead down wind under bare pole was extremely dicey as several times the boat took off like an express train. I quickly realised that many more events like this would see us being seriously damaged in a massive broach or pitchpole. We therefore tried to stay head-to-wind and would have streamed warps over the bow had we had enough length to justify their use. As it was we were not at all comfortable and once or twice we were carried astern at a rate which threatened to take the rudder off. Beam-reaching was then tried. This seemed all right for a while, but then we were knocked down twice to the point where the mast was in the water.

Finally we tried broad-reaching with the sea on the quarter, treating the J 30 just like a dinghy. Steering was extremely difficult, especially at night, when one had to listen for the breaking waves and line the hull up in order to surf down the wave at an angle. When caught by a breaker the low lateral area of the keel allowed the boat to slip sideways at a speed of, possibly, up to five knots. The method worked and, apart from one or two extremely alarming situations brought about by breaking waves from above, we were able to maintain control. I feel that a heavier and less manoeuvrable boat than a J 30 would have rounded up frequently with disastrous results. On the other hand if we had tried to lie a-hull with tiller lashed, such a light boat would have been thrown about unacceptably.'

It is worth recording that Andrew Cassell has lost both of his legs, and is well known for his tenacity and the tremendous strength of his arms.

The inquiry report revealed that, in general, owners felt it was necessary to keep the helm manned during the storm. Some of those who did not do so felt, in retrospect, that they should have done. Some 80% of yachts from which information was available had someone at the helm when hove-to or lying a-hull. Obviously it can be an advantage to be able to lash the helm in this situation, and it may be a reflection upon modern yacht design that lashing the helm did not occur more widely. Incidentally, in at least one yacht there were very few members of the crew prepared to take the responsibility of steering in the severe conditions.

Of the other active tactics used, there were several reports of yachts being knocked down with warps streamed and others which had success. It appeared to be best to endeavour to have sufficient drag to give control when surfing off the tops of waves, but not so much drag as to lose steerage way in the troughs. Use of *Lorelei*'s engine clearly was an advantage in her circumstances, and it worked well for her. Lying to a sea anchor led from the bow did not seem to work for the one boat known to have tried it, for the reason that the pull was not constant and the boat, over which there was no steering control, tended to take up an awkward angle to the seas. Besides, sudden backsliding movement would impose high loads on the steering gear. On the other hand there is an interesting and encouraging account from Chris Dunning whose 45 ft *Marionette of Wight* broke her rudder at 2.30 am on 14 August, having made 20 miles homewards from the Fastnet Rock. After lowering sails, *Marionette* lay beam-on to the sea with her crew feeling decidedly exposed to the curling breakers while they put together a very long warp – perhaps 1000 ft – composed of their two anchor lines and three genoa sheets. They led the warp over the bow, whereupon *Marionette* took up a steady and comfortable angle of 20 – 40 ° to the wind, though perhaps making up to four knots of leeway.

If the 1979 Fastnet did not give strong pointers to the best storm tactics to adopt, many significant lessons did emerge. The storm drew attention to the fact that the range of positive stability of certain yachts was such that they could remain perilously inverted for some time after a capsize. As a result, a yacht's range of positive stability has now become an important criterion in deciding a vessel's seaworthiness. The race also demonstrated that, however uncomfortable a yacht might have become during extremely severe weather, she provided the best refuge as long as she remained afloat. Thus a liferaft should have been thought of only as a last resort. Seven of the 15 fatalities occurred to crew who had taken to the liferafts, and of the 24 yachts which were abandoned, 19 were recovered afloat. Incidentally, the adage that 'your ship is your best lifeboat' emerged from naval wartime experience, and seems relevant to surviving storms in the open sea. Several shortcomings in liferaft design brought further aggravation during the proceedings. For example, entry ports being at the opposite end to the raft painter, lack of stability exacerbated by inefficient drogues, and simple lack of strength. Happily, liferaft design has unquestionably improved as a result of the Fastnet experience.

The design of safety harnesses has much improved too, though again it took many harnesses to fail in the storm, accounting for much of the other half of the loss of life, to make the point. Moreover the need for adequate strong points to clip onto, able to withstand the two ton stress that the harness should be designed for, was clearly demonstrated: the inquiry report revealed that at least six people were clipped on to attachments that broke under shock load.

There were many other events from which useful conclusions can be drawn. Unproven high tech carbon fibre rudders did not fare well in the race, nor did washboards without a form of tether, particularly those designed for a tapering type of companionway. In addition, a combination of inadequate cockpit drains with unsealed cockpit locker lids, became a cause of dangerous flooding. Many participants were impressed by the speed with which solid water will enter a small submerged opening. In this respect, a frightened man with a bucket proved his worth once again as an emergency bailer, not only aboard *Windswept* while the crew were discussing the idea of abandoning ship, but also on board many other craft in the race.

Most yachts did have the necessary charts to get into ports of refuge, but in the case of *Electron II*, when considering running for Milford Haven, the skipper found that he neither had the charts, nor did his insurance policy cover him to go there. In addition to moving ballast, others reported aerobatic batteries, which must have been equally vexing. The need to stow away or secure all movable items before going to sea, and then to keep them that way, may be difficult to implement; but it was clear from experiences in the storm that such a policy would have been well rewarded in a knockdown. In particular, the good seaman's practice of stowing away spare clothing and bedding after use, contained in a heavy duty polythene bag and sealed with a wire closure, would have prevented many instances of acute discomfort and hypothermia. Likewise, adequate handrails and an absence of sharp projections would have saved several injuries. For example, it is quite common to see yachts with protruding bolt ends below decks, which have not been trimmed off at the building stage. After the race there were reports of capsize injuries from such halyard winch fastenings on the deckhead.

Finally, it seems wise that the Royal Ocean Racing Club now requires competing yachts to have entered shorter and less demanding races as a prelude to the Fastnet, besides having a large element of the same crew together for the preliminary and final events.

Dramatic enough though the 1979 Fastnet was for all the participants, an inhospitable lee shore in otherwise similar conditions would have added yet another dimension to the situation, for which even more demanding remedies than those employed would have had to be found.

22 1984 SOUTH AFRICAN STORM

Dudley Norman

In 1984 worldwide weather conditions had probably been adversely affected by the presence of the unpredictable El Niño current along the west coast of South America. For instance a storm in Cape waters earlier than the one to be described here had caused the total disappearance of a yacht and its crew during a trip along the south-east coast of the Southern African subcontinent, and amongst other damage, seven large yachts had broken from their permanent moorings in a lagoon on the south coast. Later, an Indian Ocean tropical cyclone had moved further south than ever known before and caused floods in Mozambique and the north east of South Africa.

Most wind and wave damage in South Africa, however, was done by the southern winter storm of May 1984, when deepwater waves of over 16 m maximum height, with a significant wave height of over 10 m, were measured. These are believed to be the highest yet recorded in Southern African coastal waters. The heights were measured by a waverider buoy to the south-west of the Cape Peninsula in 200 m of water. Swells of 16 m and a wind speed of over 58 knots were recorded off Mossel Bay, about halfway between Cape Town and Port Elizabeth and further down the track of the storm. Fig 22.1 shows a synoptic chart for 16 May 1984, when the storm had reached the Cape East weather area. The normal eastabout sail passage to Mauritius entails running down an easting at about 40° south until about 50° east. A 31 ft GRP sloop of the 'Miura' class, whose underwater profile is shown in Fig 22.2, followed this route, leaving Port Elizabeth on the 13 May 1984 and encountered a gale on the following day. This developed into the main storm on the afternoon of the 15 May, when the yacht was roughly 100 miles south east of Port Elizabeth. Predicted wind speed was 50 to 60 knots, but there was no anemometer in the Miura to indicate the velocities she actually experienced. The lowest barometer reading observed was 995 mb. The gale continued virtually unabated until the afternoon of 17 May, and altogether there were many other 'ordinary' gales, of about force 8, before Mauritius was reached on 9 June 1984.

Storm encounter

The Miura had a complement of two: the owner-skipper aged 31, and the crew member, myself, aged 63. Departure from Port Elizabeth had been delayed for a week while the skipper recovered from a severe attack of influenza. After clearing customs and putting to sea, it became apparent from the shipping forecast that very heavy weather could be expected in the Cape East area. It seemed likely that the weather would have abated by the time the yacht was well out to sea so it was decided to head for the Agulhas current stream which runs southwards some distance offshore opposite Port Elizabeth. In the event the Agulhas current was crossed during the night of 13 to 14 May, under motor in a light northerly air. The stream was easily identified by an unexpectedly lumpy sea with clearly defined boundaries.

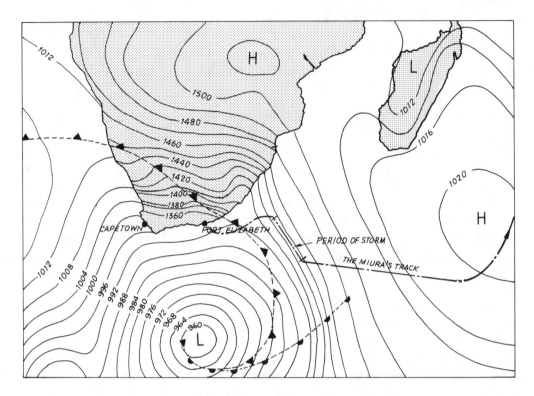

Fig 22.1. Track chart from Miura class yacht and weather chart.

By late afternoon of 14 May a wind of 35 to 40 knots came up suddenly from the south-west and the sea was soon higher than expected. Mindful of earlier forecasts, the skipper very wisely decided to remove everything possible from on deck, including the canvas doghouse, the dodgers and the mainsail from the boom. The Miura ran very well under self-steering and a small jib, surfing occasionally. The craft had a wind vane, mounted off-centre and connected to the trim tab by 10 mm stainless steel rods, which was used when conditions were too rough for the Autohelm 1000, although this latter system could generally cope when running. The Autohelm tore loose on 15 May as mentioned later, but the skipper was able to effect a repair and the instrument was used subsequently whenever possible.

During 15 May the wind veered to the north and moderated, but this was soon followed by an increase in the wind which proved to be the onset of the main storm. After a terrible night during which the wind was backing and strengthening, dawn on 16 May revealed a big swell, moving with an awful majesty. There was also a permanent banshee shriek from the wind in the bare rigging. The swells were at least as high as the truck of the Miura's mast, which is slightly over 12.5 m above water level. There was no question of self-steering of any sort. The storm jib had been long since handed, covers fixed over the cabin lights, and 80 m of 20 mm diameter warp streamed in a 40 m bight. From the troughs, this warp could be seen lying apparently lazily in the water from trough to crest, but any attempt to pull it in revealed that it

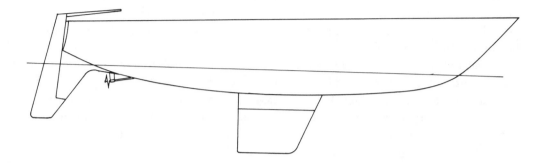

Fig 22.2. Underwater profile of the 31 ft GRP Miura class sloop.

was actually under considerable load. As the wave faces approached, little streams of water could be seen running down them under gravity, although the common optical illusion was present – namely, that waves viewed directly appear steeper than when viewed obliquely. Crests appeared to be breaking heavily everywhere but only swamped the cockpit about six times a day.

However, most waves had lesser breakers on their crests, each capable of slewing the yacht round, and these were a problem. They were not parallel to the wave crests, coming sometimes slightly from one side and sometimes slightly from the other. They could not be anticipated by any self-steering device and so constant hand steering was necessary. In spite of this, there were many near-broachings. The boat surfed down each wave face with a short sharp 'buffeting' motion.

By late afternoon on 17 May the gale had abated somewhat; the sea was becoming confused but there was still enough wind to justify remaining under bare poles until 18 May when it became possible to hoist some sail. The yacht had been driven far to the south-east, the gale having steadily backed through north-west to west. A fix on 19 May gave 33° south; 33° east.

Human element

The two 'Fs', fear and fatigue, formed a vicious circle. Whether on or off watch, normal sleep was out of the question, although there were periods of unconsciousness. On one occasion I found myself taking care not to tread on my little dog as I moved about the cabin, although the dog was at home ashore. The least disturbed area for rest was on the cabin sole amidships, and there was a natural inclination to stay in foul weather gear when off watch. This caused fits of shivering, presumably due to lack of body ventilation, particularly when I briefly elected to wear a wetsuit rather than the normal 'oilies'.

At night it was usually possible to see broken water nearby and, fortunately, the main gale peaked during daylight hours, apparently around the time of high tide in that inexplicable habit that gales appear to have. There were no 'stay awake' drugs on board, but the skipper, who stood the longest single watch of about 11 hours, did take a few sips from the ship's first-aid bottle on that occasion. The tranquillising effect of this probably outweighed any disadvantages. Watches were fairly flexible: nominally three hours, but often more or less than this, depending on the decision of the watch on deck. My age and the skipper's recent influenza provided incentives for each to call the other as late as possible.

A dangerous moment came on 15 May when the Autohelm broke its mounting and was only prevented from slipping overboard by its electrical lead. The skipper was off-watch and I was keeping watch from the cabin hatchway without my safety harness. Without thinking I found myself at the sternrail retrieving the instrument. Fortunately, the skipper was not asleep and had heard words of blasphemy, but there would have been great difficulty in motoring back upwind had I been swept overboard. With such incidents on deck in mind it is clear to me, in retrospect, that when on watch a safety harness should be worn even down below.

Ship performance

During the main gale, whenever the yacht was kicked from behind by a breaker it was not possible for it to yield except by surfing, and breakers could appear on the crests before the gravity-assisted surfing had started. With the hull near its maximum theoretical speed, the impact had to be taken by the transom and companionway washboards. The stoutly-constructed Miura survived this onslaught but it would be wise for all designers to give some thought to transom and washboard strength. The bows should be strong too, as full speed can continue through the night in such a gale, and any object such as a floating log could cause heavy damage. The Miura has longitudinal stiffeners at waterline level on the bow section, but a collision bulkhead could have been a wise addition. It is worth recording that the 10 mm stainless steel rods connecting the wind vane to the trim tab each somewhat over half a metre in length were both bent, though the vane itself was not rigged at the time.

Another point is that when a breaker hit the washboards, water spurted into the cabin around the boards even though they were a good fit. Due to the floor carpeting, a quantity of water accumulated on the cabin sole every time this happened before it could drain to the bilges and be pumped out. Easier drainage to the bilges would have been an advantage and if there had been more serious leakage the situation could have become a major handicap.

A safety harness alone was not sufficient to prevent the helmsman from being carried out of the cockpit by the breakers, and it was found that the mainsheet track across the cockpit provided a convenient additional object for the helmsman to lash himself on to.

Later in the trip, when reaching in one of the lesser gales, a breaking sea knocked the yacht violently sideways. Fortunately, there was no structural damage to the hull or coachroof, possibly because a modern short keel design, such as the Miura, does not offer much lateral resistance. Arguably, modern yachts may be less comfortable than long-keeled yachts in a beam sea, but are, perhaps less vulnerable to structural damage. In this incident I was thrown back against a projection of the cabin furniture and sustained what was later diagnosed ashore as a rib fracture, thus demonstrating the need for well-rounded corners in a yacht's interior.

Hindsight

It would have been a brave man who attempted to lie a-hull in the storm as the boat would have certainly been rolled over. A desire to exploit the marvellous research possibilities of the occasion was not uppermost in the minds of those present, but perhaps the Amercian technique of using an aircraft-type parachute streamed from the bow could have been successful. Some years ago an American visitor to the Cape mentioned using one regularly, and I had read about early American experiments from a motor yacht. I have ever since wanted to try one out (I think). Discharging oil in the small quantities that vessels of the size of the Miura are likely to carry would not have been helpful due to the rapid change in the boat's position.

Writing some time after the event my belief is that my skipper, Jean-Louis Deyne, knew his business and got it right.

23 AT SEA IN THE GREAT STORM OF OCTOBER 1987

Peter Bruce

There were several small vessels at sea during the great storm of October 1987 but with massive destruction ashore, both on the north coast of France and the south coast of Britiain, not much has been heard about them. The intense small depression brought winds of over 100 knots to the English Channel area, and though the wind speed was of hurricane strength it is, of course, strictly speaking wrong to describe the event as a hurricane. It was a storm, albeit a truly violent one.

October is late in the year for leisure sailing, but the sea water temperature is still warm; so there are still a few sailors who go out for pleasure at this time. Moreover, delivery trips continue all through the year, and the Joint Service Sailing Centre yachts, based at Gosport, have a long season. Thus, with manageable conditions during the day, and without a warning of ultra severe weather made manifest, it was not surprising that several yachts were still out at sea on the night of Thursday 15 October. A small fleet of warships from Portland naval base were also at sea for the opposite reason. In very strong winds, ships at anchor can drag onto the shore even in Portland Harbour, and as a result of a timely warning from the senior meteorological officer there, Lieutenant Commander Peter Braley, who had made his own interpretation of the impending weather, the Portland fleet was ordered to sea.

Five descriptions from the sailing vessels at sea in the storm are related in this chapter, to give a wide cross section of experiences; but before the accounts from the sea, there follows a meteorological report by Peter Braley.

Meteorology

'As early as the weekend of 10 – 11 October 1987 the Meteorological Office at Bracknell had been forecasting severe weather on the following Thursday and Friday.

At midnight on Tuesday 13 October, the depression in question was centred near 50°N 47°W, off Newfoundland, tracking east. It was expected to deepen progressively by midnight on Thursday 15, after which, explosive deepening was predicted, with storm force winds expected in the South Western Approaches during Thursday, moving up Channel and into the North Sea overnight. The prediction was reinforced and updated at 1545 on Wednesday 14 October, when the low had begun to move rapidly east. The storm warning suggested that the centre could be expected to be 150 nautical miles north-west of Cape Finisterre by 0600 on Thursday 15, with winds in its southern quadrant expected to reach storm force 10 up to 200 nautical miles from the centre. The low was then predicted to curve NNE to bring SW storm force winds to the east Channel and southern North Sea areas late on 15 October.

The development became noticeable to meteorological authorities throughout Western Europe after midnight on 14/15 October. At that time, an elongated trough was located over a distance of some 500 nautical miles westwards from Cape Finisterre with two apparently distinct depression centres – one 983 mb just north-west of

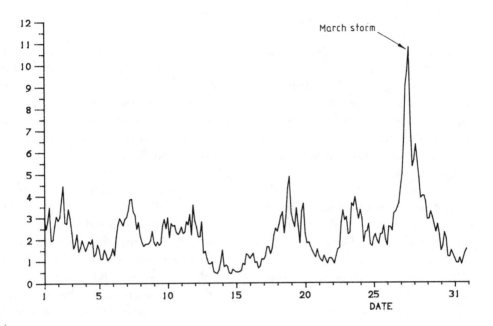

Channel LV, March 1987
Significant Wave Height (Hs)

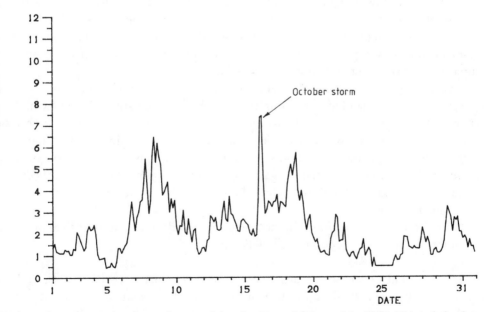

Channel LV, October 1987
Significant Wave Height (Hs)

The charts show two months of wave data recorded at the Channel lightvessel in 1987. This includes the very severe storm of October, also the one in March of the same year. Though the intense, fast-moving system of October 1987 caused havoc on land, it was not around long enough to create exceptionally high waves unlike the very large depression of the March storm.

Coruña, the other about 986 mb near 43°N 19°W, some 450 nautical miles further west. A strong pressure gradient existed on the warm southern side of the frontal trough, but with a slack gradient on the cold side. Noticeable potential temperature differences across this front were apparent from the few available observations, suggesting a sharp differential between the polar maritime and tropical maritime air masses. By midday on 15 October this broad surface trough lay from Brittany across Finisterre, and pressure in the trough was falling markedly. By 1800 the axis of the surface and upper troughs had extended north-east, bringing the frontal feature across the Channel and into southern England. The two centres were still apparent, one over Ushant the other now more vigorous centre of 963 mb near $45\frac{1}{2}$°N $8\frac{1}{2}$°W. Surface pressure continued to fall within the trough but, more significantly, marked rises in pressure had occurred over the previous six hours both south-west of the main centre and astern of the cold front, with corresponding increases in wind speed. At about 2100, a signal was received at Portland from the forecaster in RFA *Engadine* in the north Biscay area, drawing attention to the sudden falls in pressure in his vicinity, thus suggesting explosive deepening of the Biscay centre. The signal also remarked upon the pronounced cross sea and swell conditions. Soon afterwards the centre passed close to *Engadine* and winds rose rapidly in strength, occasionally to exceed 90 knots.

Thus by 2100 the depression centre had tightened considerably and begun to move rapidly north-east at about 40 knots to deepen to about 960 mb in the extreme south of sea area Plymouth. The trend continued towards midnight when a centre was indicated just south of the Eddystone with a central pressure nearing 952 mb. Notably, this feature was now a single, definable centre. A clear and fairly accurate track became discernible, making short term forecasting and timing somewhat simpler and more certain. These, together with surface analyses, demonstrated the exceptional pressure gradients now developing in a narrow swath to the south and east of the centre with surface gusts exceeding 100 knots in exposed locations.

During the next six hours the centre of the depression moved in a north-easterly direction across England after crossing the coast between Charmouth and Lyme Regis at about 0215. By 0500 on 16 October it had reached Lincolnshire and the occlusion had swung rapidly across southern England. Very marked rising pressure tendencies were widespread in the three hours from 0300 to 0600 as the depression filled: for example +25.5 mb at Portland. This later rise has now been established, by a comfortable margin, as the greatest three-hourly pressure change ever to be recorded in the United Kingdom.

Retrospective examination of meteorological records has been extensive, but it may be of interest to draw attention to a few facts:

1 The band of strongest gusts, with isolated maxima over 100 knots, extended across the Channel areas from north-west France in a band some 90 miles wide, parallel to the track of the depression centre.

2 Gusts of 70 knots or more were reported for a period of three to four consecutive hours, during which time the wind veered from 180° to 230°.

3 A double-peaked windspeed region apparently occurred over central southern England and the adjacent Channel coast. The first was associated with winds from 170 and 190°; the latter from a direction around 230° – the peaks being at around 0200 and 0500 respectively.

4 A separate area of strong northerly winds affected Cornwall and North Devon from 2300 on 15 October until about 0200 on 16 October.

5 The strongest gust over the UK was recorded at Gorleston (Norfolk): 106 knots at 0424. Over 20 other gusts in excess of 90 knots were recorded between 0400 and 0715. A higher gust was estimated at 119 knots at a station near Quimper on the Biscay coast, while a gust measuring 117 knots occurred at 0030 at Granville on the south-west corner of the Cherbourg peninsula.

6 The passage of the cold front heralded the onset of the really strong winds at the surface. It was the strength of mean windspeeds that was noteworthy, rather than individual gusts.

7 The pressure falls ahead of the advancing depression were large but not remarkable, yet the subsequent rises in pressure appear to have been exceptional. Large areas of south England experienced rises in excess of 8 mb per hour.

8 Whilst very marked, short and steep seas built up in the Channel, recorders show that no remarkable storm surge followed and no pronounced swell developed due to the short duration and small size of the maximum windfield.

A Sadler 34 off Ushant

The first sea account comes from a new shoal-draft Sadler 34 called *Muddle Thru*. She had been bought by a Canadian, Allan McLaughlan, who had come over to Poole in Dorset, England from Ontario with two other Canadians, Harry Whale and Bill Bedell, both of long sailing experience, to take delivery of his boat and sail for the Caribbean.

They set out on 13 October 1987, but progress towards Ushant was very slow due to strong head winds, and the boat was 'caught out' sixty miles to the north of Ushant in the 100 to 120 knot winds of 15 – 16 October. The persistence of *Muddle Thru*'s crew was enhanced by the fear that VAT might have to be paid on the boat if she was to return to the UK. After the great storm, another gale came, and this time the boat was rolled through 360 degrees. The effects of the capsize, which included a broken lower shroud, caused Harry Whale, the skipper, to head back to Falmouth, where, incidentally, HM Customs took a benevolent VAT view in the circumstances. Harry Whale kept a log of events by speaking into a tape recorder, and it is his account of the storm which follows.

'Winds increased to force 9 from the WSW during 14 October; and though close hauled and carrying only a deep triple reefed mainsail, we made fairly good progress in almost the right direction. To keep a proper lookout was difficult because one could never be sure in the dark if we were looking toward the horizon or merely into yet another big black wave. Certainly we did see many ships and made every effort to stay well clear, believing that under the prevailing conditions we were probably invisible despite the radar reflector on the mast.

By about 0300 of 15 October the wind had abated enough for us to add some foresail to the sailplan and, after experimenting with the furling genoa, we rigged the detachable headstay and hanked on a small flat jib. The rest of the day was spent in heavy rain, crashing along into force 7 wind, heading SW by S. In the afternoon the wind ceased quite suddenly and after an hour or so came in from the opposite direction, north-east. It then rapidly built up to force 7 again. Under jib only we were soon making splendid time directly toward our destination with a following wind, but meeting the waves which were still coming from south-west. Some waves we seemed to jump over and others we went through; nevertheless it was good finally to be able to make our course at a respectable speed. Suddenly the turn at Ushant didn't seem so far away.

The run continued until 2100 when the wind died completely for half an hour. It then came in from the south, quickly built up, and veered to the south-west. By midnight we had taken off all sail and were lying a-hull. On the tape I said 'we are in a shocking blow.' Having experimented with the helm and the windvane steering I finally came to the conclusion that the boat took the best attitude to wind and waves when left to its own devices with the wheel lashed amidships and the windvane disconnected. I feel sure this was largely due to the sprayhood and its considerable windage, which had the effect of keeping the bow into the weather.

Having seen several east-bound ships I was concerned about traffic in the area. We had heard the traffic controller at Ushant speaking to another vessel on VHF radio. I called to inform him that we were lying a-hull and virtually at the mercy of wind and sea, so suggested a *securité* call to vessels in the vicinity. This he did very obligingly, giving our position, estimated drift and course. He also told me that his wind gauge indicated windspeed ranging from 100 to 120, and that he was concerned for the safety of his radar. I did not think to ask if the windspeed given was in nautical miles, statute miles or kilometres per hour. It seems now that they were knots. Psychologically I think all of us on board felt better for the exchange, especially when he called an hour or so later for an update on our position. I provided this, and mentioned that we had noticed two or three west-bound ships passing to the north of us. I was impressed by the fact that he wanted to know how our position was obtained. By now we were starting to hear parts of one-sided conversations on the VHF and realised that something pretty major was going on by the number of distress calls, including a MAYDAY from a yacht in a French port. Two other conversations I remember involved one ship which had been driven hard aground, and another which had lost its steering.

Generally our little ship seemed to be handling things very well, although the noise of the wind and the waves breaking over the boat was extreme and the motion quite erratic. Occasionally the boat would seem to get temporarily turned broadside onto the waves and a great thumping one would hit her full on the beam. The furling jib was tightly wound about its stay, the small hank-on jib tightly lashed to the toe rail and

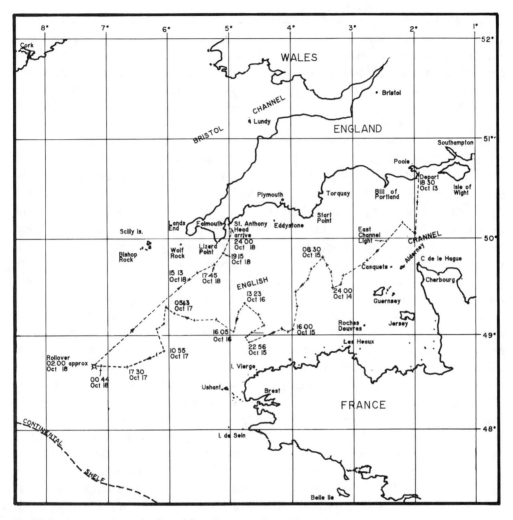

Fig 23.2. Approximate track of Muddle Thru, *October 1987.*

the mainsail lashed to the boom. A small part of the mainsail got free and proceeded to thrash so wildly I felt sure the sail would end in tatters and possibly even strain the rigging. Having determined that the loose part of the sail was below the second reefing point and therefore most of the sail should survive, I decided to let it flog rather than risk a man or two overboard trying to retie it to the boom, which would involve working around the sprayhood where the holding was least secure. As it turned out the sail fared well, though the battens were lost.

The three of us remained fully dressed in foul weather gear and harnesses; we spent our time below with just a brief check on deck every 15 to 20 minutes. With so much water constantly coming onboard the boat, one had to be careful about the timing when opening and closing the main hatch.

Eventually about 0730 on 16 October the wind abated sufficiently to start sailing again. We had gingerly worked our way up to a double reefed main and small jib by 1400, and even had glorious sunshine for 3 hours. We were visited by some dolphins, and though still on a dead beat generally felt very fortunate to be in such good overall conditions as we listened to the radio telling of the damage done during the night both at sea and ashore. At 1600 more gale warnings were broadcast – force 8 going to force 9 from the south-west. For the next 13 hours we made good course of WNW at an average speed of $3\frac{1}{4}$ knots.

During 17 October the wind direction varied from SW to WSW and we tacked accordingly, always mindful of giving Ushant a wide berth. By 2200 we were again down to bare pole in a raging gale, this time with mainsail lashed very tightly to the boom and the outboard end of the boom lowered and lashed to the toerail. Once again I experimented with the windvane steering and various positions of the helm, but came to the same conclusion as before – to lash the wheel amidships and allow the boat to take care of itself. From our perspective the main wave pattern was from the south-west; however we seemed to be constantly under attack from quite unpredictable waves from the south-east and north-west. Some of these seemed to be not so much waves as just great eruptions of water, happening without rhyme or reason.

As before, we huddled below fully dressed in foul weather gear and harness trying to catnap, with a check on deck every 15 to 20 minutes. I suppose this continued for 3 or 4 hours. I was lying down in the starboard bunk and had dozed off nicely when suddenly I realised I was airborne and on my way across the cabin, easily clearing the lee cloth and table en route. I came up short against the other side of the boat on top of Bill. As I scrambled around trying to extricate myself I remember enquiring after Al, who had been using the quarter berth. He spoke up from underneath me saying that he was OK. He had given up the quarter berth and rigged himself an accommodation on top of some cans of water we had lashed into place in the U of the dinette settee and was wedged between Bill and the folding table. About this time I heard 'we're upside down boys' from Bill in a quite conversational voice – indeed almost a mumble.

By use of the two full height teak pillars which the Sadler has in the vicinity of the galley and chart table, I was able to heave myself over the galley sink and by the time the boat was completely inverted I was standing on the deck head holding on to a pillar with each hand, facing forward with my back to the engine compartment and the main hatch boards.

The exact sequence of events and time involved is of course very uncertain, but I do have a vivid recollection of standing there and being aware that the noise and motion had ceased, other than the sound of water rushing in through various small openings, I was expecting at any moment to be struck from behind by the hatch boards collapsing inwards under the pressure. It was as well that Bill had insisted that the cockpit locker lids be well secured.

As the boat continued its roll-over I remained clinging to the pillars, negotiating them rather like climbing a large two rung ladder which is being rotated. As the boat resumed its normal upright position several impressions struck me; one was the resumption of the old noise and motion, another was the additional sound of water sloshing back and forth, and the most disturbing was what I can only describe as a 'wobbly' feeling about the boat. Several explanations flashed through my mind, the first being that may be we had sustained some sort of a longitudinal hull facture – yet there didn't seem to be much water coming in. Then I wondered if the keel had loosened, but discounted this thought on the grounds that surely it would have gone entirely or not at all, and since we seemed to be remaining right way up the keel must still be in place. The only possibility was that the hull was being wrenched around by a loose rig.

My first resolve was to try the diesel engine. This would enable us to run the electric bilge and shower pumps, and also provide motive power should we need it. When we tried the switches we got the whistle of the Calor gas alarm. We waited for some time and it didn't stop. Because the alarm had acted strangely on another occasion, and because we had always been careful to turn off the supply, I felt sure this was a false alarm, at least as far as the Calor gas was concerned. We turned off the alarm and gave the engine a try. It started. Now to check the rig.

Getting on deck to investigate was no easy matter. The sprayhood, with its stainless steel framework, was crushed down over the main sliding hatch and the hatch boards. Eventually Bill and I both got through and discovered with the help of a small torch that the port lower shroud was loose. Bill bravely went forward and took a jib halyard one turn around the mast and attached it to a stanchion base near the chain plate. He then did the same with the main halyard so that we now had the two halyards running from the spreader roots to the stanchion base. From the cockpit we did our best to sweat up these halyards, no easy matter without winch handles, which had left their pockets during the roll-over. Having thus more or less stabilised the rig we proceeded to haul inboard sheets, halyards, and other odds and ends all hanging over the port side of the cockpit.

Below, Al had been doing his best to assist the bailing process by trying to pour water into the sink with a saucepan. The outlet was not very big and sometimes, because the motion was so violent, the water was flung back at him before it had a chance to drain away. However, we certainly seemed to be holding our own because the water level was not much above the cabin sole. Bearing in mind that a large part of the volume of the bilge was taken up by canned food and built-in flotation material, the actual volume of water aboard could not have been great.

We found that the shower pump was not operating, and only one of the interior lights was working. All of the electronics except the satnav had ceased to function, ie radios, windspeed and direction, log, speedometer and depthsounder. Our spectacles, charts, dividers, hand compass, torches and spare winch handles seemed to have vanished into the heap of sodden mattresses, tools, clothing and sundry other items.

I decided that the prudent course of action was to make for Falmouth. I thought of Moitessier and Vito Dumas, who advocated that the best way to survive in heavy breaking waves was to sail with them at good speed.

Because I had stared at the chart for so many hours before the incident, I knew the course was about 050 degrees. Before shoving the gear shift into forward I tied a huge knot in the furling line of the jib, gauged to stop at the fairlead. Thus about 6 ft of it and no more could be let out in a hurry by one man, should the engine stall and force us into taking a chance on the rig.

We were off on the wildest sleigh ride of my entire sailing career. In the cockpit it was impossible to hear the engine over the roar of the wind and water, and there were no gauges of any kind to indicate RPM or boat speed. The main compass seemed to be working, though a little sluggishly. It was strictly seat of the pants stuff, aided sometimes by the vibration of the engine and propeller through the soles of the feet. The boat was completely overwhelmed on several occasions, leaving the helmsman spluttering and gasping for breath. It was the blackest of nights, and I remember thinking it would be so much better once daylight arrived.

When daylight did finally arrive I wished it had remained dark. It was an impressive scene, huge steep breaking waves everywhere and the air filled with water, either rain or spume. At one point some dolphins came along to join in the fun. Like young skiers on a mogul run they cavorted and frolicked around us, frequently leaping right out of the advancing face of the waves. Never did I feel so out of my element as I cowered at the wheel, tired and afraid, whilst admiring their wonderful performance. About the time I was beginning to worry about fatigue and hypothermia, Bill squeezed his way out of the hatch and offered to take over. This was a great relief to me, firstly because I needed the break and secondly it meant that Bill was now endorsing my strategy of heading for Falmouth, albeit tacitly. I am sure he was originally not in favour.

Later, Al also came up to be initiated and join the watch roster. Time and again it would seem she was about to broach, and be rolled over in a breaking crest; yet at the last minute a little extra power would straighten her out and away we'd go, surfing down yet another huge wave, only immediately to worry that the bow could bury and cause a pitchpole. Bill was to tell me much later that, at one point when he was steering, the boat did a complete 360° turn (horizontally this time); she seemed, he said, to sail close around a great pinnacle of water, with a cliff face on one side of the boat and a precipice on the other.

About 1900 in the failing light, the loom of the Lizard light showed 10 miles to the NNW. The satnav had done a splendid job for us. Now without tidal information or depth sounder came the task of piloting our way the remaining 20 or so miles to the harbour. As it turned out everything went well: and the wind and waves even abated a little as we came into the lee of Manacle Point and the Manacle rocks. The limited information we had on board showed the harbour entrance and not much else. We shot through the entrance, did a sharp left turn and suddenly all was tranquil, relatively speaking anyway.'

A Swan 46 in the Bay of Biscay

Had *Muddle Thru* made Ushant and into the Bay of Biscay she would not have had a much easier time, as can be appreciated from another drama going on 215 miles SSW of Ushant. On the morning of 15 October a Swan 46 fitted with a Scheel keel, on passage to Spain, hove-to in windspeeds shown on their Brookes & Gatehouse anemometer of, at times, over 85 knots. Apart from periodic look-outs the crew of four remained below, where it was dry. At 1800, without warning, the boat rolled quickly to port through 360°. Floorboards, and everything stored under them, flew through the boat. The yacht's owner, who was standing in the main cabin after returning from a look-out on deck broke a shoulder. The mast broke at deck level, probably due to a lower shroud failure.

The mast remained attached to the boat by her intact rigging, and without suitable cutting devices on board to cope with rod rigging, the crew were unable to cut it free. Around 2000 the mast broke at two other points and started to pound against the hull. At this juncture the owner decided to ask for outside assistance, and an EPIRB was activated. The Swan 46 was 'localised' in 45 minutes, and a Japanese freighter appeared on the scene at 0100 in the morning of 16 October. The freighter launched a boat but this soon capsized; so the yacht, with her engine still working, was asked to go alongside. Whilst the crew were disembarking, the yacht was smashed several times against the freighter's hull, and at one point she surfed down a wave and struck the freighter at full speed with her bow. Though the EPIRB was heard for an hour afterwards and hatches were left closed, the yacht was never found.

A significant point in this episode was the difficulty experienced in detaching the mast. Bolt croppers will not cope with heavy rod rigging: only hydraulic cutters or other special tools will do the job.

Rescue seen from destroyer bridge off Portland Bill

But for the seamanship of the rescuers, another craft might have been lost that night. The story is a little bizarre, and is best told by the captain of HMS *Birmingham*, Commander Roy Clare, who is himself a most experienced yachtsman. The ship is a modern gas-turbine powered destroyer, carrying a Lynx helicopter on board.

'HMS *Birmingham* was operating 7 miles SW of Portland Bill, turning endless circles on the calibration range. At about 1800 on 15 October my navigator showed me the latest Mufax weather chart, with isobars tightly packed over the Brest peninsula. I commented that I was glad we weren't further south, and forgot about the weather. We did not see Michael Fish, the weather-forecaster on TV that evening.

At about 0115 I was nearly thrown out of my bunk by an unusually heavy roll. I called the officer of the watch on the intercom to find out what he was up to; he answered at once, saying that the weather had suddenly freshened and that he had just heard a call on VHF for help from a yacht ... would I come to the bridge?

I quickly established that the yacht was in touch with Portland Coastguard. Her position was some 18 miles south of me, about 24 miles from the Bill. Our weather conditions were wind: SSW force 8 – 9; visibility 6 miles, less in drizzle. We set off on a course of about 200° at 22 knots; this was the maximum we could do in the prevailing conditions. We pitched heavily in the steep seas, but there was only a modest swell and we did not thump at all.

Meanwhile the yacht transmitted frequently on VHF, using Channel 67, as directed by the Coastguard. I put an experienced yachtsman on the circuit to give an informed point of contact. We started an incident log book, told the naval operations room we

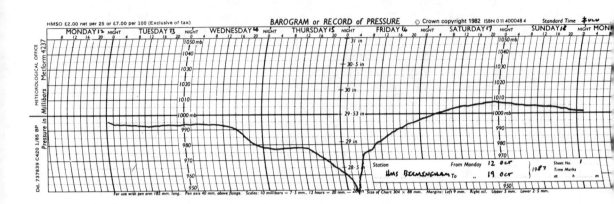

Fig 23.3. The barograph trace from HMS Birmingham *recorded during the October 1987 storm whilst standing by a distressed multihull some miles south of Portland Bill.*

were proceeding to assist, kept in touch with the coastguard, established that the Royal Fleet Auxiliary tanker *Black Rover* was also in the area and made basic preparations on board for search and rescue. Flying was out of the question; the sea state was worsening rapidly and the wind limits for helicopter operations would have been exceeded unless we had been heading out of the wind at some speed. This was clearly undesirable; in any case the ship's Lynx helicopter was not equipped for night search and rescue over a small object like a yacht, so it would only have been able to aid the search, not the rescue. Ashore, the coastguards decided that the Weymouth lifeboat should be called out. I do not know whether the shore-based search and rescue helicopter was considered; in my view it would have been of very marginal value in those conditions. In the event, search was not necessary; location was straightforward, aided by a reasonable position report in the first instance, and by continuous VHF transmissions from the yacht, which were DF'd by the coastguard stations at Portland Bill and Berry Head to give a cross bearing fix.

By 0150 we had established that the yacht was a cruising multihull of 40 ft length. The vessel was in no immediate danger of foundering, had sustained no damage and was managing to make way across the wind at about four knots on a westerly course, apparently bound for Ushant. The crew sounded terrified; they were pleading to be taken off their craft, urging rescuers to hurry before dire things happened. The voice on the radio repeated uncomplimentary things about the skipper, who was presumably out of earshot in the cockpit. The voice alleged that the skipper was a lunatic and would not turn back. The yacht was said to be shaking violently in the gusts and the voice was convinced that each moment was to be his last. In the background of each transmission, it had to be admitted, there was a fearful noise of wind in rigging and crashing waves. We did our best to reassure and console the yacht, keeping the crew informed of our progress.

By 0215 we were down to about 15 knots. The sea state was rising high, the wind gusting 75 knots. Visibility was down to a mile or two. *Black Rover* and *Birmingham* arrived on scene at about the same time – 0245 approximately. Between us we manoeuvred to windward of the yacht, whose lights were clearly visible in the driving spray.

Black Rover kept pace with the catamaran and provided a solid breakwater; I manoeuvred *Birmingham* between the two, pumping light oil from my bilges to take the edge off the sea. The yacht crew reported that our combined efforts gave them an easier ride; how much of this was due to oil and how much simply to our lee I don't know. Derek Sergeant, the lifeboat cox'n, said he thought the oil helped, but none of us felt it was a conclusively valuable contribution. I believe that the yacht was moving too fast to make maximum use of my slick; with hindsight it might have been more effective if laid to leeward and ahead of the yacht, to allow the craft to pass through the relative calm. That the oil calmed the water it was not in doubt as we could see its effectiveness in the search light beams: placing it where the yacht (and, subsequently, the lifeboat) could benefit was more tricky.

The craft continued to make progress to the WNW as the wind increased to 90 knots. My anemometer would not register above this speed, but it was clear from the roar in my own masts and aerials that the wind was still rising. There were long periods, while I lay virtually a-hull, rolling violently, keeping pace with the yacht, when the anemometer needle lay hard against the stops, and the note in the wind continued to rise steadily and unbelievably higher. On the bridge of *Birmingham* I think most of us were too busy to marvel at the strength of the storm. The ship rolled nearly 45 degrees at times, her stabilisers unable to operate as designed at these very low speeds. I used the bridge throttle controls myself as if they were Morse controls on a speed boat, using high power on each shaft, alternately ahead and astern, keeping the ship close – but not too close – to the yacht. This seamanship was as nothing compared with that in *Black Rover*, she had a single shaft, variable pitch screw and has a generally heavier and more ponderous hull. Yet she stayed in position and coped very well indeed.

Shortly after 0320, with no warning whatsoever, the wind swung from 180 – 190 degrees, 95 plus knots, to 270 – 280 degrees, 95 plus knots. This was remarkable for all sorts of reasons, not least of which was that the sea flattened out for a while. Confused is too elementary a word; the sea was gob-smacked as one of my sailors put it! The catamaran at this point began to head NNE, maintaining a beam wind and a speed of about four knots. The crew of the yacht continued to plead periodically with me to take them off. I pointed out that since they were in no immediate danger they were better off where they were. I was about 4,800 tons of crushing death if I closed with them in those conditions and I was not about to put my rigid-hulled inflatable in the water in those seas.

Meanwhile, the Weymouth lifeboat approached steadily from the NNE. They had a heavy time of it and it was to their undying credit that they arrived so quickly, at about 0420, to set about rescuing the crew of the catamaran. For the ensuing hour we in *Birmingham* had the privilege to be spectators as the Arun class vessel manoeuvred repeatedly alongside the yacht. We closed in to about 200 yards from the scene to provide the best lee. The sea was beginning to build steadily from the west. The dramatic wind shift was followed by two hours of unabated fury: the sea became very angry indeed by 0530 when the lifeboat completed her last approach, with, incidentally, no damage on either side. By then the wind was 260 degrees, 80–100 knots, with possibly stronger gusts. The seas were confused and steep with a very heavy Atlantic swell. The female member of the yacht crew had been invited to cross to the Arun first but she was in fact the last to go, showing some courage. The skipper elected to stay on board and drove his yacht to shelter in Weymouth with resolution, though nearly coming to grief on the Shambles in the process. Once in the lee of Portland Bill, at about 0800, I used the rigid-hulled inflatable to transfer an experienced yachtsman to assist the yacht into harbour.'

A Contessa 32 in the Solent

Meanwhile, not far away, a Joint Services Sailing Centre Contessa 32 called *Explorer* was on passage from Poole, Dorset to Gosport in Hampshire on the last day of an RYA coastal skippers' course. The instructor and skipper was Ray Williams, but the story is told by one of the crew, Martin Bowdler, due to take this examination the next day.

'We intended to make this a lazy passage, adding a few night hours to our week's trip. Therefore when we heard the shipping forecast at 1750, we knew we would have a lively sail. The general synopsis gave gale warnings for most sea areas, and for Wight it was cyclonic, becoming west to south-west force seven to the severe gale force nine, decreasing six later.

We cast off from the Town Quay at Poole around 1845. Our course had already been planned and plotted on the charts. All the sails were sorted out for quick access, in case the conditions strengthened. Anything that could crash around was secured or put away and our harnesses were fitted once more. Though in the week we had many a soaking we all still had at least one dry shirt at the ready to don if we were to have another wet trip. All other clothes were stowed away. One important aspect that we did not remember was to fill our thermos and make lots of sandwiches for the passage.

When we cast off the sea and sky gave no indication of the waves and wind to come. The sky was watery; dark – yes; forbidding – no. The course planned was a very simple one: from Poole Fairway across Christchurch Bay to Bridge Buoy, through the Needles Channel to the Solent and thence to Gosport.

At Poole fairway we had one slab in the main and the No 1 genoa set. The weather sharpened to force six from the WSW and a well rehearsed sail change to the working jib was executed. Our thoughts of this passage had not been to push ourselves, therefore our sail changes were always one step ahead.

By the time we were crossing Christchurch Ledge this routine was overtaken by events. The wind picked up to what we imagined to be around force eight or more, which caused the crests of the waves to froth up and turn into heavy spume. By then we had three slabs in the main and the storm jib set. We were surprised how suddenly it had become very dark; this was not your average cloudy sky at night. The wind continued to pick up; and as a result the sea was gathering strength and height. We had been able to pick out the lights of Bournemouth and Highcliffe continuously, but this became more difficult with the spume and the deepening of the troughs.

Our navigation became an increasingly odious task. As the seas increased so, too, did the emergence of a pattern of rogue waves. These would hit us on our quarter and happily break on the cockpit and coach roof. Every so often, one of these would be much larger than the others. We had the hatch shut and the washboards in except for the top one, but these larger rogues had enough force to slide the hatch forward and completely drench the navigator and chart table. I had just finished my stint at the helm and, as part of the rotation of crewing, found myself down below navigating. At last I had a chance to change my sodden shirt. This done, I felt so much better. However the first of the larger rogues swamped the boat. I know this because much of it came down my drier and warmer neck and the chart table was awash. As I appeared in the hatch this caused much amusement to the crew, who had already taken the brunt of the wave.

We found that we could be prepared for the normal series of waves by continually weaving along them to avoid the breaking crests. We could feel the normal pattern, as the boat began to lift, and whoever was at the helm would play the waves, riding through them as they broke at their crests. A Contessa may be a most forgiving boat but in these waters we had to play her firmly, constantly dumping and tightening the main. The motion gave us the overall effect of cork-screwing through the water.

However the rogues were different, in so much as we could not see them coming until the last possible moment – too late. One of the crew, armed with a torch, was detailed to watch for these rogues. Not a very satisfactory task as he would be certain for a head-on soaking. This was one of the jobs no one really enjoyed, but certainly amused the rest of the crew.

Our crew rotation proved valuable in these conditions, due to the various physical tasks on the boat. No one person could continue on a particular job for a long stretch, as certain jobs required more exertion or concentration than others.

The most spectacular seas were off the Shingles where the large waves piled up on meeting the shallows. White waves resulting from this could be made out, although the visibility was limited, while the crashing noise was almost deafening. We had the impression that we were much nearer to the Shingles than we actually were. The seas around us were very steep, broken, erratic and white. This made steering difficult, as our rudder would be as much out of the water as in. We did notice we were being blown down towards the Shingles, and had to harden up to allow for our considerable leeway. For about the next hour the heel of the boat was such that her cabin windows were almost constantly under water. When we bore away there was a moment when *Explorer* would drunkenly come upright, only to fall over again in the next gust. One of the crew found it all a bit overwhelming and he had to retire to the cabin, where he spent the rest of the trip. I cannot say that we were all scared, because there was so much to check on. Having said that, we were apprehensive off the Shingles.

The sails and rigging were being put under a great deal of strain, so the general appearance of the sails, hanks, reefing lines and sheets had to be checked so as to pre-empt any sort of accident. This was a hazardous job, but one that was enjoyable.

Explorer was fairly bucking around and very wet at the bow. But with a life-line to prevent one being lost overboard, the job felt like a fairground ride.

By the time we had passed Hurst Castle, the wind became increasingly local under the lee shore of the island. The seas had lessened, so whilst navigating I tried to take the opportunity to make a sandwich for the crew. However, odd omnidirectional gusts threw *Explorer* around; the same can be said of my attempts below – but we were all grateful for something to eat, even if crude, wet and very salty. Past Yarmouth this freak condition lessened. The wind had not yet reached its full force, even though it was screaming through the rigging. The barometer was still falling at an extraordinary rate.

People have often asked why we did not make for shelter to somewhere like Yarmouth; the simple answer being that we seemed to be safer on the water than making for land, and we had our appointment with the examiners the next day at Gosport. However, passing close by Cowes, we did have a small twinge as we could see all the various lights on. The thought of a whisky in the warm, talking about the storm, was not lost upon me.

Past Prince Consort buoy we put our final slab in the main. This did little to affect our speed, some six and a half knots. At just before midnight we came out of the lee of the Island and were once again in the full force of the storm. This was an interesting experience as the change of conditions was so quick. From the very loud wind we had experienced in the lee of the Island, the wind now roared. *Explorer* could be felt groaning under the strain, while rapid easing of all the sheets did little to improve the slamming of the boat. We fairly hurtled towards Gilkicker Point.

The 0030 shipping forecast was now nearer the mark. Wight was 'south-westerly gale force eight to storm force ten decreasing force six to gale force eight.' The storm force ten surprised us, but accounted for the noise and strong conditions.

Normally we might have taken the inshore small boat channel between Gilkicker Point and Portsmouth but to give a greater margin for error we took the Swashway route. When the moment came to turn towards the harbour entrance we tacked rather than gybed as far to windward as possible. On the new course our angle of heel was markedly increased, such that the cabin windows were constantly awash, and we noticed that we were making a good deal of leeway towards the beach. We were able to make a large allowance in our course for the leeway and thankfully no extra tacks were needed to get inside the harbour entrance.

Once we came within the protection of Fort Blockhouse, *Explorer* immediately came upright and we seemed to stop. Our handkerchief representation of sails was dropped and secured. It was just after 0100 and we had made our destination in good time, though, ironically, when day came our examiners had to call off the tests due to the weather, being blocked in their homes by fallen trees.

A Hallberg Rassy 42 in the North Sea

At much the same time some more heavy weather sailing was being experienced in the North Sea. In this case the account is by Jeff Taylor who, with two crew, was delivering a new Hallberg Rassy 42 from her builders in Sweden to Southampton.

This was a trip he had done many times before in all weather conditions.

'After a trouble-free trip across the North Sea, we called into Lowestoft on the morning of Thursday 15 October to fill up with diesel. With a forecast of S to SW force 4 – 5 for Thames and Dover, but a wind from the E to SE of 5 to 10 knots, we departed at approximately 1400 and made good progress motor-sailing at 7 knots for the first 3 to 4 hours. The 1750 shipping forecast was the first indication that there might be a strong S or SW wind on the way. At this time we were just north of east from Harwich, with a light SE wind, making good progress motor-sailing with full sail. I looked at all the options and decided to carry on, with the intention of going in to Ramsgate should the wind pick up later, as forecast. After all, even if the wind should increase to force 8 from the SSW, it was nothing that we could not cope with, as we were in a well-found boat in which I had utmost confidence.

The wind gradually increased from the south and by midnight was approximately 20 to 25 knots. We were by now motor-sailing with no genoa and one reef in the main. Our position at the time was 8 miles WSW of the Galloper light buoy and the tide was about to turn with us. The sea was starting to build up from the south. I heard a gale warning on the VHF just before midnight; the forecast was for gale force 9 from the SW, imminent. I was firming up my plans to make for Ramsgate as we were only 21 miles away, and obviously something was going to happen. The wind steadily increased up to SW 30 to 35 knots, with the sea becoming quite rough and

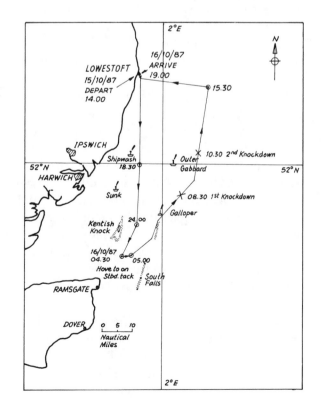

Track chart of the Hallberg Rassy 42 during the October 1987 storm.

uncomfortable. It was enhanced by the now south-going tide. There was a lot of water over the decks at this stage, and our boat speed was down to 4 knots.

At about 0130 on Friday morning I was on watch when George popped his head out of the hatch; he told me there had just been a storm force 10 given for Thames and Dover. By this time I was starting to get nervous about the situation, so I asked George to make sure everything down below was well stowed and to prepare for some very heavy weather. The next gale warning came at about 0330 and forecast SW violent storm 11, imminent.

The wind was still only 35 knots and we were making about 4 knots towards Ramsgate; but it was obvious that we were not going to arrive before the gale.

At 0430 the wind suddenly increased to 60 knots. I decided to heave-to on the starboard tack and stand out to sea. This presented me with a bit of a problem as we had no storm jib or trysail on board. In the end I found that with no sail up, the helm lashed to weather and the engine running ahead at about 1200 rpm, the yacht lay with the wind about 60 degrees off the starboard bow and was reasonably comfortable in the circumstances. We were 12 miles NE of Ramsgate at this stage. The seas were getting very big with breaking crests and spray everywhere; we started to get laid over on our beam ends as the boat rose up the face of the waves and over the tops. Visibility was down to about half a mile.

At 0500 the wind had increased further and we were approximately 14 miles NE from Ramsgate. Our radar, which worked well throughout, showed a lot of shipping in the vicinity; so I was very keen to get outside the North Falls banks, and decided to run off into deeper water. We were running downwind in generally a north-easterly direction, using our faithful Decca for navigation. The sea continued to rise and became very steep, with every crest breaking. Flying spume filled the air, reducing visibility to nil.

It became increasingly difficult to steer, due to our speed; but it was essential to keep the boat dead down the face of the waves in order to avoid a broach. I felt we really should have had a sea anchor out but we had nothing suitable on board to rig one with, and I thought it too dangerous to let anyone out of the cockpit.

We suffered our first knockdown at approximately 0830. As we surfed down a very steep wave the boat started shearing off to port, and I was unable to correct this. The wave then broke on top of us knocking the boat down to about 90 degrees. The boat righted herself very quickly, and the only problem left was a cockpit full of water; but this drained away quickly. The seas were continuing to get worse. I had sailed in winds as high as we were experiencing, but I had never encountered seas as severe and dangerous. Previously I had always been in very deep water which made the seas long, unlike these very steep short waves that were being caused by the shallow waters of the North Sea.

The storm was at its peak around 1030, which was when we suffered our second knockdown. This happened in the same way as the first one, but was far more severe. We were covered by a solid wall of water that seemed to pin us down below the horizontal with a suddenness that was quite alarming. When we came upright we noticed the liferaft, complete with its stainless steel cradle which had been attached to the deck, had been washed over the side. The liferaft, having by now inflated itself, was being towed astern by the painter. Our sprayhood had been turned inside out and the lifebuoy washed over the side. The water in the cockpit had seeped through the sides of the washboards, soaking the VHF radio and rendering it inoperable. Then the liferaft broke adrift and I started to feel very vulnerable, as without a liferaft or VHF radio, I realised that we were very much alone.

The storm started to abate around mid-day when the average wind speed dropped to approximately 40 knots; but there was still a very fierce sea running. In comparison to the average wind speed of 70 knots that we had sustained over a period of about half an hour, with gusts up to 95 knots (as recorded on the B and G equipment), it was relatively quiet. The sea had died down enough by 1530 for us to set some sail and confidently turn the boat beam onto the seas. We headed in a westerly direction for Lowestoft, from whence we had come. We were approximately 25 miles east of the port and arrived at 1900 where three very tired and relieved yachtsmen were glad to get ashore.

As a professional delivery skipper, sailing approximately 15,000 miles a year offshore, it is easy to become complacent. I felt that I had a lot of experience interpreting the UK shipping forecasts, with the time delays that one normally gets prior to a severe gale.

In this case I was wrong about the weather, and perhaps could have got some shelter in Harwich. In restrospect this may not have been a good idea, as there was much damage to yachts in this part of England. Certainly I feel that when we turned and ran, this was the only and correct decision to make.

An experience such as we had brings home the fact that we are all human and one must always have utmost respect for the sea. There was no doubt in my mind that if I had been in a less seaworthy boat we could have been in serious trouble.'

24 FORCE 12 IN THE NORTH ATLANTIC

Robin Knox-Johnston

Meteorological charts can only be relied upon to give an indication of the sort of weather that may be expected in any particular area, and it is well to bear in mind that storms can develop quite quickly where they are least to be expected. The North Atlantic has given one or two unpleasant examples of this in recent years and it behoves anyone thinking of making a crossing to obtain all possible information before departure and be prepared to alter the plan if necessary. November and December of 1989 are good examples. I was in Norfolk, Virginia in the United States and planned to sail to the United Kingdom via the Azores. The course is almost due east and the Pilot Atlas gives the maximum frequency of gales to be expected on the route at 9% in November and 14% in December, both just north of Bermuda. In this area only 19 observations gave winds of force 9 or greater out of 3,500 taken over a period of many years, and we expected to be well to the east of the 14% area by December.

The boat, *Suhaili*, is a 32 foot Bermudian ketch, built of teak in Bombay between 1963 and 1965 to an Atkins design based on Colin Archer. She is well proven having sailed for five months through the Southern Ocean on a non-stop voyage around the world, which included winds in excess of force 12. There were three other crew in addition to myself, none of whom had crossed the Atlantic before, but all had sailing experience. Philippa King had done a Fastnet, Dave Balls owned his own boat, and Chris Hipwell had raced dinghies. We set sail on 14 November the wind being a southerly force 5.

We quickly learned that the voyage was not to be quite as simple as we had anticipated when, after two days, the wind rose to force 8, gusting 10. The forecasters then admitted that there was a storm about 700 miles away, putting my mind at ease as far as the cause of our winds was concerned, but leaving us very unhappy with the effect. By daybreak we had handed all sail and had started to stream the warps. The objective with warps is to put out sufficient to extend astern at least more than one wave length, so that even if a large part of a wave is moving in unison with the boat, there will be a section of warp providing drag. Then, as the boat tends to try to surf down the front of a wave, the warp acts as a brake, and prevents it from broaching round in front of the wave. My usual method is to fasten one end of the warp to the eight inch square keel-stepped post forward, the strongest securing point in the boat, and to lead the other end down one side, over the stern and back to the post on the other side. Thus the warp is streamed in a bight, the length of which is easy to adjust, in the same way that parbuckling is easier than a direct lift.

On this occasion the seas were not overly dangerous, perhaps 35 ft in height and very steep as is to be expected in the Gulf Stream. With the stern anchored by the warp and in line with the waves, we bounced about safely all day, occasionally being swept by a breaking wave, but at least the water was warm. The storm passed quite quickly and shortly after midnight we were able to hand the warps and start setting some sail although I did not want to get up too much speed until the waves had subsided a little.

It had been a good introduction for the crew, who had come through it all with flying colours, but I hoped that this would be the last storm for the voyage; the further east we sailed the further we ought to be from the normal path of north Atlantic depressions. What was not clear at the time was that there was a high pressure system over Europe, and all the systems that usually head in that direction were being deflected north; we soon learned that this was causing them to intensify as well.

For the next couple of days the weather was, as expected, moderate southerlies, but then another low was forecast, intitially to pass about 500 miles north of us. This low quickly deepened to a storm and moved closer to our position. I altered course to the south-east to remove ourselves as far away as possible from its path, but with a maximum speed of about 6 knots, the difference this made was more psychological than anything else. It was obvious that we were in for a pasting, but having survived some extreme weather in *Suhaili* in the Roaring Forties I was reasonably confident of her abilities. The wind and seas rose well in advance of the storm, and by daybreak on 21 November, we had short steep seas of at least 30 ft in height, with the wind strength at force 8 to 9, the gusts reaching 58 knots. Sail was reduced to a storm jib I had made on the outward voyage, and we had just started to stream the warps when a wave swept the boat and fashioned them into a beautiful snakes' honeymoon. The rest of the morning was spent trying to disentangle knots that had appeared from nowhere.

All morning the waves built in height. The anemometer broke shortly after daybreak so we had no idea of the wind's strength thereafter, however it was certainly increasing. We ran before the wave, keeping our faces down wind as the spindrift was blinding. Even under the storm jib our average speed was 6 knots, about *Suhaili's* maximum, but when a large wave caught us we tended to surf at much higher speeds. This was where the danger lay, as the rudder tends to loose grip when surfing and, unless very lucky, the boat will then swing round in front of a breaking wave. As a general comment, at best this wave will roll a boat, at worst, if the wave is breaking at that moment, it can hit with tremendous force and smash the boat right over. If the wave is slightly hollow, a percussive effect can knock a hole in the hull or push portlights straight through the structure.

For *Suhaili* the only solution was to get a brake on her, and we worked steadily disentangling the warps, but as fast as we cleared one tangle, a wave sweeping over the boat from astern put in others. These waves were not dangerous to the boat herself, but all of us were carried to the full extent of our safety lines at one time or another, collecting bruises and becoming thoroughly soaked, and water found its way below through the hatches.

The first serious broach happened just before midday. *Suhaili* was picked up by a very large wave which thrust her forward. Without any control from the rudder she swung round to starboard and the wave broke, pushing us over at least to the horizontal. When the water had cleared we found that the compass had disappeared. The next broach came about half an hour later and took away the main upper port

All morning the waves built in height – the anemometer broke shortly after day break so we had no idea of the wind strength thereafter, however it was certainly increasing.

spreader. Within the next thirty minutes yet another broach occurred, and this time the helmsman, whose harness was unwisely attached to the mizzen shroud, was carried up the mast until he hit his head on the spreader. As the boat came upright he had to slide down the rigging to regain the deck. From then on everyone in the cockpit clipped their harnesses to the main sheet block deck eye.

The seas were now very large, and to compound our difficulties, were coming from the south-west as well as the west. Most boats can cope with a sea from one direction, but a cross sea is a very different and a much more dangerous matter. Every wave was breaking, but periodically three or four larger than average would follow each other. The wave period was so short that there was barely time to correct the boat's course between the smaller ones, the large ones gave no time and our situation was becoming precarious. We had about 100 ft of warp out in a bight, not nearly enough to check a surf, and I knew that if we could not get more out quickly we could be in for a very nasty broach.

The next forecast was far from encouraging. The centre was now closer and showing a pressure of 966 millibars. Although the wind speed was given as between 45 and 55 knots, this was an obvious underestimate, and the seas were certainly greater than 30 ft. In fact later it was admitted that they had reached 50 ft and that over three million square miles of the Atlantic had been affected by winds of force 8 or above.

Running with warps streamed before the dismasting.

I had just finished taking the forecast when I heard the roar of yet another breaking wave. It hit, and like a bomb burst, we were flung instantly hard over to the starboard side in a tangle of people, provisions and anything loose. As we regained our breath and cautiously disengaged ourselves to see whether we had broken any bones, our first thought was for the helmsman, but at that moment Chris threw open the hatch and yelled 'She's breaking up'. I stared at him in disbelief simply because I could not believe that *Suhaili* could ever break up at sea. She is incredibly strong, $1\frac{1}{4}$ inch thick teak planking, with 2 x 3 inch frames every 14 inches, and had survived five months in the Southern Ocean with no breakages. Although these seas were very short and steep, and to make matters worse, not always coming from the same direction, I was sure that the boat would come through.

I rushed on deck to see for myself and the initial sight was not encouraging. Both masts had been ripped out and were lying across the deck amid a tangle of rigging. However, the cabin top was still in place and there were no apparent holes in the deck or hull. Our first task was to clear up the mess, which was useful therapy for the crew anyway, so Philippa and Dave started to bale out the water that had come in, and Chris and I began to release the rigging in order to throw the masts over the side and avoid damage to the hull and deck.

It was time to take stock of our situation. There was no point in even thinking of

A view down the seas in force 8.

abandoning the boat. One of the major lessons learned from other people's recent experience is that the last thing to do is take to a liferaft. It is less easy to spot, is thrown around far more, and cannot carry all one's food and the means to cook properly. It is best to stay with the boat until there is no alternative. Strangely enough, now that she was dead in the water, *Suhaili* was lying broadside to the waves, and seemed more comfortable although we were being thrown about heavily as rolling without her masts was considerably more violent. Two hours later the water was out of the bilges, the decks were cleared, and at last the warps were streamed again, but this time from the bow like a sea anchor. We even attached the anchor to the warp and released it to its full scope of chain to add weight to try to keep us head to the seas. The effect was only to bring the bow round about 20°, but this was more comfortable than being broadside on to the sea.

We had three EPIRB's on board, two of the older 121.5MHz and a new Lo-kata one on 406MHz as well as the Argos Satellite transponder which could be employed to send a distress signal if necessary. However the situation did not seem to warrant a distress call. *Suhaili* was lying comfortably, and was evidently not leaking through the hull, although both hatches were now letting in water every time a wave swept over us. It seemed to me that we were distressed rather than in distress, and as such it would be far better to wait at least until everyone had got over the shock and the weather eased, and then see how everything looked. We ate some biscuits and then I told everyone to turn in. There was nothing we could do, and a good rest was probably the best therapy. There was little point in posting a lookout as it was almost impossible to see anything, and in any case we were temporarily without power and both radio aerials were missing. I wished that we had not lost the Firedell radar reflector though, as although any radar set close to us was probably going to lose us in its clutter, there was just a chance we might show up. Despite our tiredness nobody slept very well, the motion was just too violent, but we did manage to doze through the night.

With daylight we had the best breakfast we could manage and then made a plan. Bermuda was less than 400 miles away, but to the south-west. Faial was 1,600 miles to the east and down wind. There was no way we could rig anything that would take us to windward under sail, and although I was confident of getting the engine to work after dealing with water in the fuel system, bashing our way into the prevailing wind under engine did not seem very attractive. It was agreed that we would sail to Faial. Despite the fact that none of the crew had sailed the Atlantic before, nor encountered weather like this, and the hammer blows of the knockdowns and losing the masts were new and shocking experiences, they remained calm and totally supportive.

As soon as we felt it was safe, we put a line around the stump of the mainmast about 8 ft above the deck, and re-set the storm jib. As we were doing this a ship passed less than a mile astern but obviously did not see us. Next we got to work on the engine and were rewarded after a couple of hours by a nice healthy roar which gave some warmth and re-charged the batteries. The seas were still large and the wind gusting force 9, but it seemed that the worst had passed.

The following day, 23 November, the wind dropped to a force 4, and the time had come to rig up a more effective means of sailing, and something to give us a course to steer. Fortunately I had an old fashioned hand bearing compass on board, and this was lashed in the cockpit for the helmsman to read. None of the mast pieces had been long enough to be worth retaining, and the longest spar left was the main boom. We rigged shrouds and halyards to its outer end and then took a tackle to the bow. By dint of holding the boom up and taking in on the tackles, we managed to raise it vertically and then set up the shrouds. The staysail was our smallest surviving sail, but it proved too long in the luff to set properly, however it caught the wind so we set it forward and then took in the warps, and we were sailing again. Our speed was soon up to 4 knots as yet another system came through, but this one only reached force 8. It swept us once, breaking the staysail boom, but fortunately passed quite quickly.

By 25 November we were making good progress with the mizzen sail set on the main boom lashed vertically to the stub of the mainmast with its foot as the luff. It was not very efficient, and I planned to set a form of gunter rig as soon as conditions permitted. However just before noon a US Navy Orion flew overhead and circled round before coming in again and dropping a flare. This worried me, as we had not put out distress signals and I could not think why he wanted to attract our attention. Our main problem was one of communication since we had no working radio transmitter, however we had my Aldis signalling lamp and started to send a message in morse. Each time the plane approached I sent *Suhaili – OK – Faial* and in response the plane switched on its landing lights. However it did not go away and circled all the afternoon until just before dusk, we saw a ship heading towards us. The plane dropped a number of flares and the ship came close and slowed down. Obviously the plane had not read my signals, and the ship seemed to have no knowledge of the morse code, nor the procedures for using it with a signal lamp which, as an ex-merchant navy captain and communications officer of a frigate, I found staggering.

The ship, the *Aquarius*, came in alarmingly close to leeward, and not liking the close quarter situation I gunned the engine and shot round to his other side. There we found half its crew lined up along the main deck, wearing lifejackets and standing by a scrambling net. Over the loud hailer they told us to come in alongside and they would take us to Europe. The situation was getting out of hand. I had no intention of abandoning *Suhaili* and in any case there was quite a large sea running and it would have been dangerous to have jumped across since the boat might have crushed anyone on the scrambling net as she surged on a wave. We yelled that we were OK and heading for Faial. This seemed to astound the crew, but they eventually absorbed the information and having told us we were crazy they went ahead and slowly disappeard over the horizon. Since we had no means of communcation it seemed prudent to shun all contact with ships in future as this sort of situation was embarrassing. I had mentioned this to Chris earlier whilst we were watching the *Aquarius* approach and he had asked if I wanted to avoid her. Apparently I replied that we could hardly do that when the aircraft was dropping flares as if it was fireworks night!

The most important thing from our point of view was to discover what message might be passed back to our families. We knew that Peter Dunning in Newport was monitoring our Argos position, and must have become worried when we stopped making the radio schedules with him. That evening we heard him telling Jim Holman, who was crossing the other way to the south, that we had been sighted and our mizzen mast was missing. This was partially correct, but the aircraft had assumed that our jury mast was our normal main mast. If people were not aware of the true situation at least they knew that we were in one piece and proceeding towards the Azores.

The next day was spent making up a long boom out of the emergency steering oar, and the mizzen boom, using the dinghy oars as splints. To this we laced the mainsail. When all was ready we hoisted the boom about two thirds from its aft end, and tacked down the foreward end to provide a cross between a lugsail and a lateensail. It set a lot more sail, and also enabled us to point higher, but one or two improvements were required. As it was getting dark by the time we finished, we lowered it and re-set the mizzen on its side for the night. That was the last time we hoisted the lugsail, as the weather deteriorated, and then on 3 December we had a very serious blow and were knocked down again,and the jury boom was smashed.

This last knockdown was very worrying as we had the full warps out at the time, and *Suhaili* has never before gone over when lying to warps. Analysing the situation afterwards, I can only presume that the sea that threw us was a rogue from the south, when the main seas were from the south-west. This was a cross-sea situation again, and does demonstrate that however seaworthy one thinks one's boat is, every small boat is

Suhaili *approaching the Azores under jury rig.*

vulnerable in a large sea. Later we met one of the crew of a French aluminium 50 ft sloop that had been a couple of hundred miles north of us which had been similarly knocked about, but they had lost a cabin window and their deck had buckled.

We had been making good progress since we lost the masts, averaging almost 90 miles a day. The fuel tanks had been filled before we left, and I had no hesitation in motoring when the wind dropped. Every other day we stopped for an hour to have a swim since it is the most pleasant way of having a wash, and on a number of occasions we were joined for a short time by dolphins. But even out in the middle of the the Atlantic we were not totally cut off from signs of civilization. On 1 December we passed through a 4 mile oil slick and were always seeing items of plastic rubbish that had been thrown into the sea. The constant series of depressions, many developing to storms, were still passing about 600 miles to our north, and we had to stream the warps a number of times. Slowly the distance to Faial reduced, and at 200 miles to go I could no longer put off finding our exact position and added star sights to my regular schedule of sun observations. I had been taking a morning sight, and a meridian altitude at noon, but there is always likely to be a slight error in the noon position. The crew had wanted to learn astro-navigation, but the other problems we had undergone had kept us preoccupied and I only got as far as showing them how to take and work out the meridian altitude.

Finally, late on 8 December Faial came into sight right ahead. As night fell we began to pick up the necklace of lights along the southern coast, and at 0200 on 9 December we entered the marina and made fast. People were both helpful and sympathetic, and suggested that we lay up there and return in the spring when the weather should be more settled since no one could remember such bad conditions.

Like everyone else we used Peter's Cafe Sport as our base in Hora, and had the opportunity of seeing his scrimshaw museum which I can thoroughly recommend. However, I still wanted to get home, so on 12 December having been told that the weather was going to improve, we set sail again.

We expected the weather to be bad when we departed, but to calm down during the night. In fact the reverse occurred. By the time we were passing between Graciosa and San Jorge Islands, the wind increased and there were heavy squalls, one of hail stones as big as marbles. Quite obviously the forecasters had got it wrong, and it seemed prudent to take cover until we could find out exactly what was happening. The only place left to leeward was Terceira, so we motorsailed down to its eastern coast and sailed into Praia de la Vittoria, which has a large sea wall on its eastern side. Once through the entrance we made for two large US Army tugs and made fast. The Amercians could not have been more hospitable, particularly on the second night when another storm came through giving winds of 154 mph at the airfield and about 98 mph in the harbour. We held *Suhaili* off the tugs with a series of aircraft tyres, doubled up in places, as even inside the harbour the waves were reaching a height of $4\frac{1}{2}$ ft.

That storm was the final blow as far as I was concerned. There was clearly something vicious going on in the northern Atlantic, and in that case it was no place for a small boat that was already partially crippled. I made arrangements to haul out, and we laid *Suhaili* up for the winter. Hopefully, in the spring the weather will have settled down, read the meteorological books, and realised what it ought to average.

The main lesson learned was that one must have good forecasts, and they need to describe the whole picture. At no time was it ever explained why the depressions were intensifying and heading north instead of north-east. Had we been aware that a large stationary high over Europe was the cause of the disturbance to the normal pattern, we could have headed south 5° or 6° and still found a favourable wind, but of much less force. The American forecasts extend east to 40° West, and it is not possible to pick up the British forecasts on the radio until within 1000 miles of the United Kingdom. A weather fax would have given us the information we needed but unfortunately we did not have one.

It would also have made our lives a lot easier if we had had a fall-back communications system, one that was not dependant upon aerials attached to the masts. A hand-held VHF would have made communications with the aircraft and passing ships possible.

The final lesson is that however good your boat, the sea can always rustle up that extra large wave or cross seas which will catch you and fling the boat around like a piece of flotsam, and in those conditions only the strongest will survive – with luck!

Caption to photo opposite: Hoisting a storm trysail aboard ChaCaBoo *during the 1989 Fastnet after the mainsail slides had failed. Photo: Peter Bruce.*

PART 2
Expert Advice

25 YACHT DESIGN FEATURES FOR HEAVY WEATHER

Olin J Stephens II

Heavy weather has taken its toll among vessels of all the shapes and sizes that one can imagine, and the survivors have been just as varied. Is it worthwhile, then, to consider the design characteristics of yachts that should best survive the worst that weather can offer; or is handling the only factor? It seems clear that the boat counts too, and could be decisive, although once at sea, the action of the crew is what counts.

To declare the obvious, to survive means to stay afloat, to keep water out of the hull; and further to remain in, or at worst return to, an upright condition. Strength and range of positive stability are first requirements. In the course of this study we shall try to determine how these essentials can best be refined and combined with other characteristics to provide for the safety and comfort of the offshore crew.

A lifetime around the water has shown me the many types that have come through the extremes of weather on long cruises. I think it must have been in 1926 that my brother Rod and I spotted Harry Pidgeon's *Islander* lying in New Rochelle harbour, not far from our home. We were quick to greet him from a borrowed dinghy and to take him for a tour of the nearby countryside after inspecting the home-built 34 ft yawl he had sailed single-handed around the world. *Islander*'s light displacement and simple vee bottom form made for no survival difficulties. One was most impressed by the simplicity of her construction and equipment: no engine or electrics of any kind, no speedometer or even a patent log. We admired the man who had made it all seem so simple. Somewhat later we heard that Alain Gerbault and his *Firebird* were at City Island, so we went there. The contrast in every way was disappointing, but the older, heavier boat had made it through some very bad weather.

I had, and retain, a great deal of respect for the work of Dr Claud Worth, the owner of several yachts called *Tern* during the twenties. He must have been a thorough and meticulous student of offshore cruising. He advocated moderate beam, plenty of displacement and a long keel. I read and re-read his books *Yacht Cruising* and *Yacht Navigation and Voyaging*.

This background, reinforced over the years, had led me to believe that size and shape can vary widely though I like to avoid extremes. If structure and handling are sound the larger design demands more from the builder and crew as the loads increase geometrically with size. Big sails supported by great stability require strength and skill to control; small sails can be manhandled. Similar observations apply to hull, spars, and rigging.

Analytical studies, such as those carried out in the course of the joint United States Yacht Racing Union/Sailboat Committee of the Society of Naval Architects and Marine Engineers study on Safety from Capsizing and by the Wolfson Unit of the University of Southampton, have noted two characteristic conditions of capsize that bear on structure, ie those that occur due to the force of wind on the sails and those resulting from the jet-like force of a breaking sea. In the first condition the light structure of a small boat is not overloaded, but in the second the hull or deck of a very small boat may be smashed, destroying its ability to float like a bottle.

In this context size and displacement mean about the same thing, although the term 'light displacement' usually refers to the displacement/length ratio or tons of displacement divided by the cube of one percent of waterline length in feet. One can accept a range of 50 to 500 in that ratio over a range of 20 to 80 ft in waterline length. In geometrically similar hulls the righting moments increase in proportion to the fourth power of length while the heeling moments grow less rapidly as length cubed. Because of this small boat designs need more inherent power, i.e. beam and displacement, while big yachts with similar proportions need very large rigs. Thus smaller boats should avoid the bottom of this range and the larger boats should avoid the top. Figures 25.1 and 25.2 show a suggested range.

Displacement is determined primarily by required strength and stability, and further by comfort in the sense of motion and roominess. The yacht's total weight must provide for an adequate structure together with the weight of crew, stores and equipment, and for sufficient ballast to ensure stability. Truly efficient use of the best

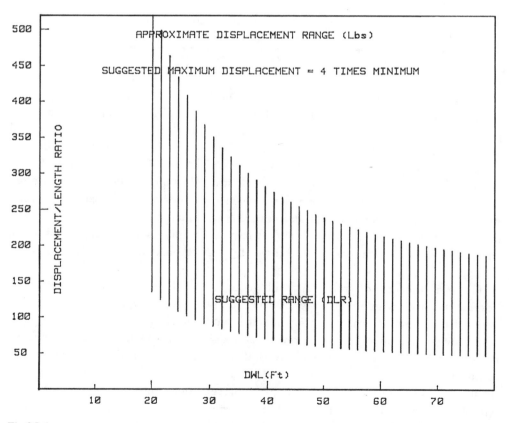

Fig 25.1.

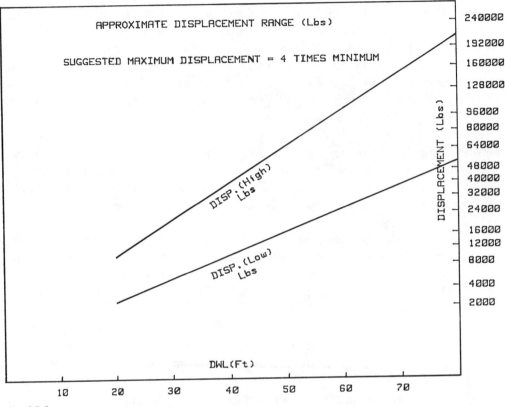

Fig 25.2.

materials can give a light hull and rig and, with enough ballast, appropriate design can provide stability, adding up to a lower limit on safe displacement. Though structural materials are not the subject of this study it should be said that, in the hands of a competent builder, sound, light hulls can be built of many different materials including wood, GRP and aluminium alloy. The high strength, high modulus materials often used as composites, such as carbon fibre, when used with care and experience offer strength and light weight. Steel and concrete are inherently heavier, especially the latter. Boats with light displacement must have light hulls so as to carry a reasonable amount of ballast for the sake of stability. In heavier boats material selection is less critical.

The vertical centre of gravity and hull geometry combine to establish the range of positive stability. A good range, for illustration's sake, over 120 degrees, will virtually assure that a capsized boat will right herself in conditions that have caused a capsize. Much lower values may assure the reverse although the larger the boat the greater the radius of gyration, accompanying a heavy hull and rig, the less the danger of capsize. It seems unfortunate that recent influences, primarily the IOR, have led to wide beam, shoal body and a high centre of gravity: the precise conditions for a poor range. The typical IOR type may show a 105 to 115 degree range while a decked-over 12-metre could go to 180 degrees, representing the full 360 degree roll-over. Moderation in beam and hull depth is the best course. Beam offers initial stability and roominess, but too much of it reduces range and results in quick motion. Depth provides easier motion, headroom, structural continuity and space for some bilge water – all desirable. Heavy ballast contributes to range but also gives quick motion. A moderate ratio of beam to hull depth seems ideal; say, a beam of not more than three to four times the hull body depth, with a centre of gravity low enough to give a positive stability range of 125 degrees.

Here one may note that Adlard Coles' three *Cohoes* fall within the range of proportions that I have recommended. Also that the damage to *Vertue XXXV* was caused by a breaking sea which threw her over and down smashing the lee side of her cabin trunk (coach roof) – a vulnerable structure, at best – due to structural discontinuity. *Vertue*'s small size probably did not permit the incorporation of the material needed for strength. A similar occurrence was the damage to *Puffin*, a design of mine and not quite so small, but hit hard at a weak point. *Sayula*'s survival of a near capsize on the second leg of the Whitbread race of 1973, with great discomfort but minimal damage, contributes to confidence in the larger boat.

The danger of weakness should not be seen as condemnation of the coach roof which is often needed to provide headroom and can also contribute usefully to range of stability by virtue of its volume. It is well to remember that all corners of abrupt discontinuities of area in a structure can be sources of weakness and should be carefully designed and built.

It must be clear to many observers that making a keel very thin or narrow at the hull juncture weakens an already highly loaded spot by even raising the loads in that area and the stress in the keel bolts, the keel and the associated hull structure. To the degree that this narrow base exists, the structure must be most carefully considered.

The power to carry sail is quite different from stability range. A naval architect evaluates both at small heel angles by measuring the vertical span between the centre of gravity and the metacentre where, in the heeled attitude, a vertical line through the centre of buoyancy cuts the centre plane. The product of this height and the sine of the heel angle gives the righting arm which, when multiplied by the displacement, gives the righting moment. We may assume displacement as constant but the metacentric height is strongly dependant on hull shape, varying with the heel angle and the ratio of beam to hull body depth. A beamy, shoal-bodied boat will have great metacentric height at small angles of heel which will be rapidly lost with increasing heel when the righting arm may become negative. Less beam and more depth will hold that factor almost constant up to angles of 35 or 40 degrees while assuring a safe range. The first example will feel stiff but must be kept upright to use the power of her rig while the deeper boat can benefit from plenty of sail power up to a wide heel angle. The beamy, light type depends greatly on crew weight to control heel: not a happy condition in bad weather.

Comfort is another characteristic related to beam. Generous beam contributes to roomy space below but too much has negative effects on motion and stability range. I am not sure that this is the place to discuss arrangement plans, but comfort can contribute to safety in heavy weather as it relates to the crew's ability to perform. Dry bunks and dry lockers, clear of either deck or bilge water, mean a great deal. I should like to refer favourably to the Dorade ventilator. Despite its appearance, and the many efforts to design something better, it still leads in supplying maximum air and minimum water.

Like other vents, the Dorade can admit solid water if fully immersed as in a capsize. Preparation for the most extreme conditions should include replacement of the cowl with a deck plate. Similar positive closure for vents of any kind should always be available.

Touching briefly on other aspects of safety and comfort we should consider strong, well-located hand rails and we should be sure that sharp corners are eliminated by generous rounding. Galleys should be arranged so that the cook can wedge or strap himself in place and, if possible, out of the path of spilled hot food. The water supply must be divided between several tanks with individual shut-off valves. This will conserve water in case of leakage and also provides control of weight distribution and reduces the undesirable effect of a large, free surface of water surging about within a

Turkish Delight, *a yacht designed to IOR, in a brisk breeze during the Admiral's Cup trials at Cork in 1987. A reef was taken in the mainsail shortly after this photograph was taken. Photo: Rick Tomlinson.*

tank. Engine exhaust systems may admit water in bad weather though careful design can eliminate that danger.

Rigs should be proportioned with ample strength for the high but uncertain loads of heavy weather. It is evident that many racing rigs lack the strength to stay in place. Recently-developed analytical methods have not replaced simple calculations based on the righting moment and the consequent rigging loads which apply tension at the chain plates and compression in the mast. Most designers use Euler column methods to predict when a mast will buckle under compression loads, often with individual assumptions on end fixity. According to such assumptions, safety factors will vary but must be generous so as to be ready for the unexpectedly severe conditions of heavy weather sailing.

Rig geometry and sail shape seem to be a matter of personal preference. Under severe conditions, however, the presence of two independently supported masts can be recommended. Strong storm sails and the rigging to set them quickly and easily are essential to a well-found offshore yacht. The storm sails should not be too large. No more than one third of mainsail area is suggested for the storm trysail and about five percent of the forestay height (squared) for the storm jib. Sail area, relative to stability, can well be considered in the light of the home port and cruising grounds, and primarily as it relates to comfort rather than to safety. Sail can always be shortened but too large a working rig means frequent reefing or sailing at an uncomfortable angle of heel. The areas of storm sails given above are very close to those defined in the Special Regulations Governing Equipment and Accommodations Standards of the Offshore Racing Council. Although drawn up for racing yachts there is a great deal of good material in these regulations which are to be found in Chapter 29, Preparations for Heavy Weather.

On the subject of hull geometry, I have stressed the ratio of beam to draft. There are other considerations, less important but meaningful. Positive and easy steering control is one such. In this day of analytical yacht design there is still no subject more deserving of intensive study than that of balance and steering control. Possibly the lack of understanding explains why there are few subjects that stir greater differences of opinion than the shape of the lateral plane including keel and rudder. Let me outline some of the problems and some partial answers.

Course stability is often characterised as that condition in which, without the adjustment of the steering mechanism, a boat sailing a given course when diverted by an external force will return to the initial course. This could be a definition of self-steering ability. Many boats can be trimmed to steer themselves under the right conditions but few will do so on all courses and wind strengths. The forces involved and the direction of their application and the tendency of the hull to turn one way or another at different heel angles and speeds form a very complex system, and one that is difficult to balance. We can accept these difficulties and yet ask for steering that is light and responsive. We should also ask that control can be maintained at both high and low speeds. Even that can be hard for the designer to ensure, but I can believe that there are a few helpful steps.

A long keel is frequently cited as the best solution. Probably it is, if all round good manners on the helm are more important than light air speed and if the length extends well aft. The disadvantage is the increased water resistance incurred by the greater wetted surface area that goes with the long keel. Such a keel seems to do several things: turning is necessarily less abrupt, and low speed control should be better; also course stability is helped if a large part of the lateral plane is abaft the centre of gravity.

Think of this the other way around and visualise a sea turning a boat with its keel area well forward. It is as though the boat were being towed from a point abaft the pivot point so that the further the new course departs from the original, the further the inertial direction departs from the course, thus causing the boat to continue to deviate further from the intended direction. Conversely if the tow point, the centre of gravity, where the force is applied, is forward of the pivot, the centre of lateral resistance, then the more the course changes the more the direction of the inertial, or tow, force is directed back toward the original course. I should add that this principle of sailing balance is not universally accepted, though to some it seems self-evident.

A small wetted surface area carries with it advantages that have resulted in the almost universal adoption of the short keel and separate rudder in racing boats and to the series production models that are too often modelled on the racing types. Comparatively it means equal performance with less sail area, especially in light weather or to windward when speeds are low. Using a short keel, the required position of the ballast dictates the location of the keel which further dictates the location of the CLR. This disadvantage can be lessened by locating disposable weights as far forward as possible, permitting the ballast keel to come aft, but such gains are limited and the best available strategy to move the CLR aft seems to be to use a large skeg and rudder. These serve the function of feathers on an arrow. Most new boats follow this pattern and, if the ends are balanced, they can behave well, exhibiting minimal loss of steering control, the ability to heave-to or other good seagoing characteristics. It is worth noting that this configuration exposes the rudder to heavy loads which have caused so many failures that designers, builders and owners should need no reminder that these rudders must be very strongly built.

Other characteristics that seem to contribute to good manners are reasonably balanced, yet buoyant, ends and moderate to light displacement. Both minimise trim change with increasing heel so that the unavoidable changes in the yacht's tendency to alter course occur gradually and the abrupt application of helm is seldom needed. Easy and positive control is valuable in a big sea. Light boats are controlled more easily when running at high speeds, though heaving-to may not be quite so easy as it is in a heavier boat.

For the sake of an easy and steady helm the pressure of the water on the hull must be evenly distributed and constant over the range of speed and heel. Long lines, minimally rounded, with relatively constant curvature make for constant water velocity, and thus constant pressure, over the hull surface. Any short, quick curve in the path of the water implies a quick change in pressure on the hull surface and, very likely, a quick change on the helm. Again, a light displacement hull with moderate beam and rather straight lines in the ends, best meets these conditions.

The motion of a yacht in rough water is probably better understood than is balance under sail. This depends a great deal on the weight and the way it is distributed. Weight distribution may be considered in any desired plane, but usually in the longitudinal and transverse senses. It is measured by the moment of inertia and is usually expressed as the gyradius which relates the moment of inertia to displacement. The former is the sum of the products of all weight elements and the squares of their distances from a chosen axis. The radius of gyration, or gyradius, is the square root of the moment of inertia divided by the total mass. It serves as a measure of the body's resistance to acceleration around the chosen axis. Thus a large gyradius, either longitudinal or transverse, tends toward easy motion in pitch or roll and is desirable in terms of comfort. In passing, it might be observed that the need for a very minimal longitudinal gyradius has become an article of faith among racing sailors. One suspects that they are right more often than not, though studies of resistance in waves mainly show that the weight distribution that results in synchronism with wave encounter is clearly bad; otherwise the effect is small.

Weight itself or, more correctly, mass also slows acceleration so that the motion of heavy displacement boat tends to be comfortable. Being heavily built, elements of mass such as framing and planking increase the moment of inertia. In this calculation the rig, due to its distance from the centre, makes a major contribution. Anyone who has experienced the loss of a mast in rough water will confirm the quick motion that followed. Thus, by resisting sudden roll, a heavy rig contributes to both comfort and safety, as studies have shown that, by increasing the transverse gyradius, resistance to capsize in a breaking sea is greatly enhanced.

Variations in hull shape are most significant as they relate to displacement stability and steering qualities, but some other effects of shape are worth noting. I have referred above to balance between the ends. This does not mean anything like true symmetry. My approach was by eye, which is a mixture of guesswork and judgement, though today it is easy to check heeled static trim with a computer, a step toward eliminating excessive trim change.

It is well to avoid flat areas in the hull. These can easily develop in the ends, expecially in a light displacement design where the fore and aft lines become pretty straight. If sections have a moderate U-shape, rather than a V, the flat area that occurs where straight lines cross in a surface will be avoided. Even some slight rounding on a long radius extends the period of impact, and so reduces the tendency to slam.

Freeboard is another characteristic on which moderation is a good guide. High sides increase stability range but present a large area exposed to the impact of a breaking sea, while a high point of impact increases the overturning moment. Low freeboard leads to immersion of the lee deck with its sheet leads and fittings. Related to freeboard is sheer. It can be argued that this has more to do with appearance than seaworthiness, but I think a good sea boat should keep her ends above water without excessive freeboard amidships. Let us agree that the beauty of the Watson and Fyfe boats of the early century was also functional.

Cockpits that can hold a great deal of water can be dangerous but may be considered in relation to the size and reserve buoyancy of the hull. It is essential that all cockpits will be self draining through scuppers which should be large. Deep cockpits offer

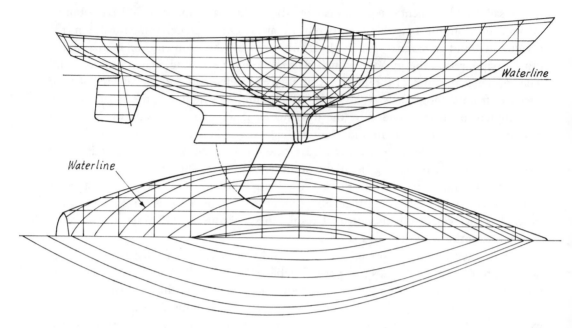

Waterline

Waterline

Fig 25.3. *The lines plan of* Sunstone, *ex* Deb. *Her present owners, Tom and Vicki Jackson, live aboard throughout the year and have achieved outstanding success in RORC races.*

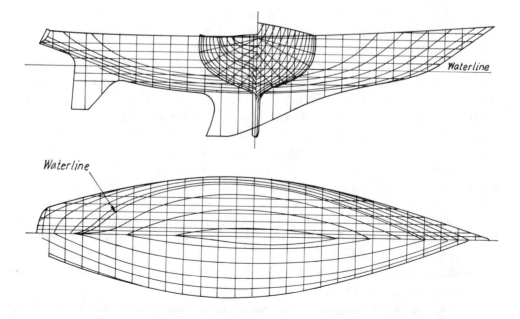

Waterline

Waterline

Fig 25.4. *The lines plan of* War Baby, *ex* Tenacious, *winner of the 1979 Fastnet Race. Her present owner, Warren Brown, has cruised extensively in her.*

protection and comfort but their effect on buoyancy must always be considered. One's priorities may play a part in selecting cockpit dimensions, but the smaller cockpit must be the safer in the end.

The conditions we are considering are less than ideal for centreboarders. They have their supporters and I have been responsible for many centreboard designs. I always tried to advise the owners that capsize was possible although, I hoped, unlikely, but in many cases the stability range was less than I should have liked. In other cases, where draft was not too restricted so that the ratio of beam to draft was not too great, and other details such as freeboard and deck structures were appropriate, the stability range seemed fully acceptable. Among S & S designed centreboarders, *Sunstone,* ex *Deb* seems a good example of a centreboarder suited to heavy weather sailing (Fig 25.3).

I hope it has been useful to consider, one by one, a number of specific characteristics. While each has an influence on the ultimate ability of a yacht it is always the combination that counts. No individual dimension means too much. Good performance can be reached by different paths and, finally, the good combinations are the ones that work (Fig 25.4). Almost every named characteristic carries with it both pros and the cons. When I think of the boat in which I should be happiest in meeting heavy weather, I visualise one that is moderate in every way, but simply as strong as possible. I should avoid extremes of beam to depth or depth to beam, either very light or very heavy displacement, or a very high rig. I should like the ends to be buoyant, but neither very sharp nor full, and neither long nor chopped right off. Though I have stressed resistance to capsize, in my own seagoing experience I have never been worried on that score, but I have occasionally been concerned about leaks or the strength of hull or rig. In the end I recommend moderate proportions and excessive strength.

26 RIG AND SPAR CONSIDERATIONS FOR HEAVY WEATHER

John Powell

Storms are inevitably a test of the design and construction of a vessel's rig. In this chapter these will be reviewed, as well as the maintenance of a rig, and choice of sail plan.

Fore and aft support of the rig

Providing lateral support to a mast presents few problems, but fore and aft control within the sail plan requires careful consideration of the options available. Either the fore and aft section of the mast can be increased to provide greater stiffness, or intermediate staying can be increased, or positioned more effectively. These aspects should be considered when choosing a rig.

The masthead sloop or cutter is still the favourite choice, being simple and efficient. A single spreader rig (Fig 26.1) allows fore and aft support at about half the height of the mast, and is well suited to a sloop. Double spreaders may be shorter, with the same shroud angles, allowing a closer-winded sail plan; but the upper two thirds of the mast lacks fore and aft support between lower shrouds and masthead, so is prone to pumping (ie oscillation fore and aft) in a head sea. This makes a good case for applying runners and an inner forestay to a spot near the upper spreader roots (Fig 26.2): all are ideal for a cutter sail plan. I believe this to be the most seamanlike offshore rig in any size of yacht; it offers greater flexibility and maximum support with less stress. The sail plan is divided into smaller units, allowing a greater choice of headsails and easier handling; although these aspects are less significant today, with the wide use of luff furling gear, now well developed for headsails and improving for mainsails.

In-mast mainsail and foresail furling gear

A good in-mast mainsail furling gear, used with a masthead foresail furler, allows easy setting, reefing and furling of the whole sail plan; thus one can set the optimum area at all times with the minimum of effort. Additionally, there are no battens to fiddle with or lose, no mainsail cover to rig or stow, no sails to hoist or flake; all are advantages, particularly in the larger yacht. However, in the case of a twin headsail rig I prefer to waive the luxury of furling gear and to set a staysail on the inner forestay in the traditional manner with the head and the foot of the sail attached to the stay with shackles to improve the set. This sail will be relatively small in weight and size and thus not too difficult to manhandle when changing, while its reliability gives confidence in storm conditions, when the failure of more sophisticated gear becomes likely. The traditional trysail should always be carried and set to complement the storm staysail.

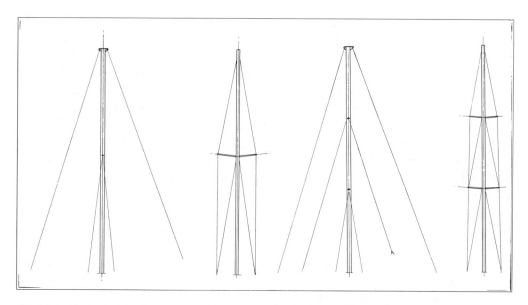

Fig 26.1. Single spreader rig.

Fig 26.2. Double spreader rig with runners and inner forestay.

Having said this, the modern, light displacement, easily driven hull, can be sailed very efficiently under bare poles, the area of which, in a really hard blow can well be more than required. In this case slowing down by trailing warps and the like may become essential.

Traditional reefing

In a traditional mainsail, with slab reefing or boom roller reefing, a row of reef points for an emergency deep reef is often advisable. By the same token, when the setting of some six feet of the top of the mainsail is quite enough sail area, it is not a bad idea to have a heavy duty cringle and patch on luff and leech at this point to enable the sail to be lashed around the boom; extreme conditions call for extreme measures and thoughtful preparation can be of great help.

Jury rigs

Thought should also be given to jury rigs in the event of spar failure; in this case short lengths of mast, boom and spinnaker pole sections suitable for sleeving, with the necessary tools and rivets, should be carried to enable a temporary repair to be effected.

Along with these tools, one should carry a pair of bolt croppers; standing rigging can be cleared quickly with these before hull damage occurs in a confused seaway from a failed and jagged mast attached alongside. In the case of high tensile rod rigging, particularly cobalt steel, hand operated hydraulic cable cutters will be required. Sophisticated gear calls for the corresponding antidote.

Rig materials and attachments

In a well-found yacht, one assumes that the mast section will be adequate, so that the most likely cause of dismasting will be failure of rigging or its attachments, with the highest proportion occurring at the masthead, ie cap shroud, forestay or, occasionally, back stay. In multi-spreader rigs the results of failure will differ; but in principle, the more frequent the support the more mast will be left; so generally one is able to set more jury rig sail area. Assuming the helmsman is unable to prevent ultimate failure of the mast (an aluminium mast will bow to a 45 degree bend without permanent deformation) that part of the mast above the next well-stayed section will fail. In a double-spreader cutter with its runners set up, this will be the top third; with the runners stowed, this will be the top two thirds. In a single-spreader mast, this will be the top half. Furling foils on headsails and mainsails can only compound the problems of sorting out a rig after a mast failure; but at the end of the day one still has to set as much sail area as possible. A halyard sheave or block at the spreaders for trisail and staysail is not difficult to provide at the design stage; a great deal of time and effort can be saved by so doing.

In choosing the section for any mast, diameter and wall thickness must be considered. Basically a large diameter with a thin wall will, for the same weight, be stiffer than a smaller diameter with a thick wall. The former in itself will produce more wind resistance, but will require less rigging to hold it up; the choice becomes complex and the degree of racing/cruising envisaged is the most important factor in making the decision. Small diameter thin walled sections can be reinforced to stiffen them but this adds weight and the balance of advantage and disadvantage requires careful consideration.

Tapering the upper section of a mast is always worth doing from the weight saving and windage point of view; the weight savings are little but worth having and come jointly from the reduced head fitting size and mast section. As the stresses become less towards the masthead such reduction is acceptable.

Aluminium masts and spars have been anodised since 1959 and this finish is hard-wearing and practical. With the coming of tougher paint finishes developed for aircraft (eg Awlgrip, Sikkens), painting has become a suitable option. It is more expensive, but allows greater scope for cosmetics, and can result in a most pleasing finish which can be touched up in the event of damage.

There is a variety of potential materials for rigging; and while 7 x 7 galvanised wire for standing rigging enjoyed a good span of life in the thirties and can still be seen today, the quality of hot galvanising in this particular field has deteriorated, shortening its life, and therefore its popularity. Stainless steel 1 x 36 aircraft wires, used in the late thirties in racing rigs, began to take over, only to be quickly followed by 1 x 19 stainless steel. This is now the norm, is very well tried and is currently the best solution for offshore work. It should be terminated with swaged terminals by experienced riggers, roll-swaged for up to 10 mm diameter wire and rotary-hammer swaged in the larger sizes. Screw terminals, properly fitted by the professionals, can also be very good and have the advantage of being easily fitted on board. Rod rigging in stainless or cobalt alloy is currently favoured by the racing fraternity as it offers greater strength with less stretch for the same diameter and similar weight. It is more brittle, so requires to be allowed universal movement at all spreader ends as well as at

British Satquote Defender *off Portland in May 1990 after sailing nearly 3000 miles under jury rig as a result of a cobalt alloy upper rigging rod failure in leg 6 of the Whitbread Round the World Race. When the top section of the mast came down, a cordless angle grinder was used to cut away the defunct rod rigging, her 18mm 'spectra' spinnaker guys were set up as a backstay, and a spare wire halliard was rigged from the top of the stump to the bow, then back to a winch as a forestay. In this photograph she is sailing under a cut-down reacher and trysail. Photo: HMS Osprey.*

upper and lower terminations. Some of the early lenticular (ie oval section) rigging had a tendency to fail from fatigue caused by wind-induced high frequency vibration, known as flutter. For example the Admiral's Cup team yacht *Prospect of Whitby* lost her backstay while on her mooring.

There are those who use rod rigging offshore for racing. Though racing encourages risk, the prudent mariner sticking with his 1 x 19 will sleep more soundly and be well pleased with his choice in the stormy southern oceans or, indeed, anywhere where extreme conditions prevail and ductility in the rigging is important.

Causes of rig failure

Causes of rigging or attachment failure are many, but are recurrent, and in the main can be prevented. Assuming again that in a well-found and designed yacht the gear was of adequate strength and appropriate material in the first place, failure must result from fatigue, wear, an incorrect component being introduced later in life, or damage from impact.

To safeguard against failure one can only inspect all components thoroughly on a regular basis, checking as follows for:

a) signs of rust at terminals (an advanced warning of fatigue)
b) loose nuts on screwed terminals
c) cracks at clevis pins and swaged terminals
d) wear at sheaves and pins
e) adequate locking arrangements to prevent unscrewing or unshipping of stay and shroud attachments
f) signs of rust, indicating fatigue, at mast tang edges or adjacent to mast tang through-bolts
g) signs of cracks in the mast at the halyard and lift exits and entrances, or in way of welds
h) adequate mousing of shackle pins or excessive wear in pin or bow.

A great many gear failures can be prevented by frequent attention to such details.

Equipment can be X-rayed or dye tested for cracks during the laying up period. Either system is good, the former being more expensive and perhaps not so convenient, requiring specialised equipment and a professional operator. Dye testing however can be carried out by anyone with common sense.

Where stainless steel is involved, and the fittings easily removed, they can be electro-polished. This serves the double purpose of revealing any hair-line cracks and restoring the appearance of the fitting at the same time. Faulty components can, in some cases, be welded, but are better replaced.

It should be remembered that mechanical polishing, while producing a very pleasing effect, will cause the material to flow over a crack and hide it instead of exposing it, as in electro-polishing. The process can also induce impurities, resulting in superficial rust spots at an early date.

Electrolytic action

Gear can also fail as the result of electrolytic action - a habitual enemy of the seafarer. Basically this is a wasting of a metal in a salt solution due to the proximity of a dissimilar metal. As in a battery an electric current is formed, and the process is accelerated in rate with increase of temperature. It can also be caused by a pure electrical leak from worn or old insulation, so mast wiring, indeed all wiring, needs to be checked with this in mind.

Aluminium must not be associated with any copper or copper-bearing alloy. A bronze inglefield clip on a burgee halyard stowed on a cleat against an aluminium mast

Yeoman 25 loses her mast over the side in a squall at the start of the selection trials held in the Solent for the British Admiral's Cup Team in 1985. The mast failed in compression due to the mainsail luff rope pulling out of the track, allowing the mast to move forward without control and out of column. Photo: Peter Bruce.

will, if left, result in a hole in the mast. A bronze toggle between rigging and an aluminium masthead can cause failure in time. Excess heat or salt accelerates this action. If dissimilar metals abound, a salt water deck wash pump used daily about the mast of a smart yacht in the heat of, say, the Caribbean is a mast killer.

Wholesome design features

Fatigue, an active enemy of seamen, can be largely prevented at the design stage. If not allowed for, its effect can be disastrous. For example, if a spreader fails it will result in the loss of a mast. While a spreader may be robust enough as a compression strut, its attachment to the mast can be the weak point due to fatigue. Failure to provide freedom for spreaders to be able to move fore and aft some ten degrees to accommodate the movement of the relatively loose lee rigging in a head sea will almost inevitably result in a fatigue break and subsequent loss of the mast. Most rig failures start on the unloaded lee side and do not always manifest themselves on the highly stressed weather side.

Tubular spreaders in tubular sockets (Fig 26.3) should also have adequate clearance between them, and a soft rubber sleeve to allow this movement. Aerofoil spreaders are best attached to through-mast bars forming lugs, which themselves are tying together both walls of the mast in the form of a diaphragm (Fig 26.4). These, profiled from aluminium plate (of similar metal to avoid corrosion), are best welded to the mast walls. The spreaders should be attached by a large fore and aft bolt through similar lugs in the spreaders. The spreaders are free to hinge up and down, thus taking care of rigging stretch as it occurs. The bolt holes should be ovalled fore and aft to allow the necessary movement in this direction. Spreader outer ends should be clamped tightly to the shrouds which will act as lifts (Fig 26.5). Spreaders so fitted have a track record second to none in yachts of all sizes.

Shroud attachments or tangs are highly stressed components and currently are usually made of stainless steel. Traditionally, these are spades or forks, through-bolted to the mast with the rigging hole, or clevis pin in the case of a fork, athwartships, thus allowing the rigging free fore and aft movement. The tang should be preset to the correct shroud angle, ie it should line up with the shroud plate at the deck. Many failures have resulted from their misalignment. In such cases the tang is pulled into line when loaded up on the weather side, only to spring back to its incorrect angle when unloaded on the lee side, and so on at every tack. This load reversal in a sea-way continues until it breaks. Shroud tangs are best designed to avoid this likelihood. Universal movement is best built-in and possible errors of the mastbuilder, boatbuilder, or those caused by mast movement in sea-way, completely eliminated (Fig 26.6). Universal movement of shrouds and stays should be provided at all times at deck and masthead by fitting toggles. Mastheads in aluminium masts are best fabricated from aluminium and welded in place for a reliable and long life.

Boom gooseneck and kicking strap attachments should be robust and well fastened to the mast and boom. They are called upon to take high loads. All tracks should be integral or well riveted. Mainsail tracks and trysail track should be double riveted, ie closely spaced at the head, over the range of the head board from full sail to third reef, to resist the considerable headboard load. Foot tracks on booms should be through-bolted or very closely riveted in way of the clew outhaul traveller. I favour an additional few turns of stout terylene through the clew cringle and around the boom.

Spinnaker tracks should be particularly well fastened to take the enormous side loads of a pole in compression or tension. Cleats are best riveted on and halyard winches should be well fastened and backed up by a cleat where possible.

It must always be remembered that a well-found rig should be capable of sustaining many a knock down and even a 360 degree roll if called upon to do so. I know of one *Robert Clark* 15 tonner, the masthead of which fouled a lifting Dutch road bridge. She was raised some twelve feet out of the water before being returned unharmed by a worried bridge operator. I believe the helmsman had his bad moments too. There are many today who would not be so lucky.

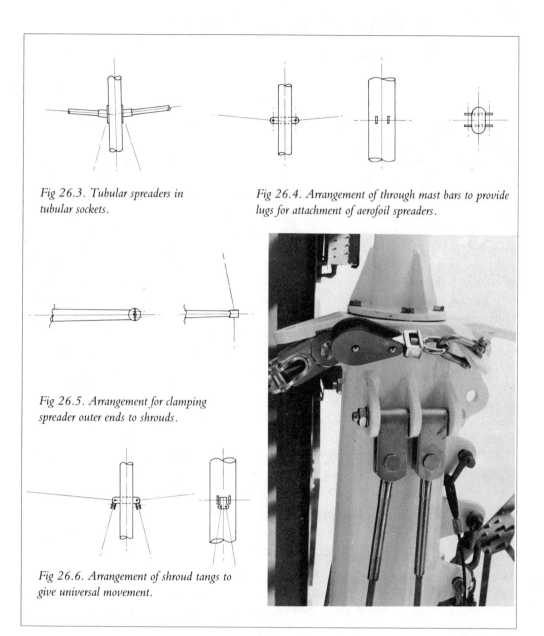

Fig 26.3. Tubular spreaders in tubular sockets.

Fig 26.4. Arrangement of through mast bars to provide lugs for attachment of aerofoil spreaders.

Fig 26.5. Arrangement for clamping spreader outer ends to shrouds.

Fig 26.6. Arrangement of shroud tangs to give universal movement.

Conclusion

In the matter of sail plan and its choice, I have concentrated on the sloop and cutter because of their popularity, but that is not to say that the ketch, yawl or schooner do not have their own charms and advantages. The mizzen staysail is a marvellous sail as is the fisherman staysail in a schooner. These rigs are most seamanlike offshore and especially suitable for world girdling; the test of time has proved them so.

27 MOTOR YACHTS IN HEAVY WEATHER

Peter Haward

My work over the past 43 years has been delivering yachts and other small craft for owners, agents and builders. It has amounted to over 700 voyages in small vessels, including 25 across the Atlantic and many others involving comparable distances. My understanding of boat construction, machinery, rigging, likely behaviour, capability, strength, stability and seaworthiness has been acquired by experience at sea, but a background of the theory and mathematics of design would be useful and at times reassuring and I must acknowledge that, when caught in winter gales, I have sometimes wished I was a fully trained naval architect.

Motor cruisers and even motor yachts (the latter expression by custom is accorded to craft of at least 45 ft) make long-haul voyages less frequently than comparable sailing boats because bunker capacity often restricts range. It is no run of the mill achievement for a 65 ft motor yacht to cross the Atlantic. Power craft of greater size will manage it routinely, with fewer problems. Fishing vessels enjoy different priorities about bunkers and even 60 ft MFVs remain at sea in all weathers for long periods.

Trans-ocean passages are possible in smaller motor yachts and cruisers but special arrangements have to be made, particularly relating to that extra fuel requirement. However these craft usually aim to replenish their bunkers frequently during their cruises. They are therefore seldom at sea for more than three or four days and they are always likely to be within range of a port of refuge in case bad weather is forecast. Compared to a similar size of sailing boat, more reliable speeds and far better windward ability reduce the chance of being 'caught out'. Nevertheless, there will be occasions when the skipper of a small motor boat will find himself confronted with severe weather and it is useful to study the effect and the ability of such power craft to survive.

Traditional hulls

The seaworthiness of fishing vessels has long been appreciated and their hull designs have been copied for yacht designs with good reason in my opinion. I have made a number of voyages in MFV yachts and can testify to their weatherliness. I remember one new 63-footer that I collected from Majorca to take to Scotland. While passing through the North Channel and onwards to east Lock Tarbert on the Mull of Kintyre, we encountered a brief but violent storm initially from the south-east and rapidly veering, force 10 or more. The performance of *Patrona* left me full of confidence. In contrast to the hard chine hull of the modern fast cruiser, true MFV type vessels have less initial stability yet possess basically stiff qualities. Rolling begins in even slight seas, but before 20 degrees of heel is reached they demonstrate an emphatic righting moment. In truth, despite their seaworthiness, they offer a pretty vigorous motion. MFV yachts benefit from having stabilisers.

A word of warning is offered. The weatherliness of MFV type hulls for motor yachts is widely known. Lately I have noticed some new designs being advertised as 'trawler yachts'. Some of these may be excellent but there are others that have underwater shapes (to say nothing about above water shapes) quite unlike any fishing vessel I have ever seen.

Despite the value of present day forecasting, there are times when events take the weathermen themselves by surprise. If the boat skipper accepts as gospel everything he has heard on the radio he will sometimes be on the blunt end of the unexpected. I am not suggesting professional weather forecasts should be lightly ignored; far from it. It is that a skipper at sea, or ready to go to sea, being aware that storms do occasionally arrive with very little warning, should build on the forecast from what he actually observes. Whether localised and of short duration or more sustained, yachts can be particularly vulnerable to an unexpected blow. The story of the loss of the *Maricelia* is a tragic example, and truly remarkable feat of endurance by her sole survivor.

I knew the *Maricelia* having had the experience of delivering this 52 ft twin screw diesel yacht from Rosneath, Scotland, to the Solent. She was a thoroughly seaworthy Brown Owl class boat, a popular product of James Silver Ltd., built just before the war. In 1964 she was purchased by James Fraser and he was cruising the North Brittany coast late in the year with his wife, his niece Alison Mitchell and two friends. Having experienced a fire in the galley they had decided to terminate their holiday and were returning to their home port of St Helier, Jersey, from St Malo, when they ran into trouble. This is a short sea crossing of some 40 miles and though the forecast told of rough weather there had been no hint at the time that conditions would deteriorate into real force 10 storm. They left St Malo at 8 am on 9 October 1964 and the weather quickly deteriorated, a SW wind remorselessly rising to gale and beyond. The course involved having the sea nearly on the port beam and the cruiser rolled heavily on her way to weather the buoys on the western edge of the Minquiers reefs. The two women were thrown several times out of their berths.

Real trouble began as the yacht approached the SW end of Jersey. The storm was due to an intense but very small depression unexpectedly careering in from the Atlantic. From a barograph reading in Jersey we learn that atmospheric pressure fell over 25 millibars in 3 hours to a minimum reading, which occurred around 2 pm, of 965 millibars. The wind must have reached storm conditions during the steep barometric fall and the RNLI still reported NW force 10 at 6 pm. A highest gust of over 90 knots was recorded in Jersey.

A sudden veer with frightful squalls must have come through at about 2 pm, just as the *Maricelia* was making her Jersey landfall, her look-outs scanning the thick weather anxiously, probably unable to see more than half a mile. Already the owner's wife had been thrown across the yacht and was suffering a broken arm. Then a heavy sea landed on the foredeck and smashed open the main saloon windows sending a great quantity of water into the ship. Efforts were made to plug the gap with a mattress and at about the same time a brief glimpse of land was reported, though there was no further sight of Jersey for another half hour. By then a very serious situation had developed. Land was very near but barely visible and not properly identified. Almost certainly the yacht was in an area of tidal overfalls that would be desperately dangerous in the force 10

The 52 ft twin screw diesel yacht Maricelia. *The picture was taken 10 years before she was caught out in the intense local storm between St Malo and Jersey. Photo: Peter Haward.*

storm that had just veered suddenly through about 90°. On top of this she was already seriously damaged and a crew member badly injured. Perhaps the skipper thought that by standing on he would identify La Corbière and then gain shelter under the lee of the south coast of Jersey. As it was, with the swell of the old south-west wind still rampant, and the rapidly rising sea from the NW becoming ever more threatening, a very violent, completely confused, utterly lethal cauldron must have existed off south-west Jersey.

Exactly what happened next probably cannot be known, except that soon one of the tide-ridden, wind-crazed seas broke on to the yacht and raised the sentry-box-type wheelhouse clean overboard and with it the whole of the upper saloon roof and coach coamings, the forward windows of which were already stove in. The owner and one other man in the wheelhouse were flung into the sea but how the others were washed from the yacht is unknown. With the whole of the ship's company in the water, miraculously together, the *Maricelia* motored on. Erroll Bruce, who is a cousin of

The Weymouth lifeboat almost submerged by a breaking sea whilst undergoing trials off Portland Bill.

Alison Mitchell, suggests the auto pilot was still working, but had this been so I suspect she would have run into the rocks near La Corbière at the extreme south-west of Jersey. In fact she went away to the east, probably her logical course not under command, and must have gained some shelter. With her heart exposed bare to the sea, another big breaker must certainly have finished her.

Two and a half hours later Capt R S Taylor and RNLI coxswain Edward Labalestier, and another, putting out in the pilot launch *La Rosière* in an attempt to render help, found her just outside St Helier breakwater, motoring round in circles. Had the crew managed to avoid going overboard it is likely that they could have been saved or might even have brought their yacht into harbour.

Only Alison survived the ordeal. I cannot conceive how she endured clinging to wreckage through the next 20 hours in the bitter, black, wild, hopeless night, but that is forever her private, incredible feat. She was washed ashore at 11.30 the next morning on the north-west side of Jersey. She climbed a 100 ft cliff, reached Lower Egypt Farm unaided and then collapsed.

This disaster illustrates a weakness in many motor yachts. The hull may be capable of surviving very heavy weather but it is the top hamper and wheel-houses which may be a source of weakness. In this respect, sailing yachts are stronger, but few, if any, could carry sail in winds gusting up to 90 knots, an occurrence which is unlikely to be encountered in a lifetime of ordinary cruising. The real danger was due to being caught out in storm force winds with exceptionally bad visibility in narrow rock infested waters. Under such conditions any small vessel whatever her type might be in peril.

Planing motor craft

The majority of modern motor craft are now designed for speed in excess of the normal waterline length formula. Speed is a safety factor for motor craft within their normal cruising range as the faster the boat the less is the risk of getting caught out in an unheralded gale or storm such as brought disaster to *Maricelia*. High speed usually requires hard chine forms, with a basically flat bottom or with a degree of V bottom, partially or completely along the whole length. Given sufficient power (now available from high speed diesels and, of course, more easily, if more dangerously, from petrol motors) and a not too heavily constructed vessel, dynamic forces can hoist her on top of the sea like a surf board. Once 'planing' she will have broken free of the speed-waterline length law. The way is open to new, interesting factors.

The craft gains an important element of stability from the dynamic forces pressing on the sea and these augment her natural stability. She will tend to have excellent *initial* stability, because the chines represent good buoyancy in the 'wings' or turn of the bilges, the position at which it will give an immediate righting moment. However, should a hard chine vessel be thrown heavily on her beam ends, this ability to come upright will be much less apparent, and it must be recognised that Rule 1 for power craft seamanship is to avoid being knocked down by a wave.

Another credit factor derives from the requirements for planing: light weight. The designer has to specify light weight and strength to cope with pounding into rough seas. Light weight is an important factor towards good buoyancy, an excellent quality towards seaworthiness. Of course it must be remembered that if a hard chine craft has to be slowed down to below planing speeds – often a necessity in bad weather – the dynamic stability goes, just as stabilizers for heavy displacement vessels do not work if speed is reduced. However, reduction of speed is more likely when punching into a heavy head sea, and that is not when stability is the big worry.

The best speed on the plane for a given power will be obtained from a completely flat bottom. The trouble is that for such a boat the sea would have to be completely flat too; a more difficult thing to arrange! Driving into even a slight sea will quickly become like driving a horse and cart over a ploughed field, not sparing the horses. Thus even the early hard chine craft had a decent V entry, diminishing as it ran aft, soon to be unrepentantly flat bottomed. Going at speed in rough weather they had to absorb dreadful punishment – and so did the crew. Wartime 'mosquito' craft, MTBs, and Air Sea Rescue boats warranted requirements where maximum speed for available power was the first consideration. I was told that some Air Sea Rescue boats which coped with UK waters had to be hastily strengthened when sent to operate in the vicious steep seas common in winter in the Mediterranean. Extra strength for an existing design will always mean extra weight – less speed for the same power – but that is what had to be.

Then came American designer Ray Hunt who introduced 'Deep V' – carrying a substantial V entry form right the way aft. You can tell a deep V hull by looking at the transom where the chine makes not a right angle turn but an obtuse angle, extending relatively deeply into the water. This simple, perhaps obvious development—all good ideas appear obvious in retrospect – went a tremendous way to improving the seaworthiness of hard chine craft. Ray Hunt's designs were adopted by Bertram Boats and Dick Bertram spectacularly demonstrated their superiority in contemporary Off-Shore Power Boat Racing. To get on the plane requires a little more power, but few of the power boat boys are squeamish about using petrol so there was no particular problem and this was more than compensated by the new sea-keeping qualities.

The advent of deep V planing boats has developed the seaworthiness at speed, and even total seaworthiness, in the same way as light displacement sailing yachts may sometimes be superior to their heavily built, take-any-punishment-we-hope predecessors. It does not mean that going to windward at speed has now become comfortable – windward work in any boat, sail or power, will always be the ultimate in seagoing discomfort – but size-for-size, and other things equal, deep V power boats are likely to make to windward faster in worse weather than other types.

A human vanity called 'styling' appears now to be having an effect on boat design, rather in the way that it holds sway in the car world. Whatever effects this whim may have on the motor trade, its influence on the boat industry has shipped in some charlatan shapes, perhaps from science fiction. Without becoming the old sailorman, I suggest that a logical shape for a projectile is not necessarily suitable for something that has to survive on the border line between two mediums, sometimes not clearly defined and neither of them always docile. Contemporary styles in boats today have complicated the problem of creating good craft. Sensible designs are tending to be at loggerheads with salesmanship to an uninitiated public. The idea of a visor, sun or whatever, extending ahead of the wheelhouse windshield, now included in so many modern designs has the effect of being an efficient lever by which a sea striking below it will more easily wrench the roof upwards.

Some years ago I was required to take a 55 ft fibreglass hulled semi-fast twin screw diesel yacht from England to Southern Portugal. The semi-fast side of the boat was scuppered before I joined because her owner appeared to wish his boat to be independent of mankind for a decade. On board were enough spares to construct two extra engines and iron rations for an expeditionary force. A feature of the yacht was her enormous storage space, which must have been a great sales point for the broker in this case. Every square inch had been used for the above-mentioned whim, mostly with heavy stuff, and I was told that the forward water and fuel tanks must be left empty because she tended to 'dip her bow'. The extra water we could do without but the fuel was necessary to promote the Biscay crossing.

In the event we enjoyed a calm trip across to north-west Spain and it was not until we were south of Lisbon with the fuel forward used up that we began to punch just a force 4 headwind. The bow dipping business was no myth. Heavily laden, the craft laboured ludicrously into the ensuing sea. I halved our speed to under five knots but we wallowed and lumbered. Thinking he was designing a cruiser with planing qualities, the designer had enhanced his magnificent sun visor roof by giving a reverse sheer forward. This was now busily engaged in scooping up ever larger samples of sea and sloshing it around the foredeck. I was occasionally cutting the throttles to discourage the habit but suddenly all our weaknesses were laid bare. She dipped into a good one, steeply. Green it rolled lazily along the short foredeck to assail the wheelhouse. The complete lump of water seemed to fit the triangular area formed by the windshield and visor roof extension and, arrested and deflected by the sloping glass, its weight came to the visor. The roof was torn from its fastenings to the super-structure, lifted 4 inches and permitted a torrent to desecrate the wheelhouse/saloon, doing an immense amount of damage, particularly to electrical equipment and switches. We crept into Villa Real the next day without even a good sea story.

In order to emphasize some glaring examples of non-seaworthiness, let me confide a yarn that similarly does not, you may think, warrant inclusion in a book on heavy weather, but it does offer a lesson for those who aim to design or sail in proper seagoing craft.

Excellently powered with twin well-proven 255 hp Mercruiser petrol motors with Z drive transmission, this particular motorcraft was 36 ft long, fibreglass, deep V, narrow gutted, low freeboard; straight sheer – but with her cabin coach roof carried right to the bow in a curve she had a reverse sheer look which would create a dynamic downward force to augment the weight of green water. As if obligatory, she had the standard visor style extension to the wheelhouse roof. Her sleek projectile effect forward seemed to have very little to do with flotation. I fear to state that she appeared not unlike some of the latest breed of off-shore powerboat racers but with a proportionally large and lightweight wheelhouse. Either there is something I have overlooked or else the ensuing experience is a complete freak.

She could cruise at well over 30 knots and on our first leg we made good 150 miles in 6 hours, then having to make port to replenish the fuel supply that vanished considerably more quickly than we had been given to understand would happen. For my purpose, the tale of teething troubles can be dismissed; sufficient to say that after some difficulties we finally left Brest bound, I hoped, for Leixoes, Portugal, with the 'standard long range' fuel tanks augmented by 44 five gallon Jerry cans and 17 empty non-returnable oil cans, 20 litres capacity. To save overall weight, we carried less than a quarter of the designed freshwater capacity; even so, for the first few hours, because of the burden of extra fuel, this lightweight, very powerful craft could not get fully on the plane. When it did, however, it was almost immediately capable of approaching its normal cruising speed. However, having made our departure from the Ar Men buoy off the Saints rocks at the northern end of the Bay of Biscay, in the interest of fuel supply, we continued at a slow speed throughout most of the night – except for various stationary periods while we manually transferred fuel. At first the wind was force 4, probably reaching 13 or 14 knots mean true velocity, and by dawn it had veered north-east and slackened to about 10 knots. The long, barely perceptible swell from the north was crossed by the new slight sea from the present wind.

Around dawn we increased speed and at eight in the morning I estimated we were making 25 knots and overtaking the seas. There was no sign of trouble – until it all happened. My crew Tim Jackson was happily asleep in the wheelhouse/saloon berth, while I was at the helm. Suddenly we seemed to be in the middle of an explosion. The boat overtook and fell over the leading edge of what must have been a rather steeper sea than usual. She dipped I suppose a little lower or a little steeper, though that was not obvious at the time. Then she was impaled steeply into the gently upward sloping trailing edge of the wave ahead – and just went on at 25 knots. Remember the unbroken water of the sea is not moving; it is the undulations of the sea that move. Only when the waves or the crests break is there actually movement of water; thus it is that when the boat dug herself into the solid water, the impact with the wheelhouse was at her true speed.

Both windscreens went to smithereens and the two heavy beams carrying the windscreen wipers were hurled aft, luckily hitting nobody. The centre stainless steel tube supporting the roof let go and was buckled; the roof itself was lifted a foot. The aft window was bent backwards and broken, as if it had let some of the sea go straight on. One starboard glass panel disappeared. A torrent of water was pouring into the forecabin nearly to bunk height and holding the fibreglass floorboards down so tightly that it was slow to find its way into the bilge.

The wheelhouse/saloon was a complete wreck, much equipment damaged and all our personal gear soaked and much of it ruined. I estimated she was about an eighth full of water. Broken 'pebble' safety glass was everywhere. I blessed this invention because normal glass would have cut me to ribbons. With this stuff I got away with 15 minor cuts and somehow a small pebble was embedded in my chest.

At some stage during the instant shambles I cut the throttles and put the engines into neutral. The boat now lay broadside on to an unbelievably slight sea. Imperceptibly she lifted to long swell, bobbing gently to the modest waves.

We had used up a great deal of our fuel and, taking into account the deliberately small amount of fresh water aboard, we were now within our design laden weight. The accident could not be blamed on overloading.

We shovelled out glass, lifted a floorboard under the water in the forecabin to let it run aft to the pumps more quickly, and we easily pumped her out with the hand and electric bilge pumps. Because of partially blocked limbers through a bulkhead very little water got into the engine compartment before we were getting rid of it at a good speed. Power units and batteries were unharmed but we had to deal with a few live wires in the circuits. We rigged the cockpit cover as a windshield, sorted out the general mess, estimated we were some 110 miles north of Coruña and set a course in that direction first at a cautious 10 knots, later to around 12, but never again allowing our selves the luxury of overtaking the waves. We arrived at Coruña that evening.

An increasingly frequent requirement is to cruise at speeds faster than that of wind rode waves. Owners are sometimes demanding not just planing craft, but really high powered boats routinely able to race following seas. Our accident was directly related to the problems these aspirations involve.

If a high speed power boat is to overtake following seas safely (ie seas running in the direction of her course) her designer must, apart from his other deliberations, balance judiciously an adequate degree of buoyancy forward with a bow shape that will not inflict excessive deceleration when encountering steep waves of manageable size. A vessel should rise confidently to the seas, but too much lift at the bows can create a hobby horse effect which will defeat the object. Too bluff a bow will rudely check a boat falling into a trough. A balance of trim is also a requirement. Without it, a projectile type bow, with or without whaleback hull or inclined superstructure, could encourage the kind of accident that befell me in Biscay.

Speed in heavy weather

Newcomers keen to purchase high performance powerboats will best be guided by designers whose boats have been successful in offshore powerboat racing. Such craft are special: racing the seas is not run-of-the-mill motor cruising. But in facing their problems it is not necessary to sacrifice weatherliness. During a bad blow, running before wind and sea is the most comfortable heading for any boat to adopt, whether sail or power. Nevertheless, emphatically there are hazards, and a significant aspect is the surprise factor. Survivors of capsize after broaching frequently claim that disaster struck without warning. During a boisterous fair wind, it is easy to succumb to elation. People begin talking of unreal ETAs. A realistic skipper should studiously cultivate his gloomy demeanour and remain aloof from the general euphoria. Few accidents at sea occur without warning, but sometimes the signs of danger are obscure. The good captain will observe and heed them. A thorough understanding of wind, weather and sea conditions and an appreciation of a boat's true seaworthiness is vital if correct action is to become instinctive.

At first viewing in broad spectrum, a gale-torn seaway may appear a disciplined formation of white-crested, spume-streaked waves, all roughly of the same calibre, each column advancing majestically and uniformly down wind like guards on parade. A more careful watch offers a different conclusion. Random ill discipline is widespread. Wave heights along the ranks differ widely and also undergo continuous change. White crest caps are thrown about unpredictably. Rather than a mere wavetop, an expanse of green water sometimes explodes along a broad front, breaking and thundering down whatever remains of the undulation, engulfing anything that may be in its way. Sometimes, part of a column may break ranks, sheering off at an angle, stumbling upon adjacent, better ordered troops and causing a rumpus. Occasionally three columns in line ahead will grow to extravagant heights, attain a fearful steepness and generally behave in a threatening manner. But it is the outside influences that really cause dangerous seas: tidal overfalls, tide deflecting headlands, wind against tide channels, cross tides, shelving of the seabed, underwater rocks and shoals, substantial changes of wind direction and a sudden slackening of wind after a sustained blow.

Providing a skipper understands the perils of wind, tides, sea and topography, he can expect to bring a well-found, seaworthy motor cruiser through rough conditions. The requirement of seaworthiness is crucial and it embraces diverse matters. Size of a craft is of particular importance, but some remarkably small boats can give a good account of themselves in heavy weather. There are other essential considerations: stability, trim, shape, superstructure, construction, watertight integrity, power, and, not least, how a ship is handled. Thus the pundit on seaworthiness and seamanship, trying to cover everything, becomes maddeningly obscure. But there is one sound, simple message, the nearest thing to a cure-all for survival at sea, which offers sage advice particularly for heavy weather and incidently for poor visibility, exacting pilotage, and close encounters. It requires only two words: slow down.

Power boats can run into trouble in a rising wind and sea by keeping up too great a speed, both when running or punching head seas. Making a dash for shelter may be counter-productive. It is true that boats with a good turn of speed can sometimes use this successfully to gain refuge from an impending blow, but it requires mature judgement. A compulsion to extricate oneself from a disagreeable, perhaps alarming, situation may inhibit clear thinking. There is a difference between going fast when a blow is forecast and speeding along when it has arrived. Acceptance of having to spend a longer time at sea is a wholesome attitude of mind and good seamanship.

I received a firm lesson regarding speed and stability during a voyage from the South of France to England in a fast TSDY built in the seventies. Based on a well tried workboat hull that was held in esteem for its sea-keeping qualities, the yacht version was lengthened slightly to approximately 43 ft and given a substantially higher and heavier superstructure, judged necessary to transform frugal living quarters into luxury accommodation. It was my second delivery voyage in her.

The relaxed French attitude that used to accept low taxed *huile domestique* as yacht fuel had come to an end and Italian *nafta* was now less expensive than diesel fuel obtainable from French marinas. Thus when joining our 20 knot vessel in Golfe Juan, we were urged by her owner to go eastward to San Remo and there take on as much fuel as we could before heading westward towards Gibraltar.

The November night brought early twilight by the time we were ready to leave San Remo. The benign head wind encountered during our morning passage from Golfe Juan had hardened into a substantial blow from the east. Early on I adopted a speed of eight knots. The boat was comfortable, but the way she lay gently to leeward when surging before the following seas, hanging there longer than she should, indicated that she was tender. That was a reluctant conclusion, despite bearing in mind the all-weather workboats from which she had been developed. I had previously brought her out to the Mediterranean from the UK and had noticed her characteristic when running down wind and sea off Portugal.

I was the second to go on watch. At midnight I handed over to my relief, having eased her down 100 rpm. We would now make about $7\frac{1}{2}$ knots. It was very dark: no moon and heavily overcast. I briefed my relief, telling him that Agay Road was just abaft the beam, 7 miles: Cap Camaret was 15 miles ahead and its light should come up within the hour, fine on the starboard bow. There was occasional traffic but no shipping in sight at the moment. If he were to think that the seas were getting worse he must call me by shouting back aft – and if he did not know whether they were getting up or not – they were getting up! Jogging his mind about monitoring the gauges, I told him temperatures and oil pressures were steady and, as always, departed with an exhortation that he maintain his primary duty of keeping a proper look-out.

I dived aft to prepare my bunk which, as is the case with many cruisers boasting the prefix luxury, had no leeboards or bunkboards but merely a retaining fiddle for the mattress. Choosing, of course, the leeward bunk, the trick is to roll items of clothing, or some lifejackets or shoes or whatever is available and lay them under the inboard edge of the mattress so that it slopes downward to the ships side. Then into your sleeping bag you go, and retained by this amalgam of semi-security you will sleep peacefully up to a certain angle of heel or sideways-induced kinetic energy.

I dozed off happily and it was perhaps an hour before both the above mentioned provisos went over limits. Suddenly I was embroiled in events beyond my control. Mattress, sleeping bag and its occupant were in free flight, an occurrence of short duration, yet enough to fill the mind with a stream of thoughts. A development of substance was taking place in darkness. Exactly what was unimportant because at the moment nothing could be done about it. Hopefully, practical decisions would be possible later. Meanwhile the only thing to do was to wait patiently. The fate of the boat was in balance and a personal hard landing could also be on the cards.

The next instant I struck the starboard side of the cabin, high above its complementary bunk. Mattress or bedding took the brunt of the blow and I seemed to roll over face first on to the bunk which had its own mattress happily in place. I was uninjured. The boat had come upright. In some haste I went to the wheelhouse. The helmsman was getting her back on course. 'I couldn't hold her' he said, with the air of Atlas, surprised at the weight of the world.

It was good to see him in one piece. I told him to ease the revs to maintain steerage way. My other crewman had been passing the ankle high chart drawer just as it had slid clean out impaling one of his legs against the opposite side of the passageway. His air was of the grumbler, rather than a mutilated veteran of war, so again I felt relieved.

Such circumstances compel within me thoughts of bilges and engine rooms and inspection showed the former dry and the main engines both running happily. A closer examination revealed a tell-tale spillage of oil at each dipstick, both trailing away to starboard. Only heeling beyond 90 degrees, combined with considerable sideways momentum could have done this. My guess is that it was the strong, high superstructure that may have denied the boat some desirable stability but was of help in preventing total capsize.

A full gale was blowing from the port quarter, though in the black night it was impossible to evaluate either wind or sea with any precision. We had no anemometer to confirm my estimate.

This is how I explain the event. Rolling high seas induce at their crests an orbital flow that nullifies rudder effect. Our moderate 7.5 knots average speed had increased to a surfboard run. With diminishing steerage the helmsman ran out of wheel and listing steeply on the sloping wave front she romped out of control. Swinging round, broadside to, further dangerous heeling was induced by the 90 degree turn. A tumbling crest exploded on her, thrusting her bodily into the trough. Her righting moment would have reduced to vanishing point – but then came the effect of the high superstructure. Gratifyingly strong and with glass that withstood being hurled against perhaps foaming but basically solid sea, the structure held fast and remained suitably watertight. Here was an extra stability factor that exerted a crucial, nick-of-time righting effect. Just when the hull itself was rolled beyond redemption, to beyond 90 degrees of heel and to certain capsize, new-found buoyancy absorbed the momentum and pushed her back upright.

If that superstructure had been less strong, and therefore probably, less heavy, the yacht's basic stability could have been more positive. The beyond 90 degrees knockdown might not have occurred. A last ditch righting moment by the superstructure would not then have been necessary. On the other hand, if she had rolled beyond the right angle, regardless of better initial stability, a light weight superstructure could have broken up – in which case nothing could have saved her. There lies a conundrum I am happy to leave to boat designers. Suffice to say I was gratified at the final outcome of that downwind drama.

As regards boat management, attempting to maintain control by using greater engine-powered water thrust on the rudder, ie increasing speed, may be successful on occasions but if, or when, it were to fail, a far more spectacular accident would unfold. The prudent course must be to reduce speed and allow threatening waves to pass more quickly. At 7.5 knots average we had been going too fast. I cut the speed to a cautious five knots and watched anxiously. There were no similarly threatening surfboard runs and a few hours later, without further trouble, we crept thankfully into Le Lavandou. The next day locals unreservedly remarked on the stormy night, the fury of the wind, the surf on the beach, the inability to sleep and other shore based criteria.

Mechanical problems in heavy weather

Notes for the engineer do not go amiss since power cruisers are often in command of a skipper-cum-engineer. By far the most frequent breakdown at sea is due to fuel failing to reach the engine. The violent motion caused by bad weather increases the chance of the debris of dirty fuel tanks being swilled into suspension in the fuel and therefore being carried to the filters and blocking them or even blocking the fuel lines on the tank side of the filter. Be prepared for both these problems. Ensure you have spare fuel filter elements aboard and can clean existing felt type filter elements. Make sure you have a way of blowing through or clearing a pipe at the tank itself. Sometimes a wire reamed through a valve is the answer. Regular attention to filters can prevent trouble at awkward times.

Some fittings currently used in motor yachts and cruisers are unseaworthy. I once brought a handsome but heavily top-hampered 64 ft boat from West Africa against a persistent headwind reaching force 7 for a considerable time. At times we were slowed to 4 knots and three port hole fittings failed in the forepeak, owing to the pins in the hinges severing through impact with seas, which gave us fun trying to devise methods of keeping out cascading water, and a great deal of use of the efficient bilge pumping system. Other mechanical problems can occur in heavy weather as the following story illustrates.

Norway has a fascination for me and it was exciting to land the challenge of taking a modern fast production boat from the Thames to Bergen in mid-winter. The January concerned happened to be an exceptionally cold one and all Britain was engulfed in deep frost.

By the time we got to Blyth there was a slow-moving depression causing northerly gales in both Utsire sea areas, but it seemed that the disturbance would have slipped away south-east before we were near. The previous two forecasts had each mentioned gales 'at first'. By starting promptly I anticipated being in the shelter of the fjords before a subsequent depression made its presence felt. With this in mind we departed from Blyth at 0130 hrs, an unsocial hour and particularly disagreeable when only two-handed.

The pre-dawn forecast came up yet again with its gale 'at first' in the Utsire sea area. By early afternoon Forties was added to the gale warnings: 'north to north-east 6 to gale 8'. Being now well into the south-west edge of that sea area, I could confirm it. The depression, if moving, was probably heading south rather than south-east.

At a judicious speed of eight knots our boat took the rising seas well. The wind was some 45 degrees on the bow, force 6-7. Incidentally I think it is insufficiently appreciated that while deep V hulls are exemplary for driving to windward at planing speeds – their original design objective – they can also offer a remarkably smooth ride at orthodox, non-planing speeds. I have known many 'displacement' cruisers, with designed speeds of little more than eight knots that would have to slow down drastically when making to windward in such conditions.

Early darkness saw no sign of the blow easing; in fact throughout the next night wind and sea seemed to be getting slowly but steadily worse. Added to this, all was not going well with the engines. As a minor irritation, the starboard engine temperature warning light began to flick on, but a check revealed the engine itself was cool. The electrical circuit was faulty. Later, however, the same engine began to lose speed. On changing the fuel filter I noted more water present. I had been confident that all water had been removed from the tanks. I drained off some more diesel oil but the engine continued to run erratically. It seemed that the filters were able to separate water from the system only up to a certain limit. Eventually, rather than stop all fluid passing through, they would permit some water to reach the injection pumps. I determined to check the new fuel filter after a short interval, and when I did so it was apparent that there was yet more water in the starboard tank. It seemed as if there was a continuing influx of water into the tank. I changed the filter bowl again and drained more fuel and water from the tank.

During heavy weather, particularly with the wind on, or near, the beam, if two tanks are linked by a balancing pipe, heeling or rolling to leeward encourages fuel to settle in the down wind tank. As it does so, further listing occurs, exacerbating the trend. Before sailing, I had therefore isolated the tanks, closing the balancing valve. It was satisfactory to reflect that the port engine was running perfectly, with no hint of fuel troubles. Water was apparent in only one tank.

There followed a wild night with the boat heeling heavily and crashing robustly over the seas, their foaming crests and brash disorder dimly perceptible in the darkness. After midnight, prudently we eased her to six knots. Her behaviour in the prevailing conditions could only be admired; but the difficulty with the starboard engine was less reassuring. My supply of filter bowls was limited and more and more water seemed to be present in the starboard tank.

Before the second daybreak from Blyth one injector had failed and another cylinder was inefficient. Carrying a couple of spare injectors is not thought over-cautious but even if we had got some it would have been pointless to use them without first arranging a water-free fuel supply. A brand new injector would quickly succumb like the others.

The distressed fishing boat Lynnmore 'dodging' in seas described as 40-50 ft high, accompanied by a mean wind speed of 60-65 knots. The term 'dodging' entails keeping a vessel's bows into the seas, using only the minimum power to maintain steerage way, and is the fisherman's preferred storm tactic.
Photo: Kieran Murray.

I took a fresh look at the starboard tank and noted the breather pipe. It went straight from the top of the tank to the ship side. There was no swan neck arrangement to inhibit sea finding its way in. The gale on the bow was causing a heavy cork-screw motion and on occasions a particularly steep wave or bursting crest would hurl our vessel bodily sideways, heeling her steeply. When this happened the starboard, leeward fuel breather outlet would be well immersed and unless there was something to restrain it, seawater would flow into the bunkers. Some builders answer the problem by fixing a mesh strainer, so fine that water is restrained from entering. Filler funnels sometimes have a similar grade of 'water strainer' filter. Personally I prefer to see a good swan neck pipe in the vent system. Clearly the mesh, if there was any, was ineffective. It is not always easy to inspect the vent if it is in a cowl. Ours may have suffered damage. Or sometime ago it could have become blocked, inhibiting the fuel flow and somebody had deliberately broken the mesh as a short term answer. At this stage, little could be done. Two at least, probably more, injectors would need overhauling and all water must be removed from the tank. The starboard engine was labouring badly. Continuing to inflict the adulterated fuel upon it could only make things worse. I shut it down and we continued more slowly under the port engine alone.

By midday on our second day, thankfully we perceived the wind to be easing. It had backed north-east, now force 6, and was blowing diagonally from the Norwegian coast needing an alteration of course to allow for extra leeway. I intended a landfall well south of Bergen, into Skudenes Fjord for some earlier, well earned shelter. A couple of hours before midnight Gjeitungen light was sighted and, as we approached the fjord, the intense cold became ever more apparent. Later I found that even the Norwegians remarked on the exceptionally cold spell. The windscreen froze inside and out. I have to be criticised for setting off on this voyage without spare anti-freeze. Diligent work with boiling water and towels helped, but frequent resort to the great outdoors was essential. We entered Haugsund 49 hours out of Blyth.

The starboard tank was carefully cleared of water and all the injectors for that engine were sent ashore to be serviced. Then we proceeded to Bergen at a good speed, mostly via sheltered fjords.

Conclusions

Problems that may occur at high speeds in heavy weather do not necessarily mean that a particular boat will be unseaworthy, if kept at modest speed. Knowing when to slow down – and when it is unnecessary – is an essential judgement of a good skipper. Nor is it too far-fetched to view it as a vital discipline.

Both traditional displacement motor cruisers and motor yachts and also some hard chine craft, given good stability, perhaps due to good beam or even a ballast keel, can survive rough weather. It may be necessary to avoid beam sea conditions – and experience will develop reliable judgement – and it will almost certainly be necessary at some stage, in deteriorating weather, to slow down when driving into a head sea.

It must be realised that the simple factor of size affects seaworthiness. Comparing two craft of similar proportions, ballast and trim, one of, say, 25 ft would be less safe in a particular rough sea than a 35 ft boat. Having said that, many boats of 35 ft and upwards can safely struggle into a force 7 to 8 wind at around 4 knots. Size will be a factor in acceptable performance. If the weather worsens to say force 9 or more it is likely to be prudent to give up trying to make to windward. Slow right down, even stop one engine on a twin screw boat if fuel supply is a problem. Keep the seas a little on the bow, presenting a nice slab of buoyancy. Just holding her own, the vessel should be fairly happy and have steerage way in a gale or worse. Of course the smaller she is the worse the discomfort and the greater the danger. Lying-to like this she will not use much fuel.

Having never experienced truly ultimate conditions, it is not clear to me whether, like sailing yachts, safety for a motor cruiser lies finally in running before the seas or lying-to. I suspect that lying-to could be the better policy in the end. It is the known tactic of fishing vessels in storms – 'dodge' is the expression – of keeping head to sea. The bow is paid off a little to present good buoyancy, and maintaining speed sufficient only for adequate steerage way.

I have heard a number of motor yacht skippers claim remarkable deviations from their course because severe weather had made it impossible even to turn round into the seas and reach a lie-to position. I suspect that inexperience may account for many of these detours rather than the prevailing weather, but nevertheless there are doubtless times when it will be imprudent to maintain a beam-to-sea course in many motor yachts.

Light displacement motor craft must be kept light if they are to give their best performance. The weight of additional features (whether at the instigation of an enthusiastic yacht chandler or simply born of an unrealistic flight of fancy) can easily add up to a substantial total. If the designer has already finalised his weight and displacement calculations, it may make a hash of the designed stability, speed, fuel consumption and range. The owner will not receive what he dreamed of and paid for.

Despite my occasional acid comments, it can be confidently stated that increasing numbers of well designed and well built power craft are coming into service today. Many are successfully combining performance, reliability, and overall excellence. Real progress continues. Economy in weight and the greater speed this makes possible are balanced with unquestioned seaworthiness, good range, modern navigational aids and safety equipment. A tempting feature also is unabashed luxury. Sometimes at variance with the rest of the package, it is a talisman that helps lure those who can dream of owning one of these prestigious vessels.

28 MULTIHULL DESIGN FOR HEAVY WEATHER

John Shuttleworth

In this chapter I shall explain some of the many factors that affect the seaworthiness of multihulls, including windward ability, stability, motion in waves, and pitching and rolling. I will describe the broad outlines of a number of distinct types of multihull that have emerged over the past 30 years, and go on to indicate how these different types of multihull can be handled in heavy weather, and then describe how the crew of a multihull can survive if the worst happens and the vessel capsizes.

Ocean going multihulls have been designed and built for at least 2000 years and it is now generally accepted that many of the Pacific islands were colonised from the west (i.e. to windward) by early navigators. It is also possible that islands to the east were colonised to windward, rather than downwind from South America as demonstrated by Thor Heyerdal on the *Kon Tiki*. Two main types of vessel are known to have been in use in the Pacific. The first were large 'double canoes', some over 100 ft long, which were the predecessors of modern catamarans. Such vessels were seen by Captain Cook and his contemporaries, though with rigs too small and inefficient for good windward sailing. The craft he saw were paddled, and probably were used either as war canoes or for ceremonies. The second type were various forms of outrigger canoe. One sort, now called a Pacific *proa*, had a main hull for living quarters and stores, and one stabilising *ama* – or outrigger – to windward. Another sort had an outrigger on either side of the main hull, as can still be seen in Indonesia, the Phillipines and East Africa. These latter vessels are the forerunners of the modern trimaran. The Pacific *proa*, which seems to have been the most likely craft to have been used for the colonisation of the Pacific to windward, differs from the general types of multihull that are being developed today, in that the mast was always steeped on the main hull, and the vessel was 'shunted' end for end during a tack. This required a rudder at each end of the vessel, the hulls to be symmetrical, and the sails to be set from either end. Making long ocean voyages aboard these craft required great skill and experience, besides a high degree of 'toughness', since for the most part, the accommodation was on deck, exposed to the elements, and weight carrying capacity was limited.

Modern yachtsmen in general require more creature comforts than the hardy navigators of the Pacific, and while some will accept spartan living as a necessary part of the adventure of going to sea, there are others to whom a fully equipped laundry, huge freezers, scuba equipment, speed boat, jacuzzi, and dog kennel, are minimum requirements for living afloat. Therefore although the basic form of the multihull is derived from the Polynesian vessels, we have had to develop a new form that can carry weight without loss of seaworthiness, and requires less skill to sail, since the modern sailor may not have a 'navigator' aboard who was born to the task with several generations of knowledge behind him.

Over the past 30 years a number of distinct types of multihull have emerged all having different sailing qualities, and seaworthiness. There has been a steady improvement in the understanding of the factors required to make a multihull both safe and fast, resulting in boats that are extremely seaworthy, as will be demonstrated in the following pages.

The basic types of multihull are reviewed below. Obviously these are the extremes, and many boats will fall between the categories. The groupings given here represent a chronological order in only a very general way. Boats having some of the characteristics of the most modern types can be found in multihulls whose designs date back over 1000 years. On the other hand boats of all types are still being designed and built. From a subjective point of view, the order given here follows my own multihull sailing and design experience closely. I first sailed across the Atlantic some 17 years ago in a type 1 trimaran, later I crossed again in a type 2 trimaran, and a few years ago in a type 3 trimaran. Recently most of my long distance ocean voyaging has been aboard a type 6 catamaran. So the order given is more applicable to my own rather than general criteria, even though most observers of the development of the modern multihull are likely to agree with the broad outline of each type.

A keel in the sense used below is a foil for resisting leeway. The keel is not ballasted as in a monohull, and may be fixed, or retractable, either vertically (a daggerboard) or by pivoting (a centreboard). A few multihulls have been built with ballasted keels, but current practice is to make the boats wider to increase stability, and to keep weight low, to improve windward performance.

Amas are the outer hulls of a trimaran, sometimes referred to as outriggers, or the smaller hull of a proa.

Type 1 Trimaran. Relatively heavy. High windage. Inefficient underwater and keel shape, often with either a fixed keel or no keel at all. Small sail area. Hard chine with high wetted surface. Poor pitching control. Medium buoyancy amas (around 110% of the displacement of the boat). Amas usually both in the water at the same time. Narrow beam (length to beam ratio = 2). Construction often of sheet plywood and glue. This was not always satisfactory as the ply tended to rot and the hulls had low long term fatigue resistance due to the low strength of the glue compared with modern epoxies.

Type 2 Trimaran. Becoming lighter. Larger sail plans. Less accommodation. Low buoyancy amas (75 to 90%). Wide beam (L/B = 1.3). Considerably reduced windage. Improvement in structural design.

Type 3 Trimaran. Light weight (due to the use of composite materials). Large sail areas. Wide beam (L/B < 1.5 to as low as 1.0 in smaller boats.). High buoyancy amas (up to 200% of displacement). Pitching very well controlled by use of different hull shapes on main hull and ama. Sailing attitude well controlled on all points of sail. Low windage. Dramatic improvement in structures due to use of computer aided design, and better understanding of composite materials.

Type 4 Catamaran. Relatively heavy. Narrow beam (L/B often over 2). Small sail area. Inefficient underwater shape with low aspect fixed keels or no keels at all. Cruising cats very heavy by today's standards. Bridgedeck saloon versions with large flat windows in coachroof causing high windage. Often prone to hobby-horsing (ie

Modern 35 ft open bridgedeck cruising multihulls designed by John Shuttleworth. These new designs have large cruising accommodation without the weight and windage of a bridgedeck cabin. Photo: Beken of Cowes.

plunging at a standstill) and pitching due to rocker (ie the depth of curvature of the hull towards the centre compared to the ends) and symmetry of hulls.

Type 5 Catamaran. Open bridgedeck designs. All accommodation in the hulls. Greatly reduced windage. Keel shapes improved. Retractable daggerboards. Large sterns and fine bows causing bow burying tendencies on a reach. Greatly improved windward performance. Pitching control still poor, some attempts to reduce pitching by using bulb bows. Wider than early designs. Larger sail plans.

Type 6 Catamaran. Open bridgedeck designs with large accommodation in hulls. Hobby-horsing eliminated by hull shape. Windage greatly reduced by rounding and streamlining deck edges. Powerful efficient rigs. Sophisticated retractable daggerboards and rudders. Minimum wetted surface hulls. Excellent windward performance. Fast easy motion through sea. Very stable with wide beam (L/B < 1.5). Similar structural design improvements taking place as for trimarans.

Type 7 Catamaran. Basically as 6 above but with very streamlined bridgedeck cabin for large accommodation and low windage. Light hull weight maintained, giving large weight carrying ability for fast cruising.

Type 8 Miscellaneous other types. Proas (Atlantic, ama to leeward, and Pacific, ama to windward) and trimaran foilers. In general these are development types almost exclusively for racing, as far as modern multihulls are concerned, and they have special problems that require particular knowledge, experience and seamanship for handling in heavy weather. Due to lack of space these types will not be dealt with in any detail in this chapter.

Although I am concentrating primarily on cruising designs, many of the design concepts have been derived from successful racing designs. Indeed the racing designs which push the limits of performance to the edge, are an excellent test bed for cruising boats, particularly racers designed for the long offshore events like the OSTAR and the 2STAR, which are both predominantly to windward across the North Atlantic. In these races, ease of handling and motion, windward ability, structural integrity, and seaworthiness are of paramount importance.

It is interesting to note that this development, resulting in a dramatic increase in seaworthiness and speed, has taken place over virtually the same time period as some monohulls appear to have been deteriorating in seaworthiness. The primary reason for this has to be the fact that the development of multihulls has taken place without the restriction of any rating rules, with the only criteria for successful design being to improve seakeeping qualities and overall performance – resulting in the development of extremely seaworthy cruising designs.

Boat motion in a seaway, and the effect on the ship's crew

There are six basic forms of motion in a seaway, which combine in various ways to give the full dynamic movement of the yacht at sea.

1 Rolling
With the exception of type 2 above, multihulls are virtually immune to rolling. This means that the boat sits on the water like a raft – following the surface of the sea, giving great crew comfort particularly while sailing down wind. When lying a-hull, cats and tris exhibit different characteristics. Firstly catamarans have a very high roll moment of inertia (Ir), because the weight of the boat is primarily concentrated at the hull centrelines. The buoyancy of the boat is also concentrated at the extremity of the hull centreline beam, giving massive roll damping. Open bridgedeck cruising cats benefit most from this effect, and low buoyancy ama trimarans (Type 2) least. In a tri the weight is concentrated closer to the centre of gravity (CG), reducing Ir, and the amas take longer to pick up buoyancy as the boat heels, thereby reducing damping. In a low buoyancy ama tri this effect can lead to capsize in waves when laying a-hull (this will be shown later) and different techniques of seamanship are required to ensure the safety of this type of multihull in a storm.

2 Pitching
Many early multihulls were prone to pitching. This was caused by too much rocker in

The 80 ft multihull Novanet Elite *setting off from Falmouth at the start of the AZAB race on 6 June 1987 in force 8. The curvature of the 27 ft whip aerial in the stern gives some indication of the wind strength. A storm staysail has been set on an inner forestay with a long tack line to keep it clear of breaking waves and there are four slabs in the mainsail. Peter Phillips, her skipper, says that at the time his boatspeed was between 15 and 20 knots. Photo: Roger Lean-Vercoe.*

the hull profile, and fine V sections both fore and aft. This can also cause 'hobby horsing', a form of exaggerated pitching motion which matches the frequency of small waves, due to lack of any means of damping vertical motion in the hull form, particularly at the bow and stern. U-shaped sections forward help to reduce pitching, however, too flat a forefoot will slam. As hull shapes improved tending towards more U-shaped underbodies particularly aft, pitching still remained a problem, because the large width of the stern sections caused the sea to lift the stern as the boat passed over a wave, driving the bow down. However we now know the pitching can be dramatically reduced by finer sections at the stern combined with the centre of buoyancy being moved forward in the immersed hull, and aft in the lifting hull. This effect can be achieved in both cats and tris, giving a very comfortable and easy motion upwind. At the same time, windward performance is also improved, because the apparent wind direction is more stable across the sails.

3 Yawing

Any tendency to yaw due to the shallow draft of the hull has been virtually eliminated in the modern multihull by use of U-shaped sections and lightweight construction and by the use of retractable daggerboards. Once the keel is removed when sailing downwind, there is virtually no chance of broaching, as long as the forefoot does not dig in. This can be prevented by firstly reducing the forefoot, and by picking up buoyancy quickly in the forward sections of the boat above the waterline. Computer simulations of the hull in different bow-down trims, and at varying waterline positions, are now an essential part of the design process to control sailing attitude properly, both on and off the wind.

4 Surfing

A modern multihull will surf very easily, allowing fast passage-making in the open ocean. Sailing downwind in winds up to 40 knots can sometimes be quite comfortable and easy depending on the steepness of the seas. This is because the apparent wind has been reduced by the high boat speed. Once the wind speed becomes so strong, or the seas so steep, that surfing downwind is dangerous, alternatively if the boat will not make progress to windward or lie a-hull - this could well be the case for types 1, 2, and 4 – it will be essential to deploy a drag device such as a sea anchor to control the boat speed. Much has been written on this subject, and it is certainly an accepted way of surviving a severe storm in a multihull. From the designer's point of view it is essential to provide adequately strong attachment points on the bow and stern.

One of the main problems when sailing at 20 knots or over in the open sea – even if this is just caused by surfing on the odd occasion – is that the boat may overtake the wave train. When this happens there is a danger that the bows could bury into the back of the wave ahead. If the true windspeed is, say, 35 knots and the boat is sailing at 16 to 20 knots as she surfs down the wave fronts the apparent wind speed will be between 19 and 15 knots. So the 'feel' on the boat will be that of sailing in a force 4. The danger is to think it is right to carry full sail, and indeed one can, until the boat buries the bow and slows right down. Suddenly the wind across the deck will increase to nearly 35 knots, and the boat will be massively overcanvassed. At this point a jib

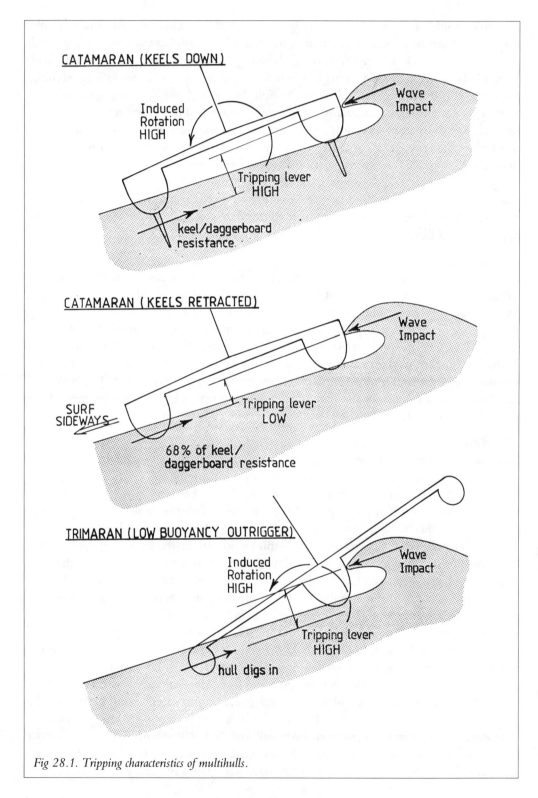

Fig 28.1. Tripping characteristics of multihulls.

sheet can be released, which may be enough to save the boat from capsize, but a mainsail cannot be depowered in the same way when sailing downwind. Therefore it is very important to be familiar with the static stability of the boat, and when sailing downwind to ensure that there is never more sail area up than if sailing with that true wind from abeam.

The bows can be prevented from burying to a large extent if there is adequate reserve buoyancy forward. This can be achieved by flaring the hull above the waterline, and by rounding the decks so that if the bow goes under, the build up of water pressure on the deck is reduced. The stern or transom should not be large, to prevent a following wave from lifting the aft end, thereby increasing the possibility of pitchpoling.

Particular attention has to be paid to rudder size and design to maintain good control at surfing speeds that may be in excess of 20 knots. Elliptical, balanced spade rudders of airfoil section reduce helm loads and drag at high speeds. Rudder stocks have to be very strong to be able to steer consistently at such high speed. I used to favour using stainless steel or titanium, rather than carbon for rudder stocks, because at least the rudder will bend if overloaded – by grounding or hitting an object – instead of shearing off. However, recently we have designed carbon rudders for some catamarans, because we have found that a bent rudder stock can jam in position, and cause more problems than if it had simply sheared off. With this in mind it is obviously important to design the rudders of a catamaran large enough so that the boat will steer with only one rudder. A couple of years ago I made a voyage from Coruña to the Canaries in a 42 ft cat with only one rudder, in wind speeds of 30 to 40 knots, without any problem.

The factor of safety for the rudder design should be at least 1.5 with rudder at 90 degrees to the waterflow at 25 knots. This situation is quite possible if the boat starts to broach and slew down a wave, and the helm is turned to full lock; ie 35 degrees. A sketch shows that the rudder is then presented at 90 degrees to the water flow.

5 Swaying

A modern light displacement multihull lying sideways to the seas with no sails up, i.e. lying a-hull, and with the daggerboards up, will surf sideways very easily in a breaking crest. It will be seen later that this is a very favourable feature in the seaworthiness of multihulls in a storm. Multihulls with fixed keels and tris of type 2 have the possibility of tripping over their keels or amas when struck by a breaking crest. Narrow beam increases the danger of capsize in this situation. In a recent tank test programme, we dragged a 65 ft cat sideways with no keels, two shallow keels, and two deep keels. The increase in sideways resistance, at 3.4 knots was:

No keels	100%
Shallow keels	123%
Deep keels	132%

Although we did not test a daggerboard, I would estimate that a daggerboard on the downwind hull would be in the order of 145% to 150%. Fig 28.1 shows the effect of the tripping action of a board or low buoyancy outrigger that has become immersed.

If sideways motion of the yacht needs to be stopped, for instance because of a danger to leeward, this can be done by either deploying a sea anchor abeam, or by putting down the upwind board (this only applies to a cat). The upwind board can act as a brake without imparting rotational momentum to the boat. If the boat does not rotate it will not capsize.

6 Heaving

This is a complex problem to define clearly for a multihull, because the two immersed hulls are at different places on the wave front at any given time. Nevertheless, they heave less than monohulls because the hulls are slimmer, allowing the yacht to cut through the water when sailing. Loss of apparent displacement at the wave crest and rotational momentum imparted to the boat by heaving on the upwind hull will assist in capsizing an overcanvassed multihull due to wind and wave action. Heaving-assisted capsize has been experienced particularly in tris of type 2, and cats of type 4. This is of particular importance and will be dealt with more fully in the next section.

Stability

Stability is generally a very contentious and little understood subject when multihull seaworthiness is discussed, and is probably the biggest fear that inexperienced sailors have about this type of vessel. While it is true that certain multihulls have capsized, there are many different types of multihull, and indeed many different ways in which they can capsize. I will endeavour to show that some multihulls are extremely difficult to capsize, and therefore can be safe in nearly all conditions. This is provided that the correct amount of sail is carried in relation to the wind and the sea state, and correct storm techniques are adopted when sail can no longer be carried. I say 'nearly all conditions' because there may be a wave out there that will overwhelm any vessel.

Stability in wind

Static stability is a measure of the stability of the boat in flat water, and is given by the following formula:

$$SF = 15.8 \times \sqrt{(0.5 \times B \times D) / (SA \times CE)}$$ Where:

D   = displacement (lbs)
CE   = height of the centre of effort above the centre of gravity (CG) in feet.
SF   = windspeed in miles per hour (mph) at which the boat has to reduce sail.
SA   = actual sail area in square feet.
B   = beam between the centrelines of the outer hulls in feet.

Note. Use of mph is a convention established in order to be consistent with wind pressure work in other fields.

This formula gives designers a measure of stability as an indication of the power to carry sail, i.e. the ability of the boat to resist capsize by wind action alone. There are two factors that can reduce SF. Firstly if the boat has a high angle of heel at the point of maximum stability (worst in trimarans of type 2, and minimal in all catamarans) the correct SF is given by replacing beam in the formula with hull centreline beam x cos

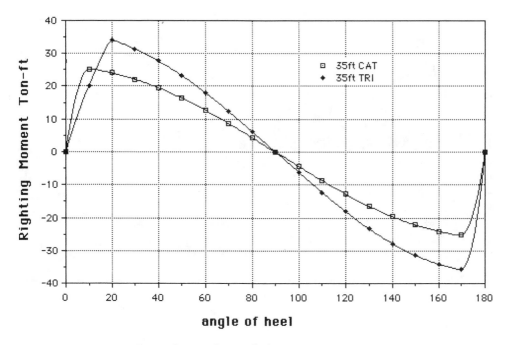

Fig 28.2. Righting moment curve for a 35 ft cat and tri.

(angle of heel). Secondly the height and steepness of waves will give the boat an initial angle of heel and possibly rotational momentum, both of which reduce stability. Thus in order to give an accurate measure of stability, SF has to be reduced as wave height increases. Typical values for SF can vary between 12 mph for a Formula 40 racing catamaran, to over 50 mph for cruising multihulls. Modern light cruiser-racers would be in the range of 24 to 30 mph. So it is clear that in addition to the different types of multihull mentioned above, the initial static stability can vary enormously.

Stability curve and stability in waves
Righting moment is the distance from the centre of buoyancy to the centre of gravity x the apparent weight of the vessel. This is basically the vessel's inbuilt static resistance to heeling. The forces that heel the boat could come from the wind and/or the waves.

Fig 28.2 shows the curve of righting moment versus angle of heel for a typical modern 35 ft catamaran and trimaran racer cruiser. The trimaran is a type 3 with high buoyancy amas and has an overall beam of 32 ft; the cat is a type 6 with a beam of 23 ft. The trimaran has less accommodation, and is lighter than the cat, but because of the wider beam it has greater maximum stability. It is important to note that the maximum stability of the tri occurs at around 20 degrees angle of heel, whilst for a cat this occurs at about 6 degrees. If the buoyancy of the ama is reduced below 100% of the weight of the boat (as in type 2 above), the maximum stability will be reduced not only in proportion to the reduction in buoyancy in the ama, but also by the effect of added apparent displacement from the downward pressure from the sails at high angles of heel. At 20 degrees this would cause a loss of righting moment in the order of 20%.

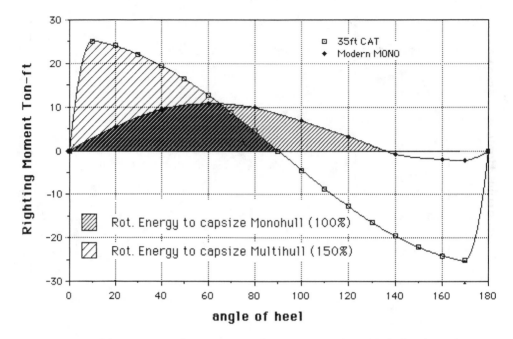

Fig 28.3. Static stability curves give the energy required to capsize a modern monohull and a modern multihull.

If the ama buoyancy was only 80% in the first place, the total righting moment would be only 60% of an equivalent trimaran of type 3.

Fig 28.3 shows the same righting moment curve for the cat versus a typical modern cruiser-racer monohull. The energy required to be put into the yacht in order to roll it from 0 degrees to the point of capsize (90 degrees in the cat and 135 degrees in the mono) is given by the area under the curve. From the graph it is clear that the energy required to roll the cat over is 50% higher than for the monohull. Of course in either case the initial angle of heel will reduce the available reserve of stability, and in the trimarans this reduction in energy resistance to roll will be greater than in a cat. The energy to roll a tri to 90 degrees is greater than a cat, provided it is of 1 or 3.

However, in all cases, in order for a capsize to occur, the energy from the wind and the waves (which is equal to the area under the righting moment curve) has to be transferred to the vessel in the form of rotational energy.

In waves alone, if the energy of the wave impact is not changed into rolling energy, the boat cannot capsize. The following table gives the displacement and dimensions of the cat, tri, and monohull shown in the graphs in Figs 28.2 and 28.3.

	Catamaran	Trimaran	Monohull
LOA (feet)	35	35	33
Beam overall BOA (feet)	23	32	10
B = beam betw. centrelines	17	29	
F = displacement (lbs)	6700	5600	10080

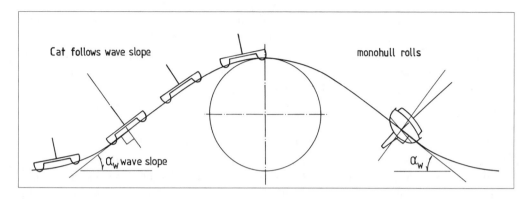

Fig 28.4. A catamaran and a monohull on a wave slope.

Let us first consider the action of the waves alone. Tank testing has shown that capsize due to the action of unbroken waves is impossible. Therefore when a vessel is lying a-hull the impact of the breaking crest is the primary means of energy entering the system which may be transformed into roll energy. A multihull follows the slope of the wave face exactly like a raft as shown in Fig 28.4. However, because the buoyancy and the weight is concentrated at the extremities of the beam – particularly in cats of type 6 – multihulls will be more stable against rolling than a simple raft.

Secondly, particularly in the case of the open bridgedeck cruising cat (type 6), the roll moment of inertia is very high because of the hull configuration. Also the inertia of the water entrained by the hulls is high. Therefore this type of cat has the least energy transferred from the wave impact into rolling energy.

On the other hand a trimaran with the board up will still have a small capsize lever, but the roll moment of inertia (Ir) is concentrated closer to the centre of gravity. Therefore more roll energy will enter the trimaran lying a-hull than for the cat.

Most modern multihulls have low draft, and if the keels are retracted, or not too deep, virtually all the energy of the wave impact is absorbed by surfing sideways. This is exactly the same effect that saves the older type of monohull from capsizing in waves, the only difference is that the monohull has to experience a knockdown before the keel is almost parallel to the surface of the water, thereby reducing the lever arm (r) and allowing the energy to be dissipated into sideways motion.

The multihull that fares worst in this situation is the trimaran with low buoyancy amas. When a wave hits the side of the boat, firstly it will roll more rapidly and to a greater extent than a cat, and if the ama immerses to the point where it digs in, thereby stopping sideways movement, all the energy will be transferred into rolling and a capsize is possible. Also, having deep fixed keels or leaving the downwind board down will greatly increase the risk of capsize in waves for all types of multihull.

The accepted technique for handling a low buoyancy ama trimaran is to keep sailing if at all possible, usually with no headsail and a very deep reef in the mainsail. The boat can be sailed at about 60 degrees to the wind, and will make slow progress to windward. The sail both keeps the tri at a more steady angle of heel, which reduces the initial rolling as the boat passes over the wave crests, and also damps rolling.

Wing masts, which are becoming more popular on performance cruisers, must not be larger in area than the storm jib. A multihull can then be sailed slowly to windward in a storm under mast alone. In this application the mast is easy to control, and can be set to a suitable angle to give the right amount of drive for a comfortable speed and optimum steadying effect on the boat. Of course the situation may arise when it may be necessary to limit the sideways drift of the boat, for instance if there is a danger to leeward. In a cat this can be achieved safely by lowering the windward daggerboard. All other types will have to use a sea anchor. Whether to deploy the sea anchor from the bow, stern or the side of the vessel depends on the type of boat, and the conditions. Several people have written on the subject including the Cassanovas, and Dick Newick, both of whom have used, and are in favour, of this method of controlling drift and rolling in a storm.

Sea anchors and drogues

There are now a number of types of sea anchor available which can be used successfully on a multihull. Whether to deploy the sea anchor from the side, the stern or the bow, seems to depend on the type, and particularly the weight, of the vessel. The Cassanovas, during a voyage around Cape Horn aboard their 39 ft trimaran – which was of relatively heavy displacement – used a parachute anchoring system on a bridle attached to the bows. They concluded that the parachute diameter should be 28 ft for boats up to 50 ft, the bridle length should be one and a half times the boat length, and the tether line – between bridle and parachute – should be 10 times the length of the boat. This system obviously worked well for them. However Nigel Irens found that when lying to a sea anchor off the bows of his light displacement 40 ft racing tri, *Gordano Goose*, the boat would surf backwards and then snub up violently on the sea anchor, to such an extent that he eventually cut it away. On the other hand, during a gale in the 1980 OSTAR, Bill Holmwood on his 31 ft trimaran, *The Third Turtle*, felt that attaching the sea anchor abeam saved his boat.

It seems that the most important lesson is that some form of sea anchor should be aboard and tried in strong winds, long before it is needed in a survival situation. Clearly proper preparation and ease of deployment and recovery are essential.

Wind and wave action

When the actions of wind and waves are combined, a catamaran is more vulnerable than a tri, because when a boat is sailing, the heaving action of the wave on the windward hull imparts rolling momentum to the boat, reducing the energy reserve left to give a righting moment. If the boat has a low static stability, and is being sailed close to the limit, with the daggerboard down, capsize will be possible in waves in a wind speed that would be safe in flat water.

Cats are also more vulnerable than tris because, in general, the static stability of the cat is less than an equivalent tri. This is the main reason that tris are considered to be safer for short handed racing – they can be sailed harder in waves with a greater margin of safety, and this is a very good reason to make cats as wide as possible to increase the static stability and thereby increase the safe sail carrying power.

Windward ability

Another area of traditionally poor performance in multihulls is their windward ability It is true that the older types of multihull (types 1 and 4) would tack through 100 degrees or more, and had very inferior pointing ability when compared to their monohull counterparts. Today, however, modern multihulls are designed with careful attention to weight saving in the structure, reducing drag of the hulls and superstructure, besides having an efficient rig and wide overall beam to give good sail carrying power. All these features combine together to give a windward performance better than any equivalent sized monohull. In a force 4 wind, a modern 60 ft racing trimaran will sail upwind at 16 knots, tacking through 75 degrees, while an open bridgedeck cruising cat like the *Spectrum 42* will sail to windward at around 10 knots and tack through 80 degrees.

Moreover, modern multihulls will sail upwind in a gale long after the monohulls have had to heave-to. Indeed this superior windward ability has been conclusively demonstrated in all the windward races of the North Atlantic and is a factor of major significance in the improved seaworthiness of modern designs. Windward ability is a very important measure of seaworthiness, and can prove vital if there is a danger to leeward in a gale.

Safety in the event of collision or capsize

Even though it has become extremely unlikely that a properly designed multihull will capsize, the possibility still exists, in much the same way as it exists for any monohull. The monohull's escape valve is that there is a chance that the boat will right itself before it sinks. The multihull on the other hand can be made into a safe raft for the crew to live on in the inverted position, provided that proper provision for this eventuality has been made at the design stage. It is desirable for watertight compartments to be built in to the amas of a trimaran, and wherever possible in a cat. A trimaran can be made virtually unsinkable by making the cross beams watertight, and by dividing the ama up into watertight compartments in such a way that, if any section is holed, the remaining volume is over 120% of the displacement of the main hull. The first 20–30 cm of the bows of all three hulls should be packed with foam, and a watertight collision bulkhead can usually be placed about 6 ft back from the bow, without affecting the accommodation. Consideration should be given to means of pumping out each individual compartment of the amas.

The structural cross beams of a cat can be designed to be watertight, with their combined volume large enough to support the whole weight of the vessel. In the unlikely event of a capsize, this will ensure that the boat floats high out of the water, which reduces stress on the structure, and allows the crew to live in the upturned hull. Escape hatches should be incorporated in all designs as a matter of course.

Construction and fatigue

During the lifespan of a multihull it is subjected to many cycles of a complex array of loads. If the boat is to survive in all conditions without damage, careful attention has to be paid to avoiding stress concentrations in the structure, and long term fatigue of the materials used to build it. By using a computer to analyse the loads at any point in the boat, and then laying appropriate amounts of fibres aligned in the direction of the stress, the stiffness and the strength of the boat can be greatly increased, while at the same time, weight can be saved by removing excess material where it is not required. This weight saving actually increases the strength of the boat, because it not only reduces the loads that the boat experiences, but it reduces stress concentrations, which are a major cause of fatigue failure. If the structural design is carried out in this way, and adequate allowance is made in the fibre stress levels in all parts of the boat to account for long term fatigue, the lifespan of the boat will be greatly increased. At present, research indicates that if a composite laminate can survive over 10 million cycles, it will last indefinitely. In general, in order to achieve this, a factor of safety of at least 10 is required. In all my cruising designs I use at least 10 as a factor of safety in areas of maximum stress. For carbon in particular the laminate is strain-limited because the material is so stiff, has a relatively low strain to failure, and an extremely high notch sensitivity (ie loss of strength due to stress concentration at a drilled hole, for example). However, the material can be very successfully used in areas where great stiffness is required, like the cross beams of a multihull. There are many racing and cruising multihulls sailing that have been designed in this way, and these have suffered no structural failure whatsoever in thousands of miles of hard ocean sailing.

Conclusion

In the past 20 years the level of understanding of the factors that affect the seaworthiness of multihulls has increased enormously. There are many different types of multihull, and though different techniques are required to handle the different types of multihull in a storm, many of the problems and vices associated with the older designs have now been eliminated.

The new generation of cruising designs have produced very exciting boats to sail, while still offering vast accommodation, crew comfort, and most important of all – safety at sea.

29 PREPARATIONS FOR HEAVY WEATHER

Peter Bruce

Whilst some heavy weather is almost the inevitable lot of the ocean voyager, the majority of coastal sailors will be in a position to avoid it most of the time, very sensibly, by the exercise of discretion and by good planning. But meteorology is not an exact science, so there is always a chance of being caught out in a blow at sea however carefully one has studied the weather and listened to the weather forecasts. If this does happen, the aim should be to come through the experience safely, without damage, independently of outside assistance and in as much comfort and good humour as the circumstances permit.

To achieve the aim, the wise skipper will make contingency plans and will practise for heavy weather, either in harbour or in calm weather, knowing that he should expect no more than simple tasks to be undertaken by his crew when it comes on to blow. For storms can creep up insidiously on the crew of a vessel at sea; in a sailing boat, this often happens via a period of exhilaration as wind and wave exceed normal expectations. There follows a period when crews show a gradually decreasing tendency to make any additional effort, save that of attending to necessary reductions of sail, adjusting to discomfort or reconciling themselves to seasickness.

Clearly, a crew's energy and resources are best reserved for matters which cannot have been anticipated: moreover a skipper with no inner doubts about the fitness of his vessel and equipment will find it easier to inspire his crew to make the special effort which may be needed in a storm. This chapter is written without a particular level of severity of weather in mind, but rather in acknowlegement of the fact that an unexpected encounter with heavy weather may occur at some time.

Harbour preparations

When planning a vessel's chart portfolio, either before a voyage or during the lay-up period, consideration should be given to ensuring that there is adequate information in the event of being blown well off course, or having to find an unfamiliar port of refuge under the stress of bad weather. For example, in the 1956 Channel Race (see Chapter 11), Air Commodore Brian Macnamara, skipper of the yacht *Beltane*, no doubt wisely kept well out to sea during the storm. Near the Le Havre lightvessel, the weather became so wild that he decided to heave-to under a backed storm jib with helm lashed hard down. Thus he remained for 24 hours while the wind took him up Channel to the limit of his charts. Happily, the weather moderated to force 7–8 and such conditions permitted him to choose this moment to get under way and make to windward, but, had this not been possible, it is easy to imagine the anguish of being forced further eastwards into an area of shoal waters without a chart. Also Adlard Coles himself, in Chapter 1, describes how he had to rely on a few discouraging notes in the *Cruising Association Handbook* to enter the shallow harbour of Walberswick in Suffolk, having arrived there under duress of a North Sea gale from a point only 37 miles short of his destination of Dover.

Safety equipment comes in many forms and will vary according to the extent of the voyage. As mentioned by Olin Stephens in Chapter 25, a useful list of seamanlike measures to cope with eventualities in bad weather is provided by the Offshore Racing Council's Special Regulations. These regulations are compulsory for yachts in most offshore races world-wide, but cruising boat owners might also do well to study them, as they have been distilled from many years of experience. A summary of the regulations is given later in this chapter. Of course, if he is to face bad weather with confidence, a skipper needs to do more than buy and mount the safety equipment that his conscience dictates or which complies with regulations. He and his crew must know how to use it through testing it in a non-threatening situation.

In addition to safety equipment, even the owner of a new vessel may need to make an occasional effort to keep on top of defects in a vessel's general equipment; in the case of an older boat, a constant effort will be necessary. For in calm weather, the adaptable human quickly reconciles himself to such things as small leaks, quirky navigation lights, a seized hull valve, a defective pump, a shortage of mainsail ties and so on; yet such shortcomings can assume enormous proportions in rough weather, creating an emergency out of something previously regarded as a trivial problem. Recording such faults in a 'defect book' is a helpful means of maintaining a boat in first class condition. In addition to commonplace failures, another potentially weak area lies with standby equipment. For example, when the jib furling gear breaks, will the alternative system be operable? When the Loran or Decca fails, will there be batteries for the old radio direction finder set, and will it work? Likewise, the reserve bilge pump, the emergency navigation lights, and so on.

As a final comment on harbour preparations it has to be said that not all new boats are fit for heavy weather on delivery. Apart from the need for inherent structural strength, owners will often have to carry out numerous modifications. For example, screws are sometimes used when bolts are appropriate, backing pieces for deck fittings can have insufficient area, anchor well lids cannot be secured and so on. A surveyor's advice can be helpful in this situation.

Planning for dire situations

A good heavy weather principle is to prepare for the worst and hope for the best. For example, Chay Blythe prepares himself on a flat calm day by staging the onset of a fictitious gale. Another method is to plan and practise for dire circumstances. Among the worst of such situations which might be considered are: one or more members of the crew going overboard; the necessity to make to windward under sail in heavy weather to clear a lee shore; capsize; or the need to abandon ship.

Man overboard
Looking at these events in turn, the man overboard situation usually comes to mind first, and, in incidents involving a fatality where the lifeboat has been called out, it is the most common cause.

Aboard the Swan 53 ChaCaBoo in the 1989 Fastnet. The damaged mainsail has been lowered and is still being secured whilst the trysail sliders are being fed into the track prior to hoisting. Photo: Peter Bruce.

There are three phases involved when someone is lost overboard: location, recovery and treatment. When the size of crew allows it, there is much to be said for these tasks to have been allocated to different individuals before the emergency has arisen so that appropriate mental preparations can be made.

At the risk of stating the obvious, location is much assisted by someone keeping the casualty in sight. Therefore it may be best to have someone permanently nominated for the task. One cannot assume that this will always be possible, so it is of equal priority to practise recording the vessel's position, both with appropriate buoyant safety gear, and by using such navigational aids as may be available. In a real situation, in addition to the measures taken on board, outside help should be summoned – when available – without delay. It might even be possible to arrange to exercise this procedure through the coastguard, bearing in mind that those who carry out safety patrols, such as the crews of 'Search and Rescue' helicopters, may be glad to take part in impromptu safety drills when not otherwise occupied.

Two different methods for the retrieval of man overboard are in favour – the 'quick-stop' and the 'reach-tack-reach' method. I adopted the 'quick stop' method with an entirely satisfactory result in my small racing yacht *Scarlet Runner* in 1973, when surfing downwind under spinnaker in the English Channel. Immediately after the man was lost the yacht, which had no engine, was turned into wind as the spinnaker was dropped. The sail backed across the foretriangle and was successfully gathered in as the boat beat back under mainsail to the man in the water. He was hauled in from the lee side at a point just aft of the shrouds.

The other method, taught by the Royal Yachting Association for many years, involves going onto a reach, followed by a tack and a reach back to the victim. Initially, it is preferable to spend time trying either one of them, rather than spend the time discussing which one to employ.

At first it may be found that an effort may be necessary to get a crew to exercise a man overboard drill, but thereafter it will almost certainly be perceived as an important evolution, worthy of practice and discussion. It could also be mentioned that the quick and seamanlike retrieval of some suitable object from the sea, without the use of the engine – as propellors so often catch a rope in a crisis – provides a source of deserved satisfaction within a crew. Equally important is to prove by practical trial that the equipment intended for quick deployment, such as a horseshoe lifebuoy with a flashing light attached to it by a lanyard, can be used effectively in stressful circumstances. Experience, supported by trials in the United States, shows that a tangle is likely at the first attempt.

Every boat should have at least one means of lifting a casualty back on board. A stern bathing ladder is not the answer in a seaway. In particular, couples who sail on their own together should consider how to cope with the situation where the stronger and heavier partner has fallen over the side. The Seattle Sailing Foundation was inspired to study this aspect of the man overboard problem by some understandable but heartbreaking local tragedies. In one case a man, an experienced sailor, went overboard on a blustery day from the foredeck of his yacht. His wife, his sole crew, who was not experienced, had no idea what to do. Frozen with horror she steered straight until she hit the beach some time later. Her husband was never found. In another case, a skipper was lost and drowned even after being brought alongside, due to difficulties in hoisting him on board.

There is now on the market some very good equipment to aid the recovery of a man overboard, such as the Seattle (or life) sling, which deserves careful consideration. At the same time one should be aware that there may be a tendency these days to place too much reliance on technology, so it is best only to buy items that have withstood the stringencies of thorough testing.

I carry a small waterproof torch in my foul weather jacket pocket as a back-up device; at night this is not only useful for finding things down below without disturbing the off-watch crew by using the cabin lights, but the torch might provide a position indicator should one ever involuntarily part company with one's vessel.

Obviously, prevention is better than cure, and, in addition to non-slip surfaces and footgear, the value of a well-designed safety harness, built to the standard Offshore

Racing Council specification and adjusted so that one will not slip out, cannot be overestimated. Foredeck crew are most exposed, but they tend to be nimble and aware of the hazard, whereas those not used to being on the foredeck can be more at risk. Less obvious moments of exposure, but ones that have claimed lives, are those when crew unclip their safety harnesses to go below or have not yet clipped on when coming on deck. With practice, education, well positioned strong points and jackstays (see special regulations) it should be possible to avoid ever being unclipped when on deck. It is worth mentioning that the harness should be clipped on in a way to prevent a person falling over the side rather than merely preventing him from being separated from his craft having done so.

In spite of more general use of safety harnesses there still remains a chance of someone being lost overboard. Two more common examples are when gentlemen are relieving themselves over the side and when crewmen are struck on the head by the boom in an unexpected gybe. Chances of survival, once in the water, are obviously much increased when a lifejacket is being worn. In the latter case, when the victim is unconscious, the facility of automatic inflation is essential. There is something to be said for foul weather jackets with built-in safety harness and lifejacket, as the decision to put them on is then directly linked with windy or bad weather, and little effort is required. However it is difficult to design built-in lifejackets with features to make them effective in rough water. For example the experience of man overboard from the yacht *Hayley's Dream* in the 1989 Fastnet showed that it is necessary to incorporate a restraining arrangement, such as a crutch strap, to prevent the lifejacket riding up over the casualty's head. Another case, which occurred in the 1989 – 90 Whitbread Round the World Race, draws attention to the value of a spray guard to prevent the casualty drowning from inhaled spray. All such features can be achieved when the lifejacket is combined with a safety harness alone, an arrangement which would appear to optimise safety and convenience.

In concluding this subject, it should be remembered that drowning, rather than hypothermia, is the predominant cause of loss of life in man overboard incidents.

A lee shore

The next eventuality to be considered is a lee shore, which was once such a terrifying prospect to the crews of square rigged vessels: but such is the weatherliness of modern craft that a lee shore is not now such a common cause for concern. Nevertheless, circumstances can come about when a yacht will only be saved from being wrecked by having the ability to make to windward under storm sails. Those who have a roller headsail system should be aware that, as the headsail is rolled away, it becomes baggy and increasingly inefficient as a windward sail. Some standard production yachts are neither supplied with storm sails at all nor provision to use them. To reduce the cost of the sail inventory, others are supplied with sails described as storm jibs which are too big for use in really severe weather.

Even when in possession of a sensibly-sized storm jib, it should be remembered that, to be effective to windward, sheet leads and the point at which the jib is attached to the foredeck will, at some stage, have needed careful thought and assessment. Ideally, the tack of the storm jib should be taken to a point somewhat aft of the stem fitting to obtain optimum balance and interaction between it and the mainsail/trysail. But to achieve this, a properly designed independent forestay, supported by running backstays, will be necessary. A tack pennant is usually advisable to raise the storm jib to match the height of the trysail or reefed main, and to give greater clearance to waves breaking over the foredeck. In addition, sheet leads will need to be positioned further outboard than for the larger headsails to give an angle of 12 – 14° from the centreline. Finally, for yachts that use a headstay system rather than hanks, consideration needs to be given to coping with an unuseable foil. These devices are tough, but can be crushed by a spinnaker pole. Moreover, plastic versions have been known to simply fragment through cold temperatures or exposure to natural ultra-violet light.

Assuming conditions warrant its use, a trysail is a most efficient sail: but it is usually difficult to set, due to the necessary height above the deck of the luff fastening arrangement – usually a slider gate – coupled with the proximity of the mainsail. Ideally, a separate mast track should be provided for the trysail. This allows the trysail to be bent on before the mainsail is lowered. Whatever system is used, a new trysail must be given at least one airing in calm conditions. Clearly, it is better to find that the sliders are the wrong size for the track in harbour, than when drifting onto a rocky lee shore in a force 10 storm. Sheeting positions should be tried with and without a serviceable boom. A good opportunity to put up storm sails might be when forced by lack of wind to make a passage under engine, or when racing has been cancelled due to bad weather.

Capsize

The importance of a strong and watertight hull has been referred to in Chapter 25 of the second part of this book. The need for these features is particularly important in the dire situation of a capsize. To be prepared for a capsize, one should try to imagine one's vessel being held upside-down in the sea, with ventilator plugs in place. Thus orientated, a well designed and soundly built craft should have minimal leaks, even through the companionway or the cockpit lockers. Clearly, heavy items such as the engine, the galley stove, the deep freeze, anchors, batteries, gas cylinders, tanks and internal ballast should be installed with total inversion in mind. This is half the battle; the other half entails educating the crew to conform to the time-honoured mariner's practice of 'securing for sea'. Adequate locker space and stowages are necessary to give this prescripton a chance, but even so it will be found that stowing and securing all loose gear before leaving harbour, and keeping a vessel tidy thereafter, does not come naturally to most mortals. Nevertheless, unexpected events at sea can be so confounding that it does not always need a capsize to prove the wisdom of an orderly ship.

Whilst on the subject of capsize it should be mentioned that a safety harness line should have a clip at both ends so that, in the event of a prolonged inversion or a vessel foundering (see *Waikikamukau* Chapter 20) crewmen have a means of releasing themselves within immediate reach.

A sinking vessel

Our final worst situation is that moment when a liferaft will shortly be the only craft left afloat, and the moment has come to pull the inflation cord, or painter. An example of such a situation took place in a force 9 gale a mile off Salcombe on 11 August 1985. The 30 ft sloop *Fidget*, built of mahogany on oak by Camper & Nicholson, fell about 20 ft off the back of a wave in a very large cross-sea, opened up and sank in about 30 seconds. The skipper, Simon Wilkinson, who habitually keeps a knife in his foul weather jacket pocket, only just had time to cut the liferaft and uninflated tender free from their lashings before *Fidget* foundered, and the crew were left swimming. It was fortunate that they were already wearing lifejackets, because there was an interval before the inflation cord, which was still attached to the yacht, came taut, and the liferaft inflated. Simon Wilkinson thinks that he and his crew might well have drowned if they had not been wearing them.

Should a vessel sink far from help, the best chance of being found these days lies with an emergency indicator beacon (EPIRB). This is a radio device, which, in early versions transmits on the aviation distress frequency, 121.5MHz. The signal can be picked up by an aircraft or a satellite passing within line of sight, but will not be passed on by the satellite unless there is a ground receiving station within its horizon at that moment. The most recent and more expensive type operates on 406 MHz in conjunction with a satellite system alone. This one has world-wide coverage, gives the identity of the casualty, a more accurate position, and is not prone to spurious activation.

The kind of event that *Fidget*'s crew experienced is not one likely to have been practised, though courses of instruction in survival exist and are rated as good value, in addition to books on the subject. Apart from those who carry a Tinker Tramp combined dinghy and liferaft, the operation of a conventional liferaft is usually taken on trust. However, an opportunity to learn about one's liferaft occurs at the time of its annual inspection, when it will have to be inflated anyway. To witness an inflation, it is best to make an appointment with the servicing agent, who should be one authorised by the manufacturer, for an 'opening day'. A garden or marina inflation in the company of the crew is entertaining, but will certainly lead to extra cost. For example, the gas bottle is not used for a test inflation, but is merely weighed and inspected, and replacement bottles are expensive to recharge and retest. Besides, it is quite easy to damage the fabric of a partially deflated liferaft when in transit back to the servicing agent. Nonetheless, it is possible for an owner to inflate the liferaft without using the gas bottle, and those who can trust their own judgement and know the technique for repacking speak highly of carrying out their own annual inspection.

The merit of having additional items to supplement the liferaft's safety equipment pack will depend very much upon the circumstances; however, the concept of a grab-bag, or calamity-bag, filled with suitable sundry items deserves recognition. For example, a waterproof VHF radio, dry clothing, space blankets, extra flares, food, water, writing materials, passports, money and crew medicines might well reward the foresight involved, come the day. *Fidget*'s crew, for example, were not to know they had been spotted, and would have been glad to have had some drinking water during the interval of forty minutes before they were picked up. Also they were short of flares, one being faulty, and the others having been used immediately.

Sea preparations for imminent bad weather

A skipper should delegate various tasks within his crew to spread the load, and to take care of the situation when he himself might be incapacitated or lost overboard. This means an organisational structure has to be formed, and though some people may go sailing to escape such duress, there is a clear need for it in times of crisis.

Likewise it is always desirable to establish and maintain a watch routine so that a crew can rest properly when off watch. With an experienced crew the watchkeeping routine will need no adjustment with the onset of bad weather. On the other hand a less experienced and well-prepared crew may find some members opting out while others are involved in one drama followed by another, leading to whole-crew exhaustion. This can be a recipe for disaster as people do not tend to make good decisions when they are cold, wet, anxious and tired.

It is suggested that all craft likely to be in the open sea could benefit from an owner's check-off list covering actions to be taken in the face of an impending storm. Such measures will depend very much on the circumstances, such as the experience of the crew, the type of craft and her distance from base. The following list may be helpful to some, whilst others may think of the suggestions given here as being standard procedure:

1 Issue seasick pills.
2 Charge batteries.
3 Insert mainsail reefing lines, and set up inner forestay for storm jib, if appropriate. Rig trysail in readiness if the mast track arrangement permits.
4 Check windows and hatches are tightly closed. Fit ventilator cover plates. Put up hatchway storm screens if fitted. Put stopper, such as plasticine, in hawse pipe.
5 Move storm sails, buckets, warps or other drag devices, to a handy position, bearing in mind that it may be difficult and dangerous to open cockpit lockers during a storm.
6 Shut WC seacocks after a good pump through. Experience may show that other cocks need to be closed during bad weather. If shutting such seacocks could cause catastrophic results, an appropriate notice should be displayed. For example, a sign saying 'engine cooling water cock closed' taped over the engine starter.
7 Plot position and, if a survival situation is anticipated, report this to coastguards. Put a dry towel, or absorbent material, within reach of the chart table.
8 Start recording frequent barometer readings, or better still refer regularly to a barograph.
9 Put on appropriate clothing, such as thermal underwear, seaboots, gloves, etc.
10 Pack a 'grab-bag'. In addition seal spare clothing, bedding, matches, lavatory paper, bread, etc. within clear heavy duty polythene bags.
11 Pump bilges. Check pump handles are attached to the boat and the whereabouts of spares.
12 Put washboards into position, and secure their fastening arrangements.
13 Secure latches of cockpit lockers.

14 Check cockpit drains are free. It may be helpful to blow through with an inflatable dinghy air-pump or sink drain clearer.

15 Check deck items, such as the spinnaker pole, anchor, anchor well lid and liferaft are properly secured. Winch handles should not be left loose, and a spare should be kept below decks.

16 Check halyards are not twisted and the free ends of the lines on deck are stowed, so that they are not likely to go overboard to foul the propeller in an emergency.

17 Check navigation lights are working; hoist radar reflector.

18 Bring dinghy inboard and stow securely.

19 Consider renewing torch batteries and changing cooking gas cylinder.

20 Make up thermos flasks of soup and coffee. Make sandwiches and place in a water-tight container.

21 Give the crew a good hot meal.

22 Secure and stow loose items below, especially in the galley area. If containers made of glass have to be brought onboard, such as, possibly, coffee or jam jars, they must be packed away with great care.

23 Instruct crew members to put on their safety harnesses and lifejackets. Ensure that they are adjusted to fit snugly over storm clothing so that the wearer cannot slip out. Also that the crew are familiar with the operating features and strap design so that they may be donned rapidly without assistance, if need be, in violent motion and darkness. Safety harnesses and lifejackets should usually be worn during the period of the storm.

24 Be ready to change down to storm canvas in good time.

25 If running, rig a heavy duty main boom preventer to avoid accidental gybes. Ideally this should be rigged round a leading block forward and led back to the cockpit.

26 Rig a rope lattice within the cockpit to give greater security to its occupants, as described by Warren Brown in Chapter 19.

27 Brief the crew on what to expect. Remind them of the importance of being hooked on to something really strong when on deck, of maintaining a good look-out, and of being aware of the position of flares, grab-bag, liferaft, a sharp knife, etc.

The well prepared vessel is much more likely to come through a gruelling storm with only minor problems. Yet storms can be a tremendous test of endurance, and despite very careful preparation, much depends upon the cool judgement, courage, physical fitness and tenacity of the skipper and crew.

The Offshore Racing Council Special Regulations are published overleaf with the permission of the Offshore Racing Council:

Categories of offshore events

Category 0 race Trans-Ocean races, where yachts must be completely self-sufficient for very extended periods of time, capable of withstanding heavy storms and prepared to meet serious emergencies without the expectation of outside assistance.

Category 1 race Races of long distance and well offshore, where yachts must be completely self-sufficient for extended periods of time, capable of withstanding heavy storms and prepared to meet serious emergencies without the expectation of outside assistance.

Category 2 race Races of extended duration along or not far removed from shorelines or in large unprotected bays or lakes, where a high degree of self-sufficiency is required of the yachts but with the reasonable probability that outside assistance could be called upon for aid in the event of serious emergencies.

Category 3 race Races across open water, most of which is relatively protected or close to shorelines, including races for small yachts.

Category 4 race Short races, close to shore in relatively warm or protected waters normally held in daylight.

The regulations below generally apply to all categories unless otherwise stated.

Basic requirements

All required equipment shall:

Function properly
Be readily accessible
Be of a type, size and capacity suitable and adequate for the intended use and size of yacht.

Yachts shall be self-righting. They shall be strongly built, watertight and, particularly with regard to hulls, decks and cabin trunks capable of withstanding solid water and knockdowns.

They must be properly rigged and ballasted, be fully seaworthy and must meet the standards set forth herein. 'Properly rigged' means that shrouds shall never be disconnected.

Inboard engine installations shall be such that the engine when running can be securely covered, and that the exhaust and fuel supply systems are securely installed and adequately protected from the effects of heavy weather. When an electric starter is the only provision for starting the engine, a separate battery shall be carried, the primary purpose of which is to start the engine.

Ballast and heavy equipment: all heavy items including inside ballast and internal fittings (such as batteries, stoves, gas bottles, tanks, engines, outboard motors etc.) and anchors and chains shall be securely fastened so as to remain in position should the yacht be capsized 180 degrees.

Structural features

The hull including deck, coach roof, windows, hatches and all other parts, shall form an integral, essentially watertight, unit and any openings in it shall be capable of being immediately secured to maintain this integrity. For example, running rigging or control lines shall not compromise this watertight unit. Centre-board and dagger board trunks shall not open into the interior of the hull.

Hatches No hatches forward of the BMAX (maximum beam) station shall open inwards excepting ports having an area of less than 110 sq in (710 sq cm). Hatches shall be so arranged as to be above the water when the hull is heeled 90 degrees. All hatches shall be permanently fitted so that they can be closed immediately and will remain firmly shut in a 180 degrees capsize. The main companionway hatch shall be fitted with a strong securing arrangement which shall be operable from above and below.

Companionways All blocking arrangements (washboards, hatchboards, etc.) shall be capable of being secured in position with the hatch open or shut and shall be secured to the yacht by lanyard or other mechanical means to prevent their being lost overboard.

Cockpit companionways if extended below main deck level, must be capable of being blocked off to the level of the main deck at the sheer line abreast the opening. When such blocking arrangements are in place this companionway (or hatch) shall continue to give access to the interior of the hull.

Cockpit structure They shall be structurally strong, self draining and permanently incorporated as an integral part of the hull. They must be essentially watertight, that is all openings to the hull must be capable of being strongly and rigidly secured.

Cockpit volume The maximum volume of all cockpits below lowest coamings shall not exceed 6% LxBxFA (6% loaded water line x maximum beam x freeboard abreast the cockpit). The cockpit sole must be at least 2% L above LWL (2% length overall above loaded water line). (Category 0 and 1 only.) The maximum volume of all cockpits below lowest coamings shall not exceed 9% LxBxFA (9% loaded water line x maximum beam x freeboard abreast the cockpit). The cockpit sole must be at least 2% L above LWL (2% length overall above loaded water line).

Cockpit drains For yachts of 28 ft/8.53 m length overall and over. Cockpit drains adequate to drain cockpits quickly but with a combined area of not less than the equivalent of four $\frac{3}{4}$ in (19 mm) diameter drains.

Sea cocks or valves on all through-hull openings below LWL, except integral deck scuppers, shaft log, speed indicators, depth finders and the like. However, a means of closing such openings, when necessary to do so, shall be provided.

Soft wood plugs tapered and of the correct size, to be attached to, or adjacent to, the appropriate fitting.

Sheet winches shall be mounted in such a way that no operator is required to be substantially below deck.

Mast step The heel of a keel-stepped mast shall be securely fastened to the mast step or adjoining structure.

Bulkhead The hull shall have a watertight bulkhead within 15% of the vessel's length from the bow and abaft the forward perpendicular. (Category 0 only.)

Lifeline terminals and lifeline material Where wire lifelines are required, they shall be multi-strand stainless steel wire as follows:

Overall length	Minimum wire diameter
under 28 ft (8.53 m)	$\frac{1}{8}$ in (3 mm)
28 ft (8.53 m) to 43 ft (13.0 m)	$\frac{5}{32}$ in (4 mm)
Over 43 ft (13.0 m)	$\frac{3}{16}$ in (5 mm)

A taut lanyard of synthetic rope may be used to secure lifelines, provided that when in position its length does not exceed 4 in and gives the enclosure at least the strength of the lifeline wire.

Pulpit and stanchion fixing Pulpits and stanchions shall be securely attached:

a) When there are sockets or studs, these shall be through bolted, bonded or welded. The pulpit(s) and/or stanchions fitted to these shall be mechanically retained without the help of the lifelines.

b) Without sockets or studs, pulpit and/or stanchions shall be through bolted, bonded or welded.

For yachts of 21 ft and over: Taut double lifelines, with upper lifeline of wire at the height of not less than 2 ft above the working deck, to be permanently supported at intervals of not more than 7 ft.

When the cockpit opens aft to the sea, additional lifelines must be fitted so that no opening is greater in height than 22 in.

Pulpits Fixed bow pulpit and stern pulpit. Lower lifelines need not extend through the bow pulpit. Upper rails of pulpits shall be at no less height above the working deck than upper lifelines.

Toerails A toerail of not less than 1 in shall be permanently fitted around the deck forward of the mast; except in way of fittings. Location to be not further inboard from the edge of the working deck than one third of the local beam.

A third lifeline at a height of not less than 1 in or more than 2 ins above the working deck will be accepted in place of a toerail.

Jackstays Jackstays must be fitted on deck, port and starboard of the yacht's centre line to provide secure attachments for safety harnesses. Jackstays shall comprise stainless steel 1 x 19 wire of minimum diameter 5 mm, or webbing of equivalent strength. Jackstays must be attached to through–bolted or welded deck plates, or other suitable and strong anchorages. The jackstays must, if possible, be fitted in such a way that a crew member when clipped on, can move from a cockpit to the forward and to the after end of the main deck without unclipping the harness. If the deck layout renders this impossible, additional lines must be fitted so that a crew member can move as described with a minimum of clipping operations.

A crew member must be able to clip on before coming on deck, unclip after going below and remain clipped on while moving laterally across the yacht on the foredeck, the afterdeck, and amidships. If necessary additional jackstays and/or through-bolted or welded anchorage points must be provided for this purpose.

Through-bolted or welded anchorage points, or other suitable and strong anchorage, for safety harnesses must be provided adjacent to stations such as the helm sheet winches and masts, where crew members work for long periods. Jackstays should be sited in such a way that the safety harness lanyard can be kept as short as possible. (Categories 0, 1 and 2 only.)

Accommodations

Toilet, bunks and water tanks securely installed.
Cooking stove, securely installed against a capsize with fuel shutoff control capable of being safely operated in a sea-way.

General equipment

Fire extinguishers At least two, readily accessible in suitable and different parts of the boat.

Bilge pumps At least two manually operated, securely fitted to the yacht's structure, one operable above, the other below deck. Each pump shall be operable with all cockpit seats, hatches and companionways shut. (Categories 0, 1 and 2 only.)

Each bilge pump shall be provided with permanently fitted discharge pipe(s) of sufficient capacity to accommodate simultaneously both pumps. (Categories 0, 1 and 2 only.)

No bilge pumps may discharge into a cockpit unless that cockpit opens aft to the sea. Bilge pumps shall not be connected to cockpit drains.

Unless permanently fitted, each bilge pump handle shall be provided with a lanyard or catch or similar device to prevent accidental loss.

Two buckets of stout construction with at least 2 gallons (9 litres) capacity. Each bucket to have a lanyard.

Anchors Two with cables except for yachts under 28 ft/8.53 m length overall which shall carry at least one anchor and cable.

Flashlights, one of which is suitable for signalling, water resistant, with spare batteries and bulbs.

First aid kit and manual.

Foghorn.

Radar reflector If a radar reflector is octahedral it must have a minimum diagonal measurement of 18 ins (457 mm), or if not octahedral must have a documented 'equivalent echoing area' of not less than 10 sq m.

Shutoff valves on all fuel tanks.

Navigational equipment

Compass, marine type, properly installed and adjusted.

Spare compass.

Charts, light list and piloting equipment.

Radio direction finder or an automatic position fixing device. (Categories 0, 1 and 2 only.)

Lead line or echo sounder.

Speedometer or distance measuring instrument.

Navigation lights to be shown as required by the International Regulations for Preventing Collision at Sea, mounted so that they will not be masked by sails or the heeling of the yacht. Navigation lights shall not be mounted below deck level. Spare bulbs for navigation lights shall be carried.

Emergency equipment

Emergency navigation lights and power source. Emergency navigation lights shall have the same minimum specifications as the navigation lights and a power source and wiring separate from that used for the normal navigation lights.

Sails The following specifications for mandatory sails give maximum areas: smaller areas may well suit some yachts.

Appropriate sheeting positions on deck shall be provided for these sails. (Categories 0, 1 and 2 only):

One storm trysail not larger than 0.175 x P x E in area. It shall be sheeted independently of the boom and shall have neither a headboard nor battens and be suitable strength for the purpose. Aromatic polyamides, carbon fibres and other high modulus fibres shall not be used in the storm trysail.

One storm jib of not more than 0.05 x IG x IG (5% height of the fore-triangle squared) in area, the luff of which does not exceed 0.65 x IG (65% height of the fore-triangle), and of suitable strength for the purpose. Aromatic polyamides, carbon fibres and other high modulus fibres shall not be used in the storm jib.

One heavy weather jib of suitable strength for the purpose with area not greater than 0.135 x IG x IG (13.5% height of the fore-triangle squared) and which does not contain reef points.

Any storm or heavy weather jib if designed for a seastay or luff-groove device shall have an alternative method of attachment to the stay.

No mast shall have less than two halyards each capable of hoisting a sail.

An emergency tiller capable of being fitted to the rudder stock. Crews must be aware of alternative methods of steering the yacht in any sea conditions in the event of rudder failure.

Tools and spare parts including adequate means to disconnect or sever the standing rigging from the hull in the case of need.

Yacht's name on miscellaneous buoyant equipment such as lifejackets, oars, cushions, lifebuoys and lifeslings etc.

Retro-reflective material Lifebuoys, lifeslings, liferafts and lifejackets shall be fitted with retro-reflective material.

Marine radio transmitter and receiver If the regular antenna depends upon the mast, an emergency antenna must be provided.

In addition a waterproof hand held VHF transceiver is recommended.

Radio receiver capable of receiving weather bulletins.

EPIRBs Emergency indicator beacon. (Categories 0 and 1 only.)

Water At least 2 gallons (9 litres) of water for emergency use carried in one or more containers.

Safety equipment

Lifejackets – one for each crew member. In the absence of any specification, the following definition of a lifejacket is recommended:

'A lifejacket should be of a form which is capable of providing not less than 16 kg of buoyancy, arranged so that an unconscious man will be securely suspended face upwards at approximately 45 degrees to the water surface.'

Whistles attached to lifejackets.

Straps It is advised that lifejackets be fitted with crotch or thigh straps to prevent the lifejacket riding up over the face.

Safety belt (harness type) – one for each crew member. Each yacht may be required to demonstrate that two thirds of the crew can be adequately attached to strong points on the yacht.

Liferafts
a) Must be carried on the working deck or in a special stowage opening immediately to the working deck containing the liferafts only.
b) For yachts built after 1.7.83 liferaft(s) may only be stowed under the working deck provided:
i) the stowage compartment is watertight or self-draining.
ii) if the stowage compartment is not watertight, then the floor of the special stowage is defined as the cockpit sole.
iii) the cover of this compartment shall be capable of being opened under water pressure.
c) Liferaft(s) packed in a valise and not exceeding 40 kg may be securely stowed below deck adjacent to the companionway.
d) Each raft shall be capable of being got to the lifelines within 15 seconds. *Note* New legislation may allow a longer interval than this.
e) Must have a valid annual

certificate from the manufacturer or an approved servicing agent certifying that it has been inspected, that it complies with the above requirements and stating the official capacity of the raft which shall not be exceeded. The certificate, or a copy thereof, to be carried on board the yacht.

Lifebuoy with a drogue or lifesling, equipped with a self-igniting light within reach of the helmsman and ready for instant use. One lifebuoy within reach of the helmsman and ready for instant use, equipped with a whistle, dyemarker, drogue, a self-igniting light, and a pole and flag. The pole shall be either permanently extended or be capable of being fully automatically extended in less than 20 seconds. It shall be attached to the lifebuoy with 10 ft (3.048 m) of floating line and is to be of a length and so ballasted that the flag will fly at least 6 ft (1.828 m) off the water. (Categories 0, 1 and 2 only.)

Distress signals conforming to the current International Convention for the Safety of Life at Sea (SOLAS) regulations to be stowed in a waterproof container or containers, as indicated:

Twelve red parachute flares (Categories 0 and 1 only). Four red parachute flares. Four red hand flares. Four white hand flares. Two orange smoke day signals. Distress signals which are more than 3 years old, or for which the date of expiry has passed, are not acceptable.

It might be added that white parachute flares, which lit up the scene brilliantly, were most valuable in locating the two men lost overboard at night from *Creightons Naturally* during the 1990 Whitbread Round the World Race.

Heaving line 50 ft (15.24 m) minimum length readily accessible to cockpit.

Grab bags

A grab bag containing the following is recommended:

Second sea anchor and line
Two safety tin openers
A first aid kit
One rust-proof drinking vessel graduated in 10, 20 and 50 cc

Two 'cyalume sticks' or two throwable floating lamps
One daylight signalling mirror and one signalling whistle
Two red parachute flares
Three hand red flares
Non-thirst provoking food rations and barley sugar or equivalent
Watertight receptacles containing fresh water (at least half a litre per person)
One copy of the illustrated table of life-saving signals
Nylon string and polythene bags

For identification of symbols and the full Special Regulations, copies can be obtained from The Offshore Racing Council, 19 St James's Place, London, SW1A 1NN, England.

30 THE METEOROLOGY OF HEAVY WEATHER

Alan Watts

Survival conditions when a major yacht race was in progress have only twice been sufficiently catastrophic for the Royal Ocean Racing Club to order an in-depth inquiry into the circumstances. These two occasions were following the Channel Race storm of 29 July 1956 and the disastrous Fastnet storm of 13/14 August 1979. These events have been covered in some detail in Chapters 11 and 21 but both of them are cogent examples of weather systems that lead to the worst kinds of heavy weather. Therefore we will describe the development of their synoptic situations as useful examples of the meteorology of heavy weather.

The Channel Storm of 1956 was a very odd storm. It would have been phenomenal even without the experiences described in Chapter 11 to back it up. Meteorologically it did not look to be much of a depression and we can pick it up in Fig 30.1 at midnight on Saturday, 29 July (0001 GMT), when its central pressure was 985 millibars and the only sizeable wind was 35 knots at Ushant. It looked a simple little low which would produce a force 8 blow in the Western Channel and nothing very much more. By 0200 it had deepened by 6 millibars, but that is not unusual, and while Scilly and Cork had 35 knots mean speed, winds in the Channel were only 25 to 30 knots (force 6–7), as forecast.

In the early stages several frontal troughs were recognizable, there being the usual veer of wind across each. During the hours 0200 to 0400 the low deepened very slowly at about 1 millibar per hour. The remarkable and significant feature of this depression's progress lies in the fact that during the time when the really excessive winds developed over the Channel it did not deepen at all. Thus some other explanation has to be found for the wind torrent which swept the English coastal region that morning. At 0400 (Fig 30.1) the low centre and its central region were coming into conflict with the high ground of Cornwall, Devon and Wales. Between lay the pass of the Bristol Channel. It might be assumed that a depression stretching its circulation many thousands of feet aloft would not be influenced by surface features of a few thousand feet. Yet this is not always the case. The depression's whole inclination was to move on its original track slightly north of east, but by 0600 the isobars were slipping eastwards very slowly and the low had developed two centres. The energy of the surface low, thwarted on its leading edge by the high ground, began to find its way into a wind corridor on its trailing side, so that incredible things happened like the wind being a few knots from the south-east at Hartland Point and 60 knots from the opposite direction in Watergate Bay, a mere 45 miles down the coast to the south-west. It will be seen that the reason the Channel Islands had a mere 25 to 30 knots and that *Lutine* reported force 4 at this time is that they lay on the edge of the wind corridor, while Portland was equally on the north of the corridor and so also had a mere 30 knots.

In the next hour the depression decided that it could not force its second centre over the barrier of the Welsh mountains and began to put all its energy back into the main centre and to swing around the mountains it could not cross, taking the wind corridor with it.

By 0800 a real torrent of 40–45 knot winds with gusts to over 60 knots had built down the after side of the thwarted low. Yet over the Channel fleet the isobars were still deceptively wide apart. However, Plymouth now had force 9, as did the exposed coasts of Cardigan Bay and the Bristol Channel.

Between 0800 and 0900 the depression took the line of least resistance by tracking through the Severn Valley, and the wind corridor, swinging down around it, forsook Wales and began to concentrate its energies on sea areas Portland and Wight. It was in this period that the hard-pressed yachts in the region of the Owers were overwhelmed. The corridor of storm-force winds was moving over its prey. By 1000 the wind's main energies were concentrated in the area between Portland and Beachy Head, with mean speeds of 50 to 60 knots (force 10–11) in the Owers region. Chapter 11 tells the sorry tale.

It is important to realize from this sequence the tremendous effect near the surface of what might appear to be minor land barriers. In the hour 0900 to 1000 the wind corridor skipped over the moors and was all in the Channel, as the 2000–3000 ft ranges to the south of Snowdon doggedly determined not to allow the isobars to pass.

By 1100 the depression finally made its way across the Welsh mountains and the winds began to become more normal, even if they did re-extend their sphere of influence all the way across southern England from the Bristol Channel to the Thames Estuary. But the worst was then over and while it still blew at 45 knots at St Catherine's Point and there was 40 knots at Boulogne and 45 at Dungeness, there is no point in continuing the saga. The storm corridor – best described as a surface jet stream – had swung around its parent centre, whose pressure was now rising. *Right Royal* managed during this time to be always within the force 8–9 sector and to run along with it, so unfortunately remaining within its influence for an extended period, even if she did miss the worst that there was.

This storm has been covered in some detail because it indicates a lot about such storms. It shows, for instance, why winds in the coastal sea areas bordering hilly or mountainous terrain (over which local depression centres have to track) may be the strongest there are. The Channel depression is obviously an entity which will intimately affect the offshore and cruising fraternity, and when it swings into the maw of the Bristol Channel (Chapters 6, 12 and 14), then the effects described for the Channel Storm are likely to occur, so that the West Channel and immediate south-west approaches become areas of severe gales resulting from the ire of thwarted and previously innocuous lows.

What the charts (Fig 30.1) also show is the extreme influence that the land has on the wind strength. For instance, on the same wind track on the 0900 chart we find the triangular storm symbol for 50 knots at Chivenor (near Ilfracombe), while at Exeter, in the same isobaric tramlines but on the other side of the moors, the wind is only 35 knots. Similarly at Aberporth, on the shores of Cardigan Bay, the wind is 30 knots, while at Swansea on the other side of the Cambrian Mountains wind speed only reaches 20 knots.

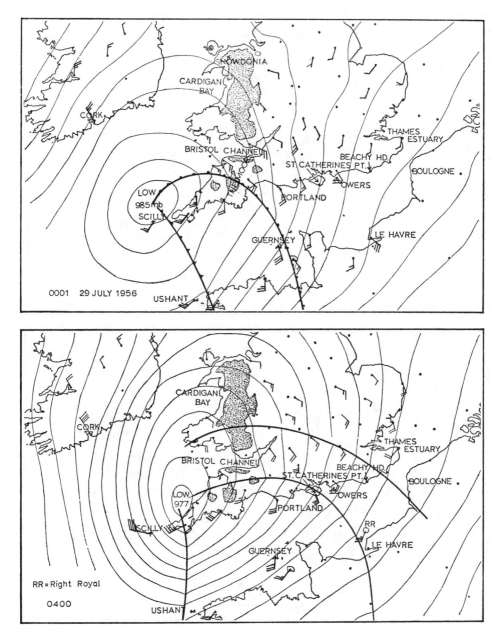

Fig 30.1(i). How the Channel Storm developed. All times are GMT and yacht and ship positions are only approximate at these times. The 1200 chart shows the path of the low from midnight 28 July to 0900 29 July and from then on the re-formation of the centre and its motion from 0900 to 1200. Isobars are at 2 millibar intervals, and full and half barbs on the wind arrows mean 10 and 5 knots respectively. Triangles mean 50 knots.

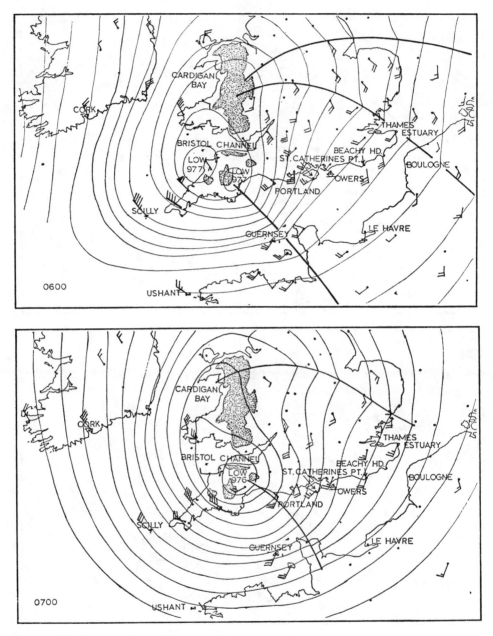

Fig 30.1(ii). How the Channel Storm developed.

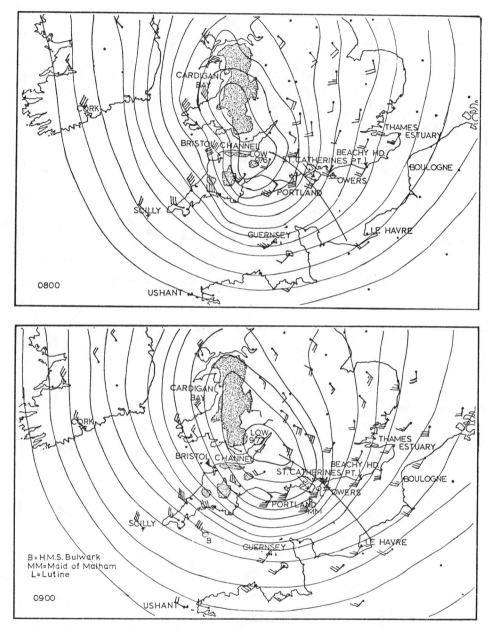

Fig 30.1(iii). How the Channel Storm developed.

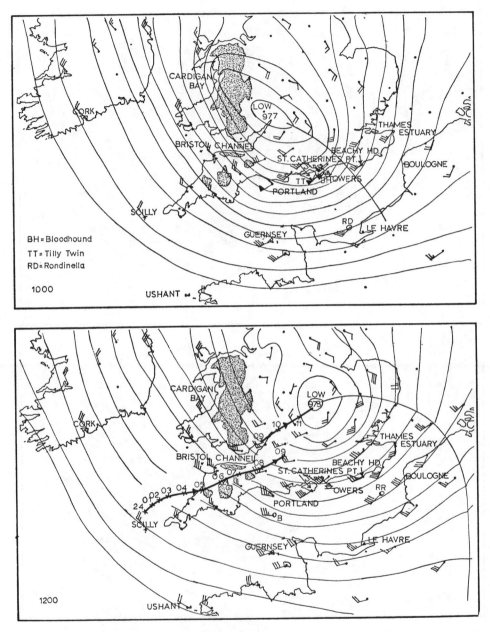

Fig 30.1(iv). How the Channel Storm developed.

Later we shall give figures culled from actual observations which will enable some estimates to be made of wind speed at sea when its value is known on land. For now we will look into the peculiarities of another very curious and deadly storm.

The Fastnet storm of 13–14 August 1979 was luckily a very rare occurrence but it may not be unique. It looked from the synoptic charts as if it was an almost copy-book re-run of a similar storm that covered the same sea area and with the same time schedule on 15–16 August 1970. But that was not a Fastnet year and so the unhappy coincidence of a storm force gale and a very large Fastnet fleet of yachts strung out between the Scillies and the Rock did not occur.

Meteorologically, the storm was a puzzle because few of the major criteria for an intense disturbance were there either before or apparently during the storm. The 1979 barograms in Fig 30.2 when compared to the Santander Race gale of 1948 (see Chapter 5), do not show at any time the massive fall of pressure that is exerienced with storms like those of the Santander Race or the storm of Jersey (see Chapter 27). These latter seem to be a distinct species of depression and the Fastnet was not one of them. It had however its own brand of meteorological eccentricities.

If you look at the official charts of the disturbance when it really began to bite around midnight (Fig 30.3), there is nothing to suggest anything unusual about the

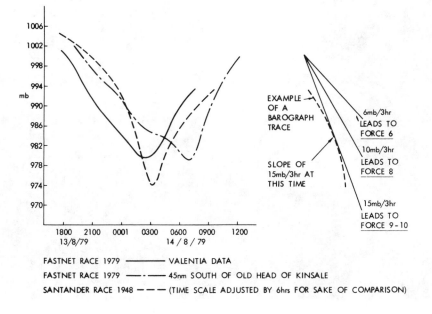

Fig 30.2. Barograph traces from a shore station (Valentia) and from the region of mid-Fastnet sea area; compared with one taken during the Santander Race of 1948. The mid-Fastnet trace is a composite deduced from the readings of yachts in the area at the various times. Slopes given on the right hand side for comparison show that at no time did the barometer fall at rates commensurate with force 10 or 11. The example of a barograph trace on the right is to show how to interpret the slopes at any time.

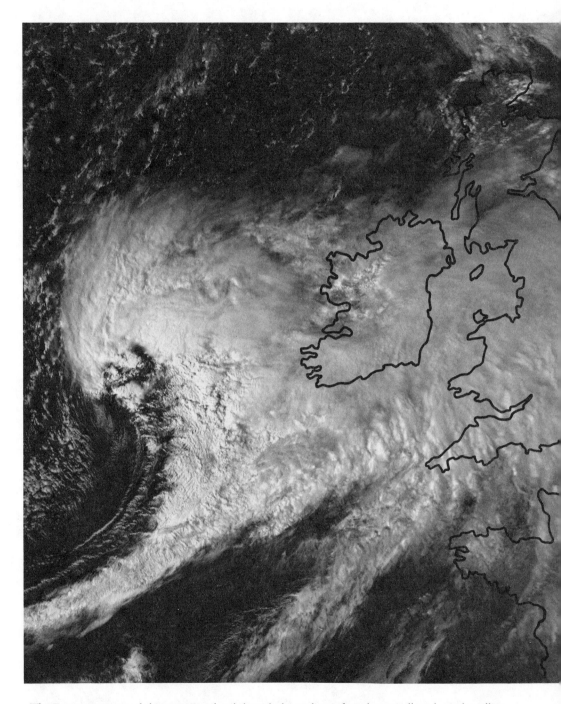

The Fastnet storm revealed in stunning detail through the medium of an electronically enhanced satellite cloud picture. The image was transmitted at 1637 on Monday 13 August 1979. Scotland can be seen in the extreme north-east corner and the Brest peninsula in the south-east corner so that the Fastnet fleet lies under the cloud mass in the centre of the picture. Photo: University of Dundee.

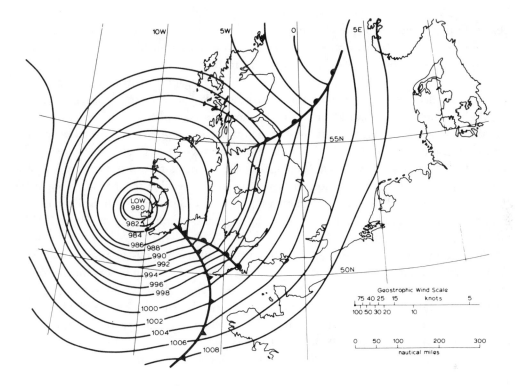

Fig 30.3. The low at 0100 Tuesday 14 August (official met sources). To use the geostrophic wind scale, take the distance apart of the isobars at any point and lay this off along the scale from the left hand end. The resulting wind speed is that just clear of surface friction (about 2000–3000 ft, 600–1000 m). The surface speed is about two-thirds of the geostrophic speed over the sea, but gusts will be up to (or may in certain cases exceed) the geostrophic value.

storm. The isobars are certainly tight and it was blowing 35 knots at Roches Point (Cork), but there is little indication that just 50 miles south of Roches Point the boats were experiencing 40–50 knots or that boats rounding the Rock itself were, within twenty minutes of the time of the chart, to be subjected to a sudden savage wind increase from a manageable gale force 35 knots to, perhaps temporarily, 60 knots and with gusts into hurricane force.

The sudden localized arrivals of winds of violent storm force, coupled with the chaotic short sea conditions in some areas, can be cited as the main meteorological oddity of this storm. From these points of view the Fastnet storm was very odd indeed and, using the observations from the yachts themselves, it is possible to bring some coherence into what could otherwise be a very confused picture.

However, we cannot assume that, because two weather charts look alike, they necessarily cover the same peculiarities of surface weather. It will become apparent that weather at sea shows localised disturbances which can deteriorate (or possibly improve) conditions in a relatively small locality. Such small-scale oddities are usually completely missed when the synoptic charts are drawn up in the major meteorological offices. This is not their fault. They just do not have a sufficient density of observations to 'see' such quirks. In fact only once to my knowledge has there ever been a chance to record the pressure variations throughout a fleet of well-found and expertly crewed yachts in a survival storm. That storm was the Fastnet of 1979 and it was Adlard Coles himself who instigated the investigation that made it possible. Wishing to elaborate on some of the questions that had been put together by the RORC Committee looking into the circumstances, Adlard sent out his own questionnaire, and this included requests for pressure readings over the duration of the storm. These needed to include the reading of each yacht as it passed St Mary's, Scilly so that their accuracy could be checked against the scientific record of pressure kept by the observers at the weather station there. Some twenty reliable records were obtained this way and enabled a complete graph of pressure readings between Scilly and the Fastnet Rock to be drawn for each three-hour period commencing at 2100z(GMT) and on throughout the night.

There really were some very queer meteorological happenings in the air over the Fastnet fleet that night as is evident when we record the following observations from a reliable source. Winds at the surface and at low cloud height can often be very different, but not when the surface wind is strong to gale over the open sea. Thus the observation of David Powell on the OOD 34 *Moonstone* at 2030 on the Monday that low clouds (estimated at 2000 ft or so) were moving over his yacht from the east at a fast rate (probably 20–30 kt), while the surface wind had risen to a 35 kt westerly, is

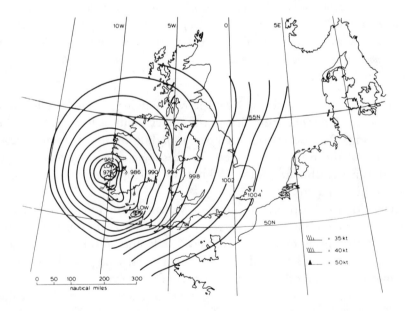

Fig 30.4. The low at 0100 Tuesday 14 August compiled from the barometric readings in the yachts themselves. It is more complex than the official version but the low of 990 mb in mid-Fastnet sea area is confirmed by no less than five boats. There are indications of other disturbances close to where 50 knot winds were reported as shown in the diagram.

very odd indeed. *Moonstone* at that time was in the area roughly occupied by the subsidiary low shown in Fig 30.4. The phenomenon continued until darkness made further observation impossible. Powell, an experienced ocean-racing man, said 'I have never seen anything like it in all my time at sea.'

Certainly the 1900 weather chart in advance of the coming depression had surface peculiarities. There was a weak depression of 1003 mb in the Irish Sea, but not so weak that most of the winds around the Irish Sea could not be bent to its influence. Some kind of trough from this feature pointed down towards the middle of the Fastnet sea area, and could have been more intense aloft than at the surface.

However, that weak feature in the Irish Sea could not have contributed a vector wind change of 180° and 55 knots within a few thousand feet of the surface. A vector wind change is the vector difference between two winds and so, in this case, is 35kt$-$ (-20kt) = 55 kt. Such an observation could be explained if a vortex of wind existed not very far above the surface – a feature that seems to have been quasi-stationary over the middle of the Fastnet sea area for a considerable time.

The yachts that were caught by the storm confirm that the seaway was often 'very strange' – not so much mountainous as curious. The outstanding feature of the seaway was its shortness and steepness with a very small wave length, which at times was of the order of a few yacht lengths. Such a sea could only be generated by intense sudden gusts of more than usual duration. A gust as normally understood will not in general have sufficient time in contact with the sea surface to produce significantly different wave formations from those already existing. In the Fastnet race, long period gusts (we might well call them jets) could have appeared when turbulence brought down the wind circulation, due to a vortex a few thousand feet up, and added it to the already strong gale force winds blowing round the main centre over Ireland.

The experience gained with the barometer readings obtained after the 1979 Fastnet race indicates that barograms from the boats that are within a few tens of miles of each other can appear very different in a big blow. You cannot tell whether there is not some local cyclonic entity that is lowering the pressure more than woud be expected in your locality while in a neighbouring locality no similar effect is occurring. Also we have to bear in mind that when comparing barograms a yacht moves, whereas a land station does not. The yacht is constantly sampling lower pressure when sailing towards a low centre, or a trough, while when sailing in the same direction as the low or trough is moving, the fall of pressure will be arrested. The yacht will not be able to sail faster than the meteorological entity but the arrival of any bit of weather will certainly be delayed when compared to the experience of a land station nearby. Thus barometric tendencies will be enhanced when sailing into a low and may make the fear of what is about to happen worse than the actual event.

When the barometric readings and the accompanying wind speeds were plotted on the rhumb line between Scilly and the Fastnet Rock, as in Fig 30.5a and 30.5b, a truly remarkable set of facts emerged. In Fig 30.5b (where the 1800 and 0300 GMT readings have been separated from those for 2100 and 0001 GMT for clarity) we see five yachts (FN, BA, TR, RV and KK) confirming that a trough existed in the area of the Labadie Bank, and that in association with that, the light to moderate wind

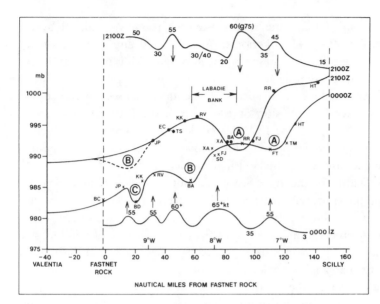

Fig 30.5a. The First Watch of the Fastnet night. The barometer readings of the yachts at 2100z (GMT) and 0000z reveal how the Pools A and B moved during this time and how a third Pool, C, had entered the area by midnight. The curves above and below indicate the speed of the wind as recorded at these times by the yachts, and they indicate how the 'wind corridors' developed either side of the pools in almost all cases. In order not to make the diagram confusing, the situation at 1800z has been combined with that for 0300z.

Key to the observers:

BA *Black Arrow*,	FN *Finnibar*,	RV *Revolootion*,
BC *Battlecry*,	FT *Fluter*,	SD *Sandettie*,
BD *Blaue Dolphyn*,	HT *Hurricantoo*,	TM *Trumpeter*,
BV *Bonaventure*,	JP *Jan Pott*,	TR *Tiderace*,
EC *Eclipse*,	KK *Kukri*,	TS *Toscana*,
FJ *Firanjo*,	PN *Pordin Nancy*,	XA *Xara*.
	RR *Right Royal*,	

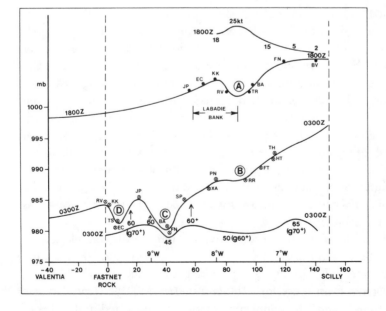

Fig 30.5b. As early as 7 pm (BST = GMT + 1 h) winds were rising in association with the development of Pool A. By the end of the night A had disappeared, B was all but rubbed out, but C and a new one D had entered the scene. However, by now the entire wind field was at least 45–50 kn. Thus the variations in the wind-speed curve are shallower than they appear in Fig 30.5a. Key as in Fig 30.5a.

Diagrams courtesy of Alan Watts' *Reading the Weather* published by Adlard Coles Nautical increased to 25 knots. By 2100 again five yachts show that this trough had remained, in the general fall of pressure across the whole area, but had deepened to some seven or eight mb below the general pressure gradient in the area. In association with this, the strongest winds developed with a mean of 60 knots gusting to 75 knots, and the first casualties of the night occurred. It is noticeable, especially on the midnight plot, that the strongest winds appeared mainly on the edges of these troughs – a situation which still obtained as late as 0300.

The meteorological reasons for these troughs are very difficult to assess and their nature even more mysterious. I have for want of a better description called them 'cyclonic pools,' but whether this is a good description, whether they were linear features stretched along the wind direction we cannot tell. The very fact that the yachts were strung out along a rhumb line, being on port tack into a generally south-west wind, meant that pressure readings outside this line were not forthcoming until later, when the fleet had been scattered.

The exact nature of the cyclonic pools on the night of the 1979 Fastnet race is not important for this text. The importance of their discovery lies in the light it throws on 'tales of the sea,' which sometimes seem unbelievable in the context of what is known about the meteorological conditions in the area where the apparently impossible weather occurred.

The loss of the British barque *Marques* sailing on the Bermuda to Nova Scotia leg of the Tall Ships race in 1984 is a case in point. She was some 80 miles north of Bermuda when she was overwhelmed in a matter of 45 seconds, and sank with only those on deck being saved – nine men in all. The rest were trapped below as she, to quote a survivor, 'fell on her side'. Then to quote Chris Phelan, a member of the British Sail Training Committee, she was 'driven under in less than a minute.' But driven under by what? It would seem by a couple of large but extremely localised waves which, according to crew members of the Polish sailing ship *Zawisa Czarny* nearby, struck the *Marques* across her bows. The first was just said to be a large wave, whereas the second was described as 'giant', after which the ship began to slip under the water. One of the survivors said that 'hurricane force winds hit us out of the blue'. Another said that the winds were the strongest he had ever experienced, about force 10 or 11.

The remarkable thing about the loss of the *Marques* was that the other ships in the vicinity only experienced relatively minor winds, and the strongest gusts reported were only between 30 and 40 knots, without a great sea running. The subsequent search for survivors was conducted in only moderate seas.

The weather map for 0600 GMT, which was just two hours before the ill-fated ship foundered, showed nothing unusual except that a ship close to the *Marques* gave a wind direction of north-west, while Bermuda's wind was south-west. On the strength of this the chart analyser drew in a trough, but there was still a large 'hole' in the isobars over the area where the *Marques* and the other attendant ships were. The reporting ship also gave a thunderstorm in progress at 0600 while Bermuda itself had light rain. Yet none of the survivors mentioned that a thunderstorm was in progress, or even that they had heard thunder near to their position at 0800. However, local thunderstorms indicated considerable instability in the local atmosphere, and one is led to ask if the hole in the isobars was as empty as the chart analyst thought. If there had been a fleet of yachts in the area at the time, would they possibly have recorded a form of 'cyclonic pool' such as was identified during the Fastnet race? It is the sheer lack of dense observational network over the sea that must miss many odd nasties as they form over deep sea areas.

In both the case of the Fastnet yacht race and of the *Marques*, we have certain conditions which appear to be necessary for the local enhancement of the wind and the sea, which could apply to many other survival situations at sea. Other related research shows that very large capsizing waves are generated by intense depressions which have cold upper troughs associated with them. The coldness aloft is a necessary condition for the development of strong convection currents, but it does not seem that these necessarily should have to produce thunder. No obvious signs may appear to the mariner that such conditions actually exist. However, it is more likely that they would occur in summer or over warm seas than over cold seas in winter. Nevertheless very cold outflows from the polar regions could produce the same effects in winter, which they have been shown to do in both the North Atlantic and North Pacific.

The survival conditions appear to occur on the southern flank of the parent depression; in the cases of the 1965 Channel Race gale and the 1979 Fastnet gale the hurricane-force winds occurred some 200 miles to the south-east of the low centre. In fact the weather maps for these two survival situations bear an uncanny resemblance. Each has a minor area of not particularly deep low pressure some hundreds of miles ahead of the low centre, the signs of a mass of relatively cold air aloft over and to the north of the depression centre, and the already-mentioned immensely strong winds 200 miles to the south-east of the centre. I am envisaging that the cyclonic pools were able to draw in to themselves winds of 10–20 knots, or more in some cases. If such winds are added to the general windfield of say 40–45 knots then sustained corridors of wind with speeds of 50–70 knots could be produced in some places on the windward sides of the pools. On the leeward sides the inflow would be opposing the general wind and so here wind speeds as low as 25 knots might be reported – which indeed they were during the height of the Fastnet storm.

These conditions may be much more prevalent at sea than previously thought, leading to locally enhanced wind speeds and attendant seas. For when winds are normal gale force or below then the frictional drag of the wind on the sea is such as to make the seas build up only over a period of time, which may be considerable. However, when winds get to near hurricane force, the frictional coupling between wind and sea surface is so strong that great breaking waves are generated almost at once. These waves have very steep faces. The speed of their crests exceeds that of the wave as a whole and so, just like shoreside breakers, they must curl over at the crest and break.

Wind at sea and at coastal stations

As is well known, the wind increases with height in the extreme conditions discussed above, so not only the position but the height of any anemometer readings must be taken into account. The standard height for wind observations is 33 ft (10 metres), which is luckily not far above, or below, the masthead height of an average yacht. When the 'effective height' of an anemometer is about this value, then the winds recorded are the mean ones to be expected in a yacht. However, the winds as plotted on the weather charts are the actual ones read at the time of the observation. Often this does not matter, but in the case of anemometers like the one on the Portland Bill lighthouse (now with an effective height of 155 ft) the height makes for stronger winds being recorded there than would be felt at 33 ft above sea-level. In this case a whole 25 per cent has to be subtracted from the Portland wind speed to make its value comparable with a level of 33 ft in a yacht skirting Portland, or farther out to sea, apart from local influences such as a wind corridor and the fact that at sea the wind is sometimes stronger than it is close to the land.

When we take the actual winds of the summer months of 1965 and compare them at Portland and Hurn Airport (inland from Bournemouth), both by day (0800–1700) and by night (1800–0700), as recorded by the anemographs (chart-drawing anemometers), then we find the following. With winds which blow in off the sea at Hurn (120–240 degrees) the Hurn wind needs to be multiplied by the following factors in order to get the wind offshore in the vicinity of Portland Bill. All the winds are reduced to the same standard 33 ft.

Wind force at sea when coastal station winds are onshore	Factor by which to multiply coastal station wind to obtain wind over the sea	
	By day	By night
4	1.1	1.7
5	1.3	1.6
6	1.4	1.8
7	1.3	1.6
8	1.3	1.6
9	no figures	1.6

It will be noticed that the factors are bigger by night than by day and that the winds of force 6 are the most affected by the friction of the land. These figures are representative of those by which a prudent skipper multiplies the wind he has in harbour (or obtains from a local met. office) to assess the wind outside, though in some harbours he may be so close under the lee of high land or buildings that the wind outside is difficult to guess at all.

A further complication is the wind direction. We can divide wind directions around a coastal station into three sectors: (1) winds off the sea (2) winds off the land, but excluding easterlies (3) easterlies. Of course, an easterly on the East Coast is a wind off the sea, but on the South and West Coasts these three divisions serve to separate the different winds.

Again comparing Portland and Hurn for winds which were between south-west and north-west (ie routed over land, but from the west), we find the following:

Wind force at sea when coastal station wind is routed over land	Factor by which to multiply coastal station wind to obtain wind over the sea	
	By day	By night
5	1.2	1.6
6	1.4	1.9
7	1.6	1.6
8	1.7	1.8
9	no figure	1.8

We see that the effect on the force 5–6 winds of flowing over land is not much different than flowing off the sea, but that the effect on the strong to gale force winds is greater. They are, in fact, slowed up more over the land than is the case with winds off the sea.

Finally the easterlies. Easterlies in Great Britain are notoriously variable and gusty and it must be pointed out that these figures are obtained from analysis of the mean wind taken over a series of full hours, so that gusts must be added to the figures.

Wind force at sea when winds are easterly	Factor by which to multiply coastal station wind to obtain wind over the sea	
	By day	By night
4	1.5	2.1
5	2.0	2.5
6	2.0	2.5
(Easterlies above force 6 mean speed are rare)		

The Meteorology of depressions

It is obvious from these figures that what may look on a weather map like a coastal station need not be unfettered by the local terrain. Cliffs may rise on one side or the other. Only a certain sector of direction may give a flat inroad for the wind. For example, the winds at Scilly and Lizard compare until the wind swings into the northerly quadrant (315°–045°) from which directions Lizard is sheltered. We have already mentioned the fact that not all anemometers are at the same height, and for real comparisons one really needs to know their exact location. An example was the anemometer on the top of the old Headquarters Building at Thorney Island in the middle of Chichester harbour. On the map it looked completely surrounded by nothing, but in actual fact its effective height of 40 ft was bedevilled by the buildings which clustered about it and more immediately by the trees which soften the view. In that position Thorney's winds were never representative of those in the open sea area, but the anemometer was later moved to the middle of the airfield well clear of all obstructions.

Only comparatively few anemometers whose readings appear on the charts are therefore representative of the wind offshore. Those on lightvessels will be, and those on lighthouses may be; the pictures in the pilots will often give a good idea of the extent to which they are well exposed. However, when winds recorded on the coasts do not measure up to those experienced (and accurately assessed) at sea there may also be other reasons, not the least of which is the sort of wind corridor effect described for the Channel Storm and for the 1979 Fastnet storm.

Other depressions

Let us now turn to the other depressions which have made the experiences of this book possible. They fall into three distinct groups, plus a small miscellaneous collection.

The first group can be called 'deep lows'. The weather maps (Fig 13.2 in Chapter 13) for 6–7 September 1957, are classic examples of a deep low. The centre below 970 millibars is really deep by summer standards and is very unusual for September, which according to statistics is the time of minimum wind over the southern half of Britain. This depression we find becoming a slow-moving whirl of air sweeping its fronts on out of its circulation, and then sticking. It stuck all the following week, as I have reason to remember, being in the act of introducing my non-sailing wife to the joys of a yachting honeymoon. The low was not the only entity which did not move that week. We decided to stay on our moorings in Wootton – and had a very restful week!

It is noticeable that many of the lows which produce blows are not many millibars below the magic number of 1000 millibars (normal pressure is 1013 millibars) and such depths need not be parts of depressions at all. But when the barometer falls to below 980 millibars, then that is a deep depression. This is not a definition of the term 'deep', but a practical figure which tends to fit the facts. Such pressures must be surrounded by dartboards of isobars as the pressure climbs up to the highs within which the lows exist. They must, therefore, produce blows and these are often long-drawn-out periods of strong if not gale force winds. As well as this unusual September gale, we find other examples of deep lows covering the Santander gale of 1948 (Chapter 5), La Coruña gale of 1954 (Chapter 9), and the Fastnet gales of 1957 (Chapter 12).

The latter had a reflection of the Channel Storm in it, in that the Saturday evening chart (10 August) shows another of those Bristol Channel depressions with the strongest of winds along the West Channel and a great no-man's-land of slack pressure and light winds up the Irish Sea. We can call such depressions, whether they track through the Severn pass or not, 'land-aggravated lows'. Their characteristics have already been covered at their worst in the opening paragraphs. It only remains to show how prevalent they are in the examples of this book. We find two examples in Chapter 6, the Belle Ile race of 1948 and the Wolf gale of the following year, while the Fastnet race of 1957, the Channel Storm of 1956 and the Fastnet gales of 1961 add to the list. There is one chart which is both deep and land aggravated and that is of the La Trinité gale of 1964 (Chapter 16). The rest are relatively shallow.

The third main division is the *multifront* system typified by the chart sequence covering La Coruña Race of 1960. To those brought up on the classic frontal-depression drawings found in textbooks, the double frontal system does not make much sense. However, weather maps as drawn by the professional meteorologists are seldom textbook and like them we must accept what nature gives us. Fronts separate air with significantly different temperature regimes and a front has the same name as the air behind it. For example, the air behind a cold front is colder than the air ahead of it. So if we start off in the cold between the lows on the 1800 (second) chart for 10 August 1960 (page 143) and move south we cross a warm front into warmer air over the Channel Islands and Brest. Then we cross another warm front into even warmer air over the Bay. Then turning westwards across the south-west Approaches we cross a cold front into cooler air and then another into cold air. As a depression whips up its wind energy by virtue of the difference in temperature between its coldest and warmest parts so such systems have a strong tendency to develop and blow hard.

We have signs of a similar situation in both the Channel Storm chart for Saturday, 28 July (Chapter 11), and the 1960 Bermuda gale (Chapter 15), where a low makes in towards an already waiting cold front and again the cold air to the north of it is drawn into the circulation and the energy of temperature contrast finds its way into a real blow. In the Saturday evening chart for 10 August 1957, covering the Fastnet Race of that year (Fig 12.2), the same kind of consideration obtains. The occlusion bending down across France separates cool and cold air, but behind a sharp cold front (from Cork towards the south-west) must come even colder air compared to which the air behind the occlusion is relatively warm. Whatever the synthesis of fronts that follow, this sort of chart signifies that air which is potentially cold is close by, and when drawn into the circulation of the already extant low must lead to its deepening at some later time. There are other examples of this principle in other weather charts, eg Belle Ile (Fig 6.2). Sometimes careful attention to shipping and other forecasts may suggest the multifront system, which is likely to lead to a real blow.

Finally there are other systems which do not fit these categories. For example, we find a curious pair of lows in a col producing a tightening of the isobars over the waters along the south coast in Chapter 2 page 17 but the main divisions fit nearly all the examples, except of course for the hurricanes.

Hurricanes

A hurricane is fundamentally different from the most potent of depressions and a cross-section of a typical one is shown in Fig 30.6. What makes a hurricane is not so much what happens at the surface (although that is important) but what happens at 6–8 nautical miles aloft. The incipient hurricane is probably not much different at the surface from any ordinary embryonic depression, but it breeds great deep cumulo-nimbus clouds over the warm oceans and these spread anvils towards the periphery of the developing system. This happens below the tropopause (the permanent inversion which divides the troposphere from the stratosphere) which is some 50,000 ft aloft over the tropics in summer. Such spreading at great heights leads to a rotation in the same direction as the winds nearer the surface. Thus the winds have a solid rotation in the same direction at all levels.

The depression, on the other hand, has upper winds which blow across the surface wind rotation and this has the effect of exerting a control on the build-up of the surface winds. There is no similar control over the winds of the hurricane, added to

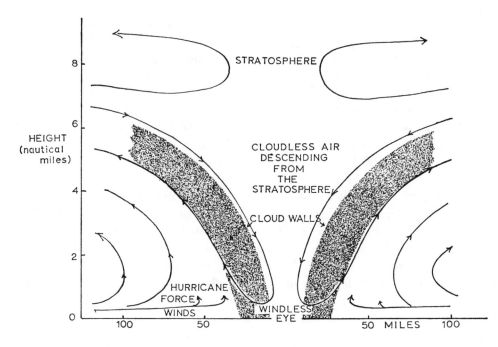

Fig 30.6. Cross-section through a typical hurricane. The way the winds descend in the eye and ascend elsewhere is characteristic of the hurricane

which there is the feeding of the cloud-free eye of the hurricane with air drawn down from the stratosphere. Depressions do not communicate with the dry air of the stratosphere in quite such a direct way and do not usually have a central area curiously and ominously clear of cloud, although the 0600 chart of the Channel gale sequence (Fig 30.1) shows that such depressions can have light winds (5 knots) in their centres and 60 knots only tens of miles away.

The general tendency of hurricane tracks, their main locations and the avoidance courses are all adequately covered in any Admiralty Pilot and do not need reiteration here. The all-seeing eye of the weather satellites has made the tracking of hurricanes very much surer than before, and the hurricane warning services of the U.S. Weather Bureau provide early warning.

When old tropical cyclones reach British shores, then they have been demoted to depressions, and this is not just a way of describing their diminished intensity. It also means that they have lost the direct link with the stratosphere that they had before and will now have winds aloft which, although strong, exercise control over the winds at the surface. For example, the Low D, which had been hurricane *Debbie*, lashed Ireland and West Scotland with winds approaching 90 knots in September 1961; and hurricane *Carrie* which sank the *Pamir* on 21 September 1957 in mid-Atlantic, finally crossed Cornwall and went into northern France on 25–26 September as a depression. Yet the full hurricane is most unlikely to ever be experienced even in the south-west approaches to the English Channel. It is worth noting, however, that *Carrie* was still considered to be a hurricane to within a couple of hundred miles of the Scillies.

Hurricane *Charley* of 1986 brought the British summer to an early end with some very strong late summer winds and a great deal of rain that lasted a long time. *Charley* came to a halt in the North Sea which meant the end of August and beginning of September were cold and unsettled. *Charley* was early but *Flossie* was more on time in 1978 bringing some very tropical weather to Britain before tracking out across the North Sea on 16 September.

September is the peak of the hurricane season in the Caribbean and of the 'old hurricane' season here. We do not get many old hurricanes, but if we do they are usually vigorous lows and should be allowed for especially off the western seaboard in September.

Tropical revolving storms, of which hurricanes are the most potent example, originate in the doldrums (between 7 and 15 degrees of latitude) and in the North Atlantic they may be observed by satellite to have started to develop as far east as the Cape Verde islands. In general they move off on a heading of between 275 and 350 degrees, but the most prevalent direction lies in the arc between 275 and 305 degrees. Somewhere around the 25 degrees north they usually begin to recurve away from the equator to travel north-easterly by the time they reach 30 degrees north. However many storms do not recurve but instead go ashore on the American mainland, where they rapidly lose their impetus.

Hurricanes do not usually move very fast. They start off at about 10 knots and do speed up a little, but rarely exceed 15 knots before they recurve. After recurving they increase speed to 20–25 knots and in rare cases have made 40 knots. While the above is the norm, it must be borne in mind that there are erratic paths which move equator-wards or even form loops. Such odd storms travel very slowly.

Typically at 200 miles from the eye you need not expect more than force 7 and even at 100 miles it is rare for winds to exceed force 8. Hurricane force is to be expected within 75 miles of the eye and within 50 miles gusts as high as 150 knots (175 mph) have been recorded. The best warning of the approach of a tropical revolving storm is via the hurricane warning service on the radio. However it is as well to itemise other criteria in case of radio failure or anomalous propagation conditions:

1 Make preliminary preparations if the barometer falls below 1002 mb and full preparations if it falls below 1000 mb.

2 Keep an eye on the sea surface for swell. Storm signs are a long swell, with frequency between two and five crests per minute. The storm lies in the direciton from which the swell comes.

3 You cannot outrun a tropical revolving storm. The outrunning swells will cut your speed to around five knots; and never attempt to 'cross the bows' of the storm towards the equator.

4 If the wind veers with time then you are in the dangerous semi-circle, and you must make as much speed as possible close-hauled on starboard tack. If the wind backs then you are in the navigable semi-circle and the action to take is to run off with the wind on the starboard quarter. If the wind does not change direction but increases in speed

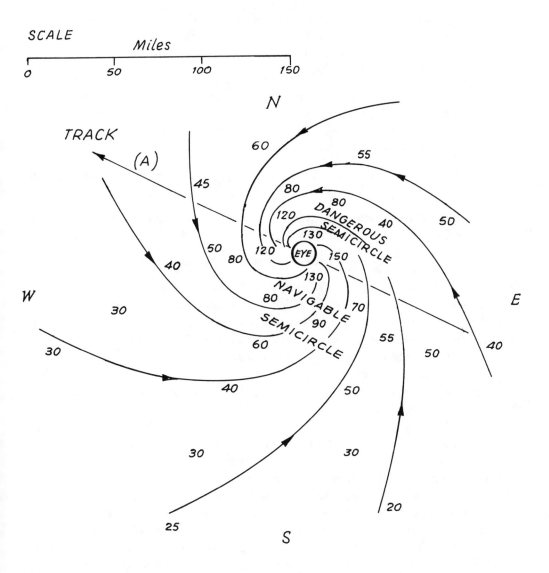

Fig 30.7. *Wind directions and speeds for a typical tropical revolving storm.*

with a falling barometer then you are in the path of the tropical revolving storm. However if the wind conditions are similar but the barometer is rising, you are astern of the eye. The reasons for this advice will become evident from Fig 30.7.

5 One way of finding the direction of the eye is to add 115 degrees to the true wind you have. For example if you lie on the track of the tropical revolving storm (A in Fig 30.7) your wind could well be northerly, so adding 115 degrees to North will mean that the eye lies on a true bearing close to south-east. When the total adds to more than 360 degrees then subtract 360 from the result.

6 Radar will observe the intense rainfall in the region of the eye, so will give its bearing.

7 When behind the centre and making way northwards, do not forget that the hurricane will probably recurve, and you could cross its track again at a later date.

Tornadoes and waterspouts

Tornadoes are another matter. Tornadoes in the British Isles have been, in recent decades, products of inland areas and form under intense thunderstorm areas, particularly where the air can get an upward push over rising ground. Once formed they may proceed out to sea, pick up a great writhing spiral of water and may deposit it in the cockpit of some unfortunate yacht. At least one liner at sea has had its crow's nest at 70 ft above the sea surface filled with water from a tornado storm spout. That was in the Atlantic. Cowes had a tornado in the last century, so it might happen again, although such things are rare.

Far more likely is the waterspout which will often form under vigorous cumulonimbus clouds over the sea. The funnel of the spout is not usually solid water and so it is not likely to swamp a yacht as a tornado storm spout may do. However the circulating winds can be temporarily dangerous and if spouts are sighted then it is prudent to shorten sail. Tornadoes and spouts come under the general heading of funnel clouds and the latter need not extend from cloud to the surface for their effects to be felt. I have only once experienced the effects of a funnel cloud and that was off a windward shore in landlocked waters. It was not a spout, as it came in an easterly wind off the land mass, having no time to pick up water. Still, the boom swung violently from side to side and the sails shivered and trembled as if they had the ague. The rotation of the wind turned us until we were heading on a course opposite to our original one, by which time I had the mainsail down. We then had time to observe the writhing pendant of cloud twisting down from the thundery cloud mass aloft and could watch it lose itself over the farthest shore and disappear.

Fog in heavy weather

Fog is not usually considered to be a phenomenon associated with heavy weather. However, the official limits of visibility for fog (ie below 1000 metres or 1100 yards) lead to fog limits being reached in rain, drizzle, and especially snow, even with force 6–7 wind. The rain needs to be heavy and the drizzle thick, in association with low amorphous cloud. This cloud is often due to the rain, which moistens the lower layers into which it evaporates as it falls, until they are sufficiently saturated to condense into cloud with no definable base. The most likely parts of depressions where such conditions can obtain is under active warm fronts and occlusions. Temporary fogs may occur under cold fronts, but they are of short duration – tens of minutes rather than hours. Snow with its high reflectance lowers visibility out of all proportion to the intensity of precipitation, and visibilities near or below fog limits will be the rule rather than the exception with snow showers if not more continuous snowfall. However, the showers pass and tend to conform to the idea that in high winds fog does not persist for very long. Only under the slow-moving parts of quasi-stationary fronts, and especially ones which were moving one way and have stopped and are now being pushed back the other way, as happens sometimes, will high wind and persistent fog be likely companions.

Poor visibility at mist limits (2000 metres or 2200 yards) still makes navigation by DR difficult and from this point of view mist might be said to define the limit below which a yacht's safety is impaired. These limits are likely in tropical maritime air (warm sector type air) whenever the cloud is thick enough to produce rain or drizzle and the air is routed over decreasing sea isotherms. This is usually the case with such air from the southern quadrants over the open sea or in the south-west quadrant when in more sheltered waters such as the English Channel.

Forecasting

What has been said up to now is hindsight. What about the immediate forecasting of the bad stuff blowing in to harry the yacht at sea?

The most potent early-warning sign is the banner-like cirrus bands which stretch across the sky and appear to converge towards the horizon. They are usually orientated from the north-west. If moon, stars or sun give a reference point, they can be seen to be moving very fast across the sky. If the very high cirrus can actually be seen to be in motion, even from the moving platform of a yacht, then the winds aloft are 80–100 knots or more. Such winds, and this cloud form, tell of a jet-stream.

A jet-stream is a high-speed wind river, usually some 6–8 nautical miles aloft, which is part and parcel of the development of vigorous depressions. Of course, every one knows that most strong wind comes with the onset of the circulation of a depression, and weather lore tells that the jet-stream cirrus banners signify the impending arrival of a depression of the stronger kind. Jet-stream cirrus must be assumed to mean that a gale is a real possibility some six to ten hours later.

Cirrus of some kind will always appear ahead of active and advancing warm fronts. These need not be part of a vigorous depression, but when the cloud thickens into grey banks of altostratus through which the sun struggles vainly not to be obscured, then some sort of front is on its way. If the barometer tumbles as well, then its rate of fall is related to the strength of the wind and, coupled to the cloud signs, state of sea, etc., will tell something of the wind to come. However, the Met Office will only issue a gale warning on a single reliable observation if there is a tendency of 10 millibars or more in three hours.

The barographs of the passage of the Santander Race gale of 7–8 August 1948 at

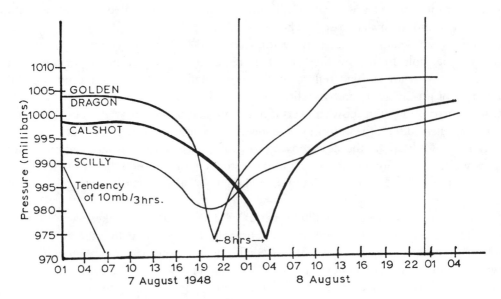

Fig 30.8. Barograms Scilly, Calshot and Golden Dragon during Santander storm.

Scilly, Calshot and at sea in *Golden Dragon* (Fig 30.8) show the relation between tendency of the barometer and wind. A line of slope, *10 millibars in three hours,* has been drawn on this figure and steeper gradient than this usually means severe gales are inevitable, if only temporarily, while lesser tendencies can mean gales, but need not necessarily do so.

At Scilly, some 40 miles north of the line of passage of the centre, only once did the tendency come close to the magic 10 millibars in three hours and that was between 1700 and 1800 GMT. The wind at Scilly at this time was a mean of about 26 knots, gusting occasionally to 35-40, but by 1900 it had risen to a mean of 30 knots, with occasional gusts to 44 knots and one gust to 48 knots.

At Calshot (200 nautical miles eastward and eight hours later 'as the low flies') the barometer plummeted, at a gradient of 10–12 mb/3 hrs between 0200 and 0400 on the Sunday morning and rose more rapidly, at gradients between 20 and 10 mb/3 hrs from 0400 to 0600.

The barograph from *Golden Dragon* shows a quite phenomenal fall, whose gradient for half hour attained the astonishing rate of 45 mb/3 hrs, and a slower but none the less exceptionally rapid rise. The position of *Golden Dragon* with reference to the passage of the low centre can be gauged from the account in Chapter 5. It seems inevitable that the report of force 11 from the Clan liner was correct, even though the very abruptness of the barometric tendency meant that such wind strength could not be sustained for long. By contrast, the far more gentle barometric tendencies associated with the Channel Storm (Chapter 11) were followed by winds of force 10 and 11 for an extended period. The old adage of 'short forecast soon past' still applies!

However, tendency is no help to a yacht at sea in foretelling what is going to happen, say, four hours ahead – you only know what the barometer has done, not what it will do. It may sink fairly steeply and look as if it will eventually knock its own bottom out, but then the fall is arrested as the trough goes through and nothing very bad occurs.

The barometer at sea is not a very reliable guide, because seeking shelter is a question of decision several hours ahead of the real blow. It must be the sky which commands the most attention.

The sky which goes rapidly through the well-known sequence of cirrus, and cirro-stratus (haloes about sun or moon) into alto-stratus (sun or moon as if seen through ground glass) presages a fairly short and not necessarily violent blow. The long build-up of these clouds signifies a bigger depression and so more prospect of a gale later. To which must be added the proviso that the ill-famed secondary depression can blow up shortly and sharply and give a real force 7–8, or more even in the summer months, as evinced by the Santander Race gale already referred to.

The cloud sequence cirrus, cirro-stratus, alto-stratus then thickens into nimbo-stratus and with this cloud comes the rain. There will be pannus (scud) beneath it and a sight of the latter will tell of the immediate onset of rain and therefore when to reach for oilskins. Normal warm fronts (which the above cloud sequence accompanies) will rain gently, but persistently, and the rain will become moderate, but not normally heavy, before the cloud-line clears. It usually does so with a low ragged sweep of dirty cloud lying athwart the wind. The latter must be watched for, since a sharpish veer will come with the passage of a well-defined warm front, as well as some gustiness. Once the front has passed the cloud may be just as low as before, but rain will usually give way to drizzle. It becomes humid and muggy, for the warm tropical part of the depression has arrived. The wind will now be blowing at its strongest and nasty unpredictable eddies must be allowed for, especially at night in what would seem to an otherwise homogenous and not very gusty airstream. These will be most severe with winds above force 7–8, because the size of the eddies in warm sectors only increases to catastrophic proportions when winds are above this strength. What happens is that super-sized eddies bring very fast wind to the surface from much higher up and may deposit it with full fury on a yacht already hard pressed by the wind anyway.

The next hazard, after it is realized that the influence of the warm sector has been felt for a reasonable time – and that time may not be very many hours – is the arrival of the cold front. However, cold fronts have a habit of producing strongly ascending currents, and what goes up must come down somewhere else. Some currents often come down ahead of the front and this temporarily breaks the cloud, so that the cold front proper is presaged by a temporary clearance. Any real clearance in a cloud type should always be suspected. It may mean worse to come.

The 'worse to come' of a cold front is the chance of very strong veers of wind, accompanied by equally vicious gusts. The veer may be 45 degrees or more and a craft beating on port tack might then be put in irons. If on the starboard tack there is a chance of being blown down as the gusts strike from the beam as the front goes through. Yet it is easier to make up to meet the veering wind on starboard than to bear away on port. With storm sails set this might not be important, but with less strong winds, when carrying as much canvas as practicable, it could be a useful point to bear

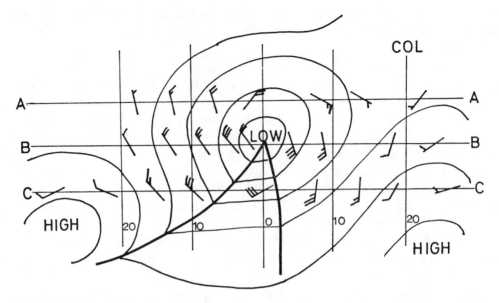

Fig 30.9. Typical wind strengths and directions when a depression moves to the south (A), passes directly over (B) or passes to the north (C) of the observer. A full barb = 10 knots mean speed; half barbs = 5 knots. Figures against vertical lines are typical times in hours ahead and behind the line of lowest pressure for a big depression moving at 30 knots from west to east.

in mind especially at night. There will also be the nasty cross-sea which comes when a sharply veering wind spreads across an already established seaway. However, careful observation of the longer, lower, faster-moving waves, running under the immediate seaway can foretell this wind change.

The sharp cold front can be recognized by the curious way in which the lower cloud elements cross one another. It may well look as if quite low clouds are scudding in from the right of the existing wind. This is no illusion, as they show the new wind direction and give a warning of an impending veer.

The cold front usually brings a climb in the barometer. The stronger every aspect of a depression is, the more likely it is that the barometer will plumb down and shoot up again.

This rise should steady out after a couple of hours, and if it fails to do so, then any improvement is likely to be short-lived. The real clearance usually comes after a couple of days of gradually rising barometer, as the low moves away. During this post-cold front stage the winds will normally be cool, squally north-westerlies, accompanied by frequent showers at first, but gradually moderating in all their aspects. Very strong and persistent winds accompanied by strong gusts and squalls can occur in this phase of the passage of a depression. The barometer may rise, but the wind increases. This should be immediately interpreted as sure evidence that the invisible depression, which was thought to have passed, is sticking or has gathered some more strength and is deepening again. The incredible wind which sank parts of the East Coast of England and much of Holland and flattened four million trees in Scotland on the night of 31

January/1 February 1953, arose in the corridor between a deepening depression slow moving in the North Sea and a galloping tendency to high pressure over the eastern Atlantic.

A yacht alone on the ocean does not normally have the benefit of weather maps, so any prognosis based on such maps will find low priority in the list of gale-forecasting methods. It may, however, be worth noting that when the charts before the big gales are studied, then a particular pattern appears. It is typified by Fig 30.1 of the Channel Storm sequence. Over the area ripe for the most trouble is a slack area of low pressure – a sort of bed into which a rogue low is about to tumble. This prepared position with a developing depression to fill it is the most suspect of situations and the sight of such a chart before putting to sea might make it worth thinking twice about going unless it were imperative. We find a somewhat less obvious example on page 88 in connection with the Cork Race. The low that caused the trouble ate its forebearer and used the energy to flay the life out of poor mariners sailing upon their lawful occasions.

It is almost the same again for the 1957 Fastnet. The Saturday-evening chart on page 116 shows a low deepening into an already low pressure area. Again the chart (page 126) for 1800, 7 August 1961 is classical. The path of the depression to deepen is already laid. It can hardly help deepening.

Wave lows and troughs

While great primary lows have secondaries in their circulations which provide sharp deteriorations, there are even smaller entities of the same general kind called 'wave lows' or just plainly 'waves'. A 'wave' is a small incursion of warm air into cold or

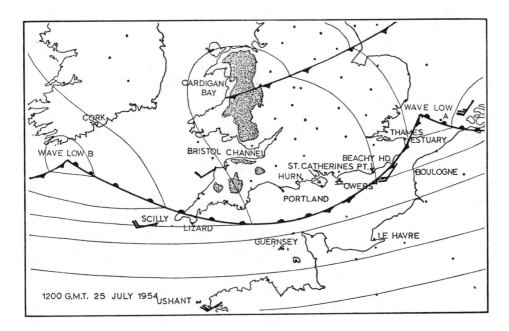

Fig 30.10. Wave lows 1200, 25 July 1954.

cool air. It mostly occurs along a cold front where it produces local deteriorations, but not usually any very strong winds. Meteorologically 'waves' are like very small depressions and the chart for 1200, 25 July 1954 (Fig 30.10), which is reproduced here and precedes the ones shown in Chapter 9, Fig 9.1 shows two waves within a relatively small distance of one another. As it requires something over 1000 miles of undisturbed front between two such waves before they both have a chance to develop into proper depressions, one or both of these embryonic lows is due to die. Before that happens, they will prevent clearance of rain and low cloud along a cold front by pulling it up northward again, rather than letting it go through to the south. They can be understood by analogy with a rope secured to a stanchion and flicked into waves along its length. The frontal waves move along the front like those in the rope, sometimes at 60 knots.

A quick glance at the chart two hours later 0600, 26 July, (Fig 9.1), shows why neither of these waves developed. All the depression-forming energy was being imbibed by the bigger wave to the west of Ireland. Both A and B died out, leaving another, not normally mentioned, entity – the sympathetic trough – to hold the stage. The inquiring mind might legitimately ask how between 1800 and 0600 next morning the wave west of Ireland grew in occlusion towards the main low south of Iceland. Often, when a more than average-sized wave looks at a parent low like this, then it grows a great line of showers and heavy rain, with squalls of some severity, in sympathy with its parent and between the two centres. Such a line can only be described succinctly on the charts by the symbol for an occlusion, so one appears. The actual weather under such troughs is often of the foulest, with cold, driving showers of rain and hail, accompanied by much gustiness. The original tip of the wave may deepen considerably, as in this example, and produce a crop of very close isobars over the areas to the south of it, also as in this case.

Then there are thundery troughs, mainly to be found in the sailing months and particularly in the worst of the summer months – August! The sooner we forget about August as a holiday month the better – everything meteorological gets worse in August – and better in September!

Perhaps the worst of thundery troughs is the damage they do to morale, expecially at night. There, heaving down over the horizon, is a fine display of lightning which stretches across the wind – what there is of it – and you have to go through it. The wind begins to build in those fitful gusts which is worrying. Then as the thunderclouds arch overhead the wind may whip up to 30 to 50 knots in the worst cases. The air becomes very cold and torrential rain roars on the coachroof and the thunder cracks and rolls while the lightning illuminates the details of the satanic clouds above. But once it's over, that's that. Normally the gusts and squall pass in half an hour to an hour and the general increase in wind is nothing more than to force 4 or 5. It is not really a heavy weather phenomenon.

Yet at the same time there are old, extra-tropical cyclones which gain new leases of life on approaching the shores of the British Isles. Hurricane *Debbie* was one and the Wokingham storm of 9 July 1959 was another. In that storm people were washed out of their homes in more than one place, while hail the size of tennis balls fell elsewhere.

Similar storms are rare but one occurred across parts of Wiltshire and Gloucestershire on 13 July 1967. There was a trail of damage due to violent winds and hailstones were up to three inches across. Like most such storms this one was inland but the South Coast hail storms of 5 June 1983 occurred across the creeks, harbours and coastwise waters from Weymouth to Chichester Harbour. Here the hailstones were variously described as between 'walnut and golfball size'. The sudden squalls capsized a large number of yachts in the Solent region, one reason being that the wind savagely reversed direction through 180° in a minute. Such storms are very localised and while the south of the Isle of Wight remained dry, at Southsea over an inch of rain was recorded. In all, six distinct storms were involved moving from west to east, each separated from the next by an interval of just over an hour. These examples give an idea of what is to be expected from severe local storms but in my experience no two are alike in their patterns of rain, hail and electric effects. Continuous lightning which keeps the cloud-tops illuminated is a sign of such storms and they are usually products of September. The line of the Caribbean hurricane jingle 'September Remember' is also true on a lesser scale in Britain, especially in the south-west approaches.

Other effects of thundery situations on the wind are to be found in my book *Wind and Sailing Boats*, including the curious fact that gusts due to thundery activity can occur tens and hundreds of miles from the actual thunderstorm area.

Rain on the sea

Finally a word about the effect of rain on the sea. There is no doubt that the meteorological conditions accompanying heavy rain calm the sea. We say the rain beats the sea down. Well, perhaps it does, but that is not all the story.

It is often noticeable that when rain starts fairly suddenly, then the wind also increases. This is easy to explain, because descending currents are pulled down with the falling rain, bringing stronger wind from higher up on to the suface. The more intense the rain, the stronger the downdraughts. They are called precipitation downdraughts and they account for the very big gusts under thunderstorms and the still sizeable gusts under other less-violent shower clouds. Likewise on a lesser scale they explain why the rain and the wind come together.

They may also explain the efficiency with which heavy rain appears to beat down the sea, because not only do we have the weight of water falling on to the surface but there is also the weight of air between – which is not inconsiderable, covering as it does a much larger area than that of the drops.

In the foregoing, some of the meteorological aspects of heavy weather have been discussed. They are not exhaustive and no seaman worth his salt can get by without some knowledge of the face of the sky and what it portends. Knowledge coupled to experience is the recipe for keeping clear of trouble.

31 WIND WAVES

Sheldon Bacon

Lay out a length of rope on the ground; take hold of one end, and give it a sharp flick. A travelling bight, a wave on the rope, will move away from your hand.

You have put energy into the rope and the energy has travelled down the rope as a wave, but you have not caused any material to be transported from one end of the rope to the other. Waves (to a first approximation) transport energy, not matter. This chapter aims to describe the ways in which energy is transferred from the wind to the sea; how the sea then behaves; how the sea is further influenced by the land, the sea bed, by currents; how the sea surface changes within years, over many years, and over the whole of the globe.

Understanding wave activity

In order to understand and describe waves, it is necessary to set out the measures used to characterise individual waves. Fig 31.1 shows a few waves together with a fixed post to be used as a stationary reference point.

The wave length, L, is the distance (usually measured in metres) from crest to crest, or equivalently, from trough to trough. Wave height h (in metres) is usually taken as crest-to-trough height. Wave period T (in seconds) is the 'duration' of each wave: for example, the time between crest A passing the post in the diagram and the following crest B passing similarly. Wave steepness is defined as the ratio of wave height to wave

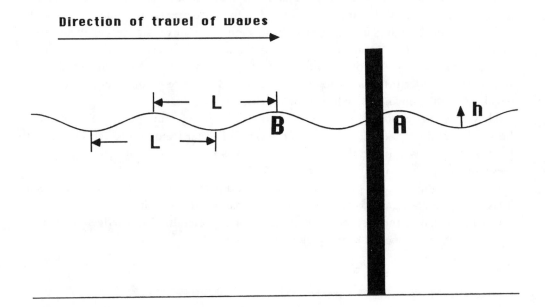

Fig 31.1

length, $\frac{h}{L}$. The term wave slope is reserved for the slope of the sea surface at any point: ie, the sea is flat (zero slope) in the troughs of the waves, and has greater slope on the wave faces. The speed of individual waves is called, correctly, the phase speed, to distinguish from group speed, which describes a complication arising when there is not just one wave train of one wavelength, but many, all of different lengths travelling over the sea. This will be dealt with later. The units of measurement to be used throughout this article are nautical miles (nm) for distance, knots for speed, metres for depths, and metres and kilometres (km) for lengths.

Gravity and capillary waves

For a wave motion to exist in a medium such as water, there must be a restoring force which responds to an initial displacement or disturbance of the medium. The two forces responsible for the propagation of wind waves on the sea are gravity, which produces waves of length from about 2 cm up to about 1 km, and surface tension, which produces ripples (also called capillary waves), of length less than 2 cm. Consider two mechanical oscillatory systems, as analogies for gravity and capillary waves: a rigid pendulum, and a trolley tethered by a spring. Set the pendulum swinging, and the trolley bouncing back and forth. When the pendulum has swung to its furthest point of travel in one direction, the downwards pull of gravity acts to set it moving in the other direction. When the trolley has reached its furthest point of travel to the left, the pull of the stretched spring acts to set it moving to the right. Similarly, gravity waves are kept moving through the influence of the force of gravity, and capillary waves by the 'springiness' of surface tension. The term 'restoring force' is used because the forces are continually pulling or pushing the substance or object in motion towards its equilibrium (or rest) position.

Other forces

The title of this chapter refers to the generation mechanism responsible for the classes of waves to be discussed here. Some other mechanisms, the effects of which will not be discussed in detail here, are: sub-sea earthquakes, which produce tsunamis (also incorrectly called tidal waves); the relative motion of the sun and moon about the earth, whose gravity fields produce the tides, which are planet-scale waves; the rotation of the earth itself, which produces an apparent force, the Coriolis force, responsible for phenomena such as the rotation of weather systems; the motion of vessels through the sea, producing wakes; and the combination of tidal currents with suitable topography, which can produce bores.

Wave movement

We know that a wave on a rope makes the rope move up and down as the wave passes; but sea wave motion is more complex. To a first approximation, the water particles travel in circular paths. At the top of the crest, the water is travelling straight forward; at the bottom of the trough, straight back. At some point on the front and rear faces of the wave, the water travels straight up and straight down. This motion is illustrated in Fig 31.2.

Figure 31.2 is a sequence of five diagrams A–E which follows one water particle (represented by the black dot), and shows how the passing of a wave can be 'supported'

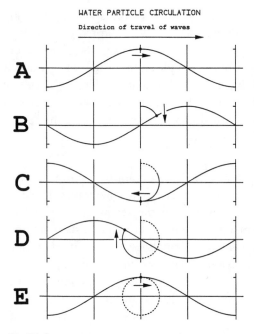

WATER PARTICLE CIRCULATION

Direction of travel of waves

A

B

C

D

E

Fig 31.2.

by water particles following circular paths. This sequence represents the passing of one wave, so the actual time taken by the particle in moving from its starting position (A) round the circle and back (E) is the wave period, T. In practice, water particles do not circulate in closed loops but advance slightly with each cycle, a drift which can produce a current of a fraction of a knot.

If the water particles at the surface are moving, the particles below the surface must also be moving, but to a lesser extent, so that there is a smooth transition between the region that is in motion at the sea surface and the interior of the sea, where the influence of the waves does not reach. In fact, the drop-off motion with depth is very rapid, as Fig 31.3 shows.

The depth to which the wave 'penetrates' depends on the length of the wave, so the vertical axis of the graph is in fractions of a wavelength, for any wave of length L, and the horizontal axis is the ratio of particle speed at depth to surface speed. Consider, for example, a 20 m long wave: at 5 m depth ($\frac{L}{4}$), the particle speeds are 20% of their surface values, and at 20 m, only 0.1%. This becomes very important in regions of shoal water: if there is sea bed in the way at a depth where the wave 'wants' to influence water motion, the wave can 'feel the bottom'. This is discussed later.

Wave measurement

A wave is described completely by four measures: its height, length, period and speed. The average yachtsman will tend to estimate most accurately

a) the length of shorter waves, since they can be compared with the length of the yacht, and

b) the period of longer waves, through the 'feel' of the rise and fall of a yacht.

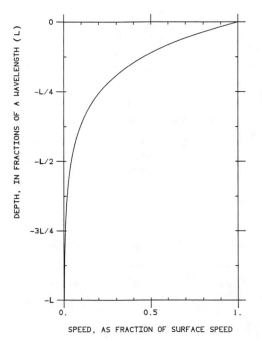

PARTICLE SPEED AT DEPTH

DEPTH, IN FRACTIONS OF A WAVELENGTH (L)

0

-L/4

-L/2

-3L/4

-L

0.   0.5   1.

SPEED, AS FRACTION OF SURFACE SPEED

Fig 31.3.

Estimates of height are notoriously inaccurate, and I suspect that an appreciation of wave speed is generally lacking. It is therefore interesting that, for deep water gravity waves, three of these four quantities are uniquely connected. Such a wave can take any height (up to a certain limit), but for any one length of wave, there is only one possible speed and one possible period. Long gravity waves travel faster than short waves, in deep water. These waves are 'dispersive', because if many waves all of different lengths are produced in one place, which occurs during a storm, the longer waves which are generated will travel out of the region of generation faster than the shorter ones. That is, they spread out or disperse according to wavelength. This is the origin of swell: fast-moving, long waves which may arrive in advance of an approaching storm. Fig 31.4 shows the relationship between wave speed, wave length and wave period.

To translate length to period (and vice versa), read from one axis to the other: a wave of length 100 m has a period of about 8 seconds. Use the curve to turn this figure into a speed: that same wave travels at a speed of about 25 knots. A wave of 10 m length (a similar length to a 30-foot yacht) will travel at about 8 knots, whilst a 500 m swell travels at 55 knots, with about 18 seconds between crests!

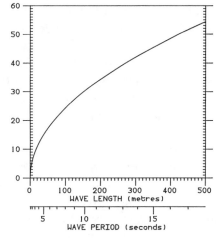

DEEP WATER WAVE SPEED

Consider now the difference between water particle speeds in a long wave and a short wave, both waves being of the same height. Let the long (fast) wave have a period of 20 seconds and the short (slow) one 5 seconds, and that both waves are

Fig 31.4.

one metre high. In twenty seconds, one long wave and four short waves pass. In each wave, the water particles are following circular orbits of one metre diameter; in the short wave, the water particles are completing four orbits to the long wave's one. So the water particles are travelling four times faster in the short wave than in the long wave, even though the long wave itself is travelling faster (twice as fast, in fact) than the short wave. At any point on a wave, a yacht will be pushed in the direction of motion of the water particles, so on the crests the yacht will be forced in the direction of travel of the waves; in the troughs the yacht will be forced in the reverse direction, and it will experience the maximum upwards and downwards accelerations about half way up the front and half way down the rear faces of the waves, respectively. Also, since water particle circulation speeds decrease with depth, the difference between surface and sub-surface speeds can cause a yacht, particularly a deep-keeled one beam-on to very steep waves, to tilt over towards the crest on both sides of a passing wave, which can produce a sudden and severe roll. Deeper draft yachts would be more affected in this way than shallow draft ones, which would follow the surface, being less influenced by the deeper, slower water circulation. This suggests that deep draft yachts should neither lie nor sail beam-on to high steep seas.

Sea state and wave behaviour

In order to make the description of the sea surface approach more nearly the 'real thing', it is necessary to know how to describe the sea state when there is not just one length of wave present, but very many. The simplest example of this occurs when a light breeze blows over a sea which is calm but for a long swell: there will be ripples generated which ride on the swell. This can be expressed graphically as a sum as in Fig 31.5.

This 'arithmetical' process can be worked backwards, and for any sea surface. The wave field can be decomposed into its many component wave trains, all of different frequencies and travelling in different directions.

The above example of swell plus ripples was chosen because the two component wave trains are of very different frequencies. However, if two wave trains of similar frequencies are added, another important phenomenon arises. The example shown in Fig 31.6 represents the result of adding two wave trains, one of period 10 seconds, the other of period 11 seconds, travelling in the same direction. Both are of height one metre. If, when superimposing the two wave trains, the crest of one coincides with the crest of the other, the resulting height is two metres, but if a crest coincides with a trough, the two waves cancel to give zero height. The most important consequence of this process is the production of a modulation of the resulting wave train: there is now a long-wave envelope around the short-wave basic wave train which groups the individual waves into 'packets'.

Now the two component wave trains will travel at different speeds according to their lengths: the longer train faster than the shorter one. This will cause the envelope to travel forward as the crests and troughs move in and out of phase, so that the envelope has a forward velocity of its own, called the group velocity.

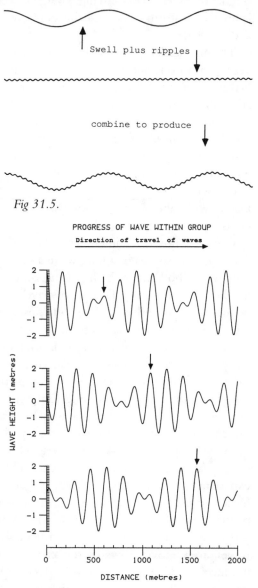

Fig 31.5.

Fig 31.6.

Within each resultant packet or group, the individual waves rise at the back, move forward through the group and finally disappear at the front of the group, while the group as a whole is moving. The three diagrams in Fig 31.6 show the resulting wave groups. The second diagram in Fig 31.6 shows how the waves and groups have moved to the right, a few seconds after the first diagram; and the third, similarly, is the sea surface a few more seconds later. One wave is labelled for us to follow its progress.

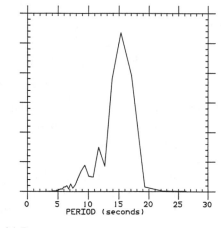

Fig 31.7.

If such a group of waves were travelling at 5 knots but the individual waves within the group were travelling at 10 knots, a yacht running downwind at 5 knots could find itself stuck in the middle of the group experiencing rougher seas than average as it is passed by the waves within the group.

In the open ocean, waves generated by the wind have different wavelengths and directions; they do not usually appear as regular trains of waves, but as part of what appears to be a random sea surface. The most important means of describing the sea surface in reality is the energy spectrum, which shows how the waves which make up the sea surface are distributed with frequency. The example in Fig 31.7 is from Ocean Weather Station Lima, measured by a ship-borne wave recorder mounted on the weather ship *Cumulus* approximately 300 nautical miles west of Ireland on 26 December 1988.

The spectrum shows the distribution of energy in the wave field; but before considering the details of the spectrum, it is necessary to describe what the spectrum represents as a whole. How does it relate to wave height? The more energy there is in the spectrum (the larger the area under the curve), the greater the wave height. The commonest measure related to mean wave height is called significant wave height, H_s an old definition of which is the average of the highest one-third of the waves. This quantity is numerically very close to a simple (and unexaggerated) visual estimate of 'average wave height', and it is useful as such because it is very close to what the practised eye would estimate as the average height of the waves in some sea state. Significant wave height is now defined (by the statistician) as four times the standard deviation of sea surface elevation, which is related to the area under the spectrum. The Lima oceanic example is of a rough sea state, of $H_s = 10.8$ m (about 30 ft).

The spectrum is often strongly peaked; that is, most of the wave energy and most of the waves are concentrated around one or two periods for each particular spectrum. For Fig 31.7, the peak periods are about 14 to 17 seconds, or 300 to 450 metres wavelength. There is considerable energy down to about 5 seconds, (40 m) and a little at 3 seconds (15 m). Of course there are also waves present shorter than these, down to ripples, but they are of too short a period to be measured by ship or buoy.

No mention has been made of directionality; a wind blowing over the sea generates waves which propagate not just directly downwind, but also a little to either side; there is a degree of spreading of the waves. The interference of waves travelling in slightly different directions is responsible for waves being short-crested.

The surface of the open sea

Idealised and simplified, some of the most important basic aspects of the behaviour of wind waves have been described; they can all be seen or felt in any sea. This section of the chapter will describe the ways in which waves grow (acquire energy) and decay (lose energy), the fundamental factors which control the height to which they can grow, and the rate at which that height can be achieved, beginning with the simplest condition of the unrestricted open sea and progressing to consideration of restrictions imposed by the proximity of land, shallowness of water and the presence of currents.

Wave growth

Your yacht is far from land, becalmed on a glassy sea. A start time is chosen, and the wind is switched on. The speed of the wind is U, measured in knots; what now happens to the surface of the sea? If the wind (of whatever speed) were a perfectly steady and uniform airflow and the sea surface were perfectly flat and smooth there would be no waves generated. The friction between air and sea would drive a current in the sea, but some irregularity in the wind is needed to disturb an initially smooth sea surface. The real wind is gusty, blustery and turbulent; there is an atmospheric boundary layer about 100 m high above the sea surface within which the air flow is directly affected by the presence of the sea surface, which induces turbulence on scales from millimetres to tens of metres by means of surface friction 'tripping' the wind (also, the air/sea temperature difference can induce circulation in the wind up to 1000 m or more above the surface). The turbulence is carried along (or advected) by the overall flow of the wind, and it is this turbulence which starts up the wave motion on the sea by resonance; that is, wave components are excited with the same phase speed as the passing turbulent wind motions. Resonance is the 'pushing' of a mechanical oscillatory system at its natural frequency, like pushing a garden swing. This is the initial phase of wave growth, and it is relatively slow. Once there are some waves present a very rapid phase of wave growth takes place, in fact exponential wave growth, due to positive feedback: the more waves there are, the rougher is the sea surface. This generates more turbulence in the wind, which produces still more waves and so on. However, this process cannot continue forever. For any given wind speed, low or high, a state of saturation is reached when energy input to the waves by the wind is balanced by dissipation of energy from the wavefield, by waves either travelling out of the generation region or by breaking. The sea state reaches a steady state which will not change until the wind changes its speed and/or direction. If the wind should increase, the waves will continue to grow; if it decreases, dissipation will remove more energy than the new lower wind speed is injecting, and the waves will become lower. If there is a change in direction, potentially the most dangerous possibility in extreme conditions, a new sea will develop on top of the declining old one resulting in a cross-sea, and it will develop rapidly since the sea is already rough and the wind turbulent.

This simple description, from calm sea to saturation, omits the phenomenon of

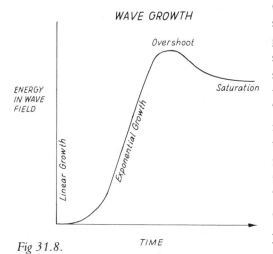

Fig 31.8.

over–shoot, which occurs before the sea settles into saturation. The exponential growth phase does not tail off to the steady state; it gets 'carried away' and for a short time, energy is pumped into the waves which the waves cannot hold. This pumping is actuated by the complex mechanism of interactions between the waves, so that energy is moved from a-round the peak frequency of the spectrum to both higher and lower frequencies. The whole process of wave growth is depicted schematically in Fig 31.8.

Breaking waves

In extreme conditions, the overshoot phase is the most dangerous for yachts. The waves will have grown over-steep and the wave field will have to lose energy by wave breaking at a greater rate than during the saturation phase. There are two distinct classes of breaking waves: spilling and plunging breakers. The former is more common in the open ocean, and occurs when the crest of a wave does not form a jet, but 'topples' down the front face of the wave. The latter, plunging, is more common on shelving beaches and occurs when the crest of a wave forms a plunging jet in front of the wave. With regard to wave breaking, it is physically impossible for a stable travelling wave to exist whose steepness is greater than about 1 in 7; if it tries to get steeper, it will break either by spilling or plunging, and the water speed in the jet of a plunging breaker can reach up to three or four times the wave phase speed. Wave shape changes as the wave becomes steeper: long, low swell waves are rounded as in the example in Fig 31.5, but as steepness increases they become more peaked at the crest.

Wave height

The time taken for the sea to become fully developed depends both on the wind speed and the existing sea state: the higher the wind speed, the longer the time taken (and the greater the ultimate height); but if there is swell present (as there nearly always is, in the open ocean) interactions between waves increase the growth rate, which makes the growth rate hard to predict. However, in the absence of swell, D = U is the time for a sea to become fully developed: U is wind speed in knots and D duration in hours. For example, for U = 20 knots, D = 20

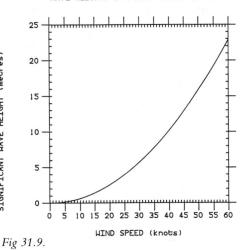

Fig 31.9.

339

hours. If swell is present, this is an upper limit on sea state development time. Fig 31.9. shows significant wave height for fully developed seas (for D greater than U) as a function of wind speed. For example, after a day of 20 knot winds, the significant wave height is 2.5 m.

Fig 31.9 is a useful estimator of wave heights, but its value is limited by factors mentioned above, and also by the fact that the wind speed and direction are rarely constant for the durations required for seas to become fully developed. Fig 31.15 (in the section on climate change) shows some examples of sequences of wave height measurements from Seven Stones LV in which can be seen the rapidly changing nature of the sea state as it responds to the passing of weather systems (northern hemisphere depressions). The typical pattern of wind change is to start with 'light' south or south-westerlies followed by increasing of wind speed with veering to westerly then north-westerly, and then decreasing wind speed. This process takes from three to six days, as can be seen with unusual regularity in the February 1974 data, with seas taking about half that time to reach their greatest height. However, some specific examples show that seas can get up very high very quickly: in the storm around 16 January 1974, the sea goes from 4 m to 10 m in only 15 hours, and during the Fastnet storm (14 August 1979) it rose from 2 m to 8 m in just 9 hours. A more recent example is the 22 to 23 November 1986 storm which took a day to grow from 3 to 11 m.

Your yacht, with the aid of these winds, has progressed to within range of land. Many more influences on the formation and propagation of waves now become apparent. These effects can be considered in four groups: offshore winds, shallow water effects, the effects of currents on waves and, briefly, sub-surface influences.

Offshore winds

These occur where fetch is a major limiting factor. The reduction, by the presence of land, of the distance over which the wind can raise the waves restricts the height to which they can grow. If a northerly wind blows off a south coast, the waves in shore will be lower than those further out. Fetch-limiting extends a very long way out from a coast, but the effect is most noticeable much closer to the coast. For example, a westerly wind of 20 knots is blowing off an east coast: the regime of fetch-limited waves ends about 150 nautical miles from the coast, after which the waves take the same height as for fully developed waves. However, at 60 nautical miles from the coast, the waves have reached 75% of their final height, so one need not progress far from the shore in such circumstances to lose much of the benefit of the sheltering. For a wind of 40 knots, fetch-limiting ends about 500 nautical miles out. Fig 31.10 shows how wave heights increase with fetch for offshore winds of 20, 40 and 60 knots up to 100 nautical miles offshore.

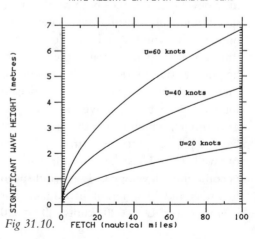

Fig 31.10. WAVE HEIGHTS IN FETCH LIMITED SEAS

Even under these circumstances, duration can be a major limiting factor, as the chances of getting a steady 40 knot wind blowing in the same direction over 100 nautical miles for some hours is extremely small.

Shallow-water effects

It has already been shown that water circulation associated with wave motion extends to a depth in water of $\frac{1}{4}$ to $\frac{1}{2}$ of the length of the wave. When a wave is in water shallow enough for it to feel the bottom, it is slowed down. This reduction in speed means that the wave also becomes shorter and therefore steeper. As the depth decreases, the wave is slowed and steepened further, ultimately breaking. This is a fundamental difference between deep-water waves and shallow-water waves; the speed of the latter depends not on their wavelength but on the depth of water in which they are travelling. Shallow-water waves are non-dispersive: waves of all lengths travel at the same speed. Short, slow deep-water waves must get very close to the shore in very shallow water before they are slowed enough to steepen to breaking; but long, fast deep-water waves (like oceanic swell) can reach considerable heights as they are slowed. For example, Hawaii is good for surfing since it is a coral atoll with no continental shelf on which swell can break, so they break close to the shore on the coral reefs. Conditions which make a lee shore dangerous present not only the danger of being driven aground in itself, but also that of destruction by the violence of the inshore seas.

Shallow water is not only encountered in the run-up to shores; shoal waters and areas of banks will generate wave breaking, but this is not the only result of wave slowing; there are also directional effects. Refraction is the change of direction of propagation of waves (light waves passing through a lens, for example); water waves are refracted by their travelling at different speeds in different depths of water, faster in deeper water, slower in shallower water. Waves are slowed and turned as they pass into shallower water; they are turned into the direction of decreasing depth. Waves approaching a shore at an oblique angle, end up rolling in parallel to it. Waves in a complex region of shoal waters such as the Thames Estuary or the Friesian Islands in the North Sea will be continually refracted by the constantly changing depths, resulting in confused cross-seas. As examples of refraction, the synthetic aperture radar images (Figs 31.11 and 31.12) taken in 1978 from the US satellite SEASAT show (Fig 31.11) easterly propagating swell curving round shallow water off the southern end of the Shetland Islands, and (Fig 31.12) the effect, in the southern North Sea, of the Goodwin Sands, the South Falls, Sandettie and other banks on the surface waves. The sketch accompanying Fig 31.12 identifies these banks. Scale bars are shown on these figures, but as an idea of scale, the Shetland image is about 13 nautical miles 'tall' and the North Sea image about 30 nautical miles 'long'.

The effect of currents on waves

Currents include coastal tidal streams (oceanic tides are generally quite weak), so-called western boundary currents (the best known are the Gulf Stream, Kuroshio and Agulhas) and wind-driven currents. Waves are affected by complex interactions with currents, which have a familiar effect on the sea surface: waves travelling in the same

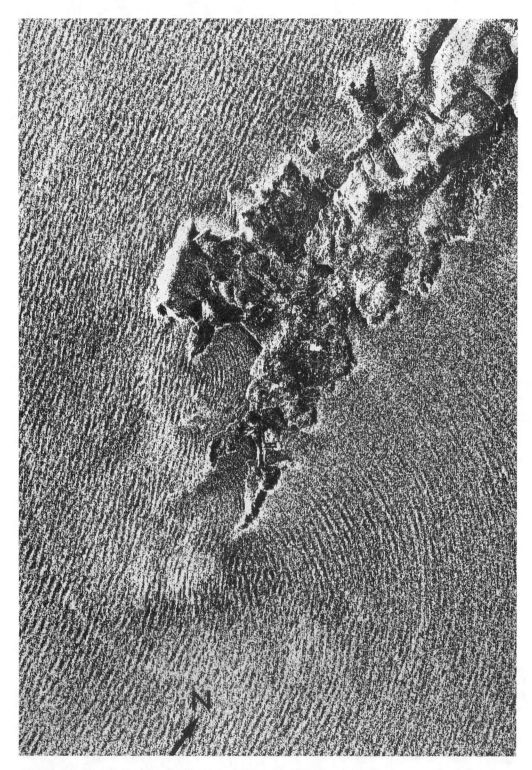

Fig 31.11. A synthetic aperture radar image showing the refractions of waves as they pass the shallower waters of the southern end of the Shetland Islands.

direction as the current are flattened, those opposing the current are steepened. If waves are travelling into an area of opposing current of increasing strength, the waves will steepen until they break, and a very choppy sea will result. If the current is sufficiently strong, the waves will not be able to propagate against it, leaving the downwind area of fast current untroubled by wave motion, an effect used in hydraulic breakwaters. This is the reason for some regions of very fast-flowing water being strangely smooth, and can be seen off Hurst Castle at the western end of the Solent in central southern England during the west-going ebb when a westerly wind is blowing.

Sub-surface influences

Yachtsmen often report that the character of the sea surface over the continental shelf is different from that in the deep ocean, that surface waves over the shelf are sometimes steeper than those in the ocean, and that the water itself is differently coloured. Now there are many classes of wave motion which can move energy around the oceans other than surface waves. It is likely that one of these classes, called internal waves, is responsible for these effects. It is well known that oil, which is less dense than water, floats on top of water. The oceans themselves are layered in this way because the density of seawater is governed by temperature and salinity, and these properties change with depth (and location) in the oceans. Put simply, a less dense, warm layer of water about 50 to 200 m thick floats on top of the rest of the ocean's waters, which are cooler and more dense. It is possible for waves to propagate at depth along the interface between these two layers; and in fact, the best regions for generating these waves are the continental shelf edges. Internal waves travel much more slowly than surface waves (up to only one or two knots); and they are much longer than most

Fig 31.12. A synthetic aperture radar image of the North Sea showing the influence of shallow water banks on surface waves.

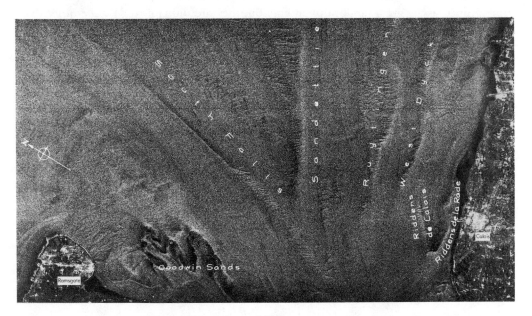

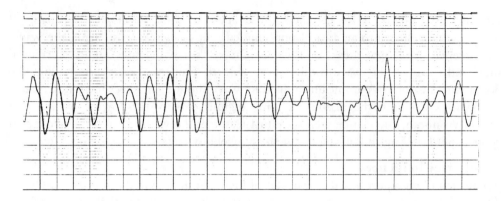

Fig 31.13. OWS Lima chart record of waves predominantly of period equalling 10 to 12 seconds (wave length from 160 to 230 metres).

surface waves (from around one to fifty km). They are responsible for two effects of interest. One is that they cause mixing of the upper layers of the ocean with deeper, nutrient-rich waters. This mixing promotes plankton growth. These tiny plants occur in great abundance in the spring and give colour to the water. This is a different effect to colouring by river-borne sediments or stirring of shallow-bottom material by surface wave action. The other effect of internal waves is that they give rise to surface currents which are 'banded' over the same scales as the lengths of the internal waves producing them. Surface waves entering such a region can be alternately 'stretched' and 'squashed', with appropriate changes in steepness, as they travel across the affected area.

Wave climate

So far the discussion has centred on individual waves and the sea state measure called significant wave height. It is also useful to know about all waves around at any one time. Since waves are a random phenomenon, they are difficult to describe other than by probability, so this section will describe the statistics of wave distributions. This is

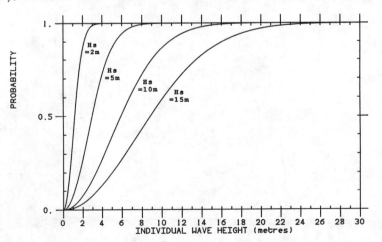

Fig 31.14. Probability of seeing individual waves within a sea state of given significant wave height.

considered in two parts: short-term statistics, meaning the types of waves found 'within' a sea over a short period when the sea state is approximately constant (usually taken as three hours) for which a value of significant wave height can be calculated; and long-term statistics, meaning the behaviour of the sea state (as measured by signficant wave height) over periods of time from months to decades. With knowledge of these probability distributions, it is possible to predict or to estimate extremes: that is, the likely highest wave conditions in a given length of time.

Short-term wave statistics

Fig 31.13 is a 4-minute-long extract from a chart record of sea surface elevation from Ocean Weather Station Lima, recorded at the same time as the spectrum in Fig 31.7. It illustrates that, as well as the range of wave lengths/periods found in any sea state (as described by the spectrum), there is also a range of wave heights. It is usual to describe this range using a probability distribution, either a mathematical one or one derived from data which shows, out of all waves occurring in a given period of time, the chance of seeing individual waves either greater or less than a given height. Useful ways of interpreting such information are in terms of probability (p) levels, such as the 10% (p = 0.1), 50% (p = 0.5, or median) and 90% (p = 0.9) levels which give the proportion of waves which will be lower than the associated height. The distributions of wave heights within a sea state of given significant wave height are shown in Fig 31.14. for H_s = 2, 5, 10 and 15 m, and the curves show individual wave height versus

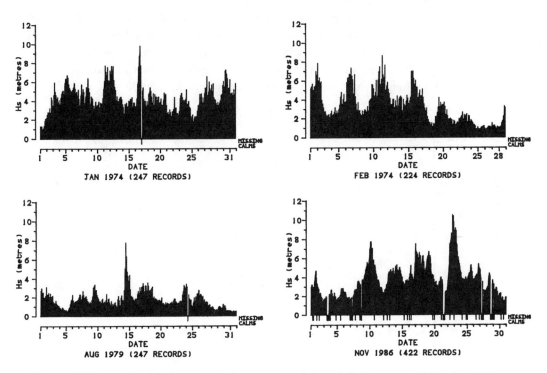

Fig. 31.15. Seven Stones Lightvessel significant wave height records for January and February 1974, August 1979 and November 1986.

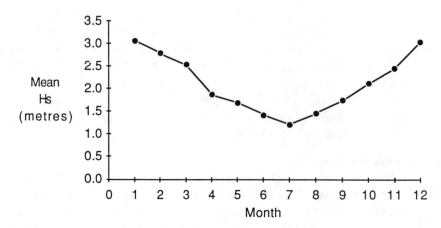

Fig 31.16. Seven Stones Lightvessel data for significant wave height, 1962-86.

probability. The probability values mean the chance of waves lower than a certain height being encountered.

The figure is of limited use because the curves are based on a distribution derived for a narrow-band sea, meaning one with a highly restricted range of frequencies present in the spectrum. In the ocean, a broad range of frequencies are present; nevertheless, these distributions serve as fair approximations to and illustrations of the wave height distributions found in reality. As examples of the use of probability levels, with the H_s = 10 m curve, 50% (p = 0.5) of the waves will be lower than 5.9 to 6.0 m and 90% (p = 0.9) will be lower than 11 m (which also means that 10% will be higher). Note that in a sea state of some value H_s, approximately 85% of the individual waves will be of height H_s or less, but also that appreciable numbers of waves (15%) will be of height greater than H_s. The distributions of sea waves are long-tailed: that is, most waves are of low or average height, but there will be some waves in a wide range of above-

Fig 31.17. Seven Stones Lightvessel annual mean significant wave height.

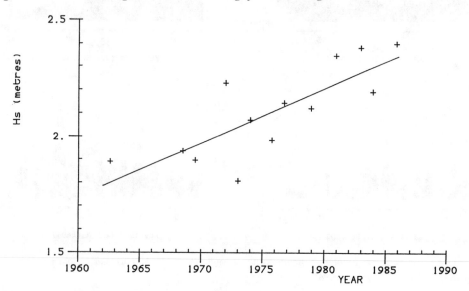

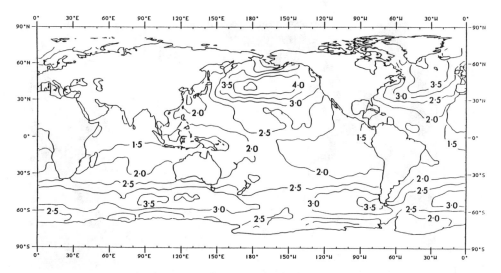

Fig 31.18. GEOSTAT global mean significant wave height for January to March 1987 contoured in metres.

average height. In fact, the most likely highest individual wave in three hours will be approximately 1.8 times H_S in height, so if $H_S = 10$ m, you are likely to encounter an 18 m wave during a three hour period.

It is important to remember that it was stated earlier that significant wave height is close to visually estimated mean wave height. It can be seen that it really is the high waves in the sea which heavily influence one's estimate, and that there is a large proportion of lower waves present, together with the higher waves which are relatively fewer in number.

Long-term wave statistics

Wave conditions are continually changing. As an illustration, Fig 31.15 shows measurements of significant wave height taken once every three hours from two particularly stormy months (January and February 1974), together with August 1979 and November 1986, at Seven Stones Light Vessel off Land's End, the south-western tip of the UK. The passing of severe storms can be seen in the rising and falling of the sea state; the average significant wave height over the whole of each month is 4.4 m (January 1974), 3.4 m (February 1974), 1.8 m (August 1979) and 4.1 m (November 1986), with individual measurements of significant wave height varying between about one and eleven m.

As well as varying within months with the passing of weather systems, the sea state varies from month to month throughout the year, from winter storms to relatively calm summers, with intermediate transitional conditions during spring and autumn. Fig 31.16, again from Seven Stones LV data, shows this annual cycle using the overall average value of significant wave height for each calendar month: the January value (month 1) is the average of all January values from all years, etc.

This timescale of climate variation is caused in the North Atlantic by i) the poleward summer migration of the mid-latitude atmospheric front between warm tropical and

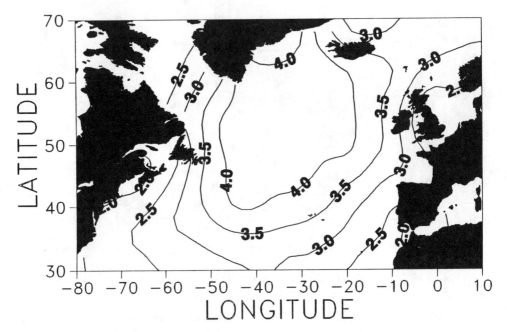

Fig 31.19. Mean significant wave height (m) GEOSTAT December 1986–January 1987.

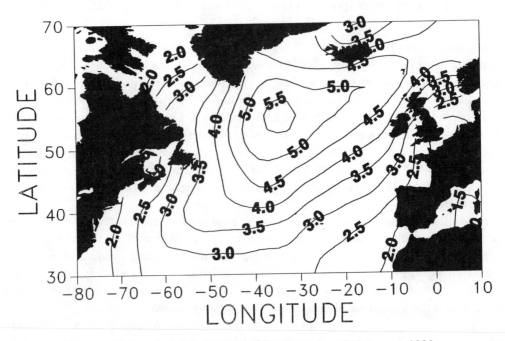

Fig 31.20. Mean significant wave height (m) GEOSTAT December 1988–January 1989.

The Channel Lightvessel whilst on tow to her station in December 1978. At the time the towing vessel was hove-to off the Lizard in an easterly force 9-10. Photo: Ambrose Greenway.

cold polar air masses, which determines depression tracks which run, for example, over mid-Europe in winter but over Iceland and further north in summer, and ii) the summer warming of the Arctic, which decreases the mean wind strength of mid-latitude westerlies. However, changes in wave conditions are not limited to the monthly or seasonal; mean conditions also change over years. The Seven Stones LV data set is well suited to illustrate this, being (currently) the longest instrumental wave data set in the world, covering 1962 to 1986 with some gaps in recording, and comprising some 50,000 three-hourly H_s measurements and incorporating 13 complete years of data with many incomplete years. Fig 31.17 shows the average value of significant wave height calculated for each of the 13 complete years.

Over the timescale of from three to five years, the average wave height varies considerably, from 2.2 m in 1972 to 1.8 m in 1973, for example, as European weather responds to global weather changes. Underlying these large inter-annual changes, however, there is a long-term change in mean condition at Seven Stones LV; the straight line on Fig 31.17 is the best fit straight line to the data, and it shows a 25% increase in severity of wave conditions between 1962 and 1986. Other evidence derived from both instrumental measurements and visual observations suggests that the wave climate in the whole of the North Atlantic has been increasing in mean severity by over 1% per year, from the earliest available data (early 1950s) to the latest (1988). We are starting to uncover changes in wave climate which are not understood; maybe wind strengths have increased, or depression tracks changed. This is part of a hitherto unknown scale of variation in wave climate. It also gives the lie to the 'it was worse in my day' comment.

As a footnote to the illustration of seasonal changes (Fig 31.17), this type of cycle is not relevant to all areas of the globe. For example, areas under the influence of monsoon conditions will have two monthly mean wave height maxima per year, coinciding with maximum monsoon winds; and the Southern Ocean, which is influenced by globally circulating westerlies, only shows a weak annual cycle with relatively little difference between its 'winter' and 'summer'.

Measurements of wave conditions have been based mainly on instruments mounted on ships or buoys and producing time series of values of sea surface elevation at a fixed position; but the advent of satellites equipped with radar altimeters and synthetic aperture radar promises a vast increase in the quantity and spatial coverage of wave data in the future. As examples of the potential of this technique, Fig 31.18 shows the distribution over the whole world of mean significant wave height as measured by the American satellite GEOSAT over three months January to March 1987. Particular features to note are the effects of the northern winter in mid-latitudes in the North Atlantic and Pacific Oceans, and even during the southern summer the belt of high significant wave height circling the globe at about 60° S (the Southern Ocean).

The next two figures (Figs 31.19 and 31.20) show how mean conditions in the North Atlantic have changed over the past three winters, 1986/87 to 1988/89, when wind and wave conditions have actually been worsening steadily, as can be seen in the measured significant wave height values (but wave heights over part of the north Pacific have decreased over these years).

Note the uniformity of the distribution of wave heights in the central North Atlantic in Fig 31.19, where average heights are all around 4 m, compared with Fig 31.20 the maximum mean height has increased to 5.5 m together with evidence of well-established storm tracks leading from south of Greenland to the Faeroes.

Unusually high waves

A term commonly used to describe unusually high waves is 'freak waves'. Having some appreciation of the distribution of wave heights, it can be seen that very high waves are a normal aspect of the behaviour of sea waves and most 'freak' waves encountered by yachtsmen should more correctly be termed extremes. This section will attempt to describe both.

Extreme waves

There is a difference in meaning between probability levels in short-term and long-term measurements: in the former case, the 90% level means 90% of the waves are below the associated height; in the latter, 90% of the time the sea state is lower than the associated significant wave height. This meaning is applied in the useful concept of the return value: to illustrate, the 50–year return value of significant wave height is defined as that value which will be exceeded on average once in 50 years. This quantity is used in offshore engineering. For Seven Stones LV, the highest measured significant wave height is 11.1 m, from a storm in October 1982; but the worst value likely in 50 years is 13.6 m. The corresponding value for the mid–Atlantic is over 20 m. From south to north in the North Sea, the 50–year return values of H_s run from 8 to 16 m. Within these sea states, you are likely to meet individual waves nearly twice as high.

Freak waves

These may be defined as waves of a height occurring more often than would be expected from the 'background' probability distribution. One area with a particularly bad reputation is the Agulhas Current of South Africa, where high north-east-going waves from the Southern Ocean run into the strong south-west-going current, with dramatic effect. Another rather unrepresentative example is that of the waves produced by the Lisbon earthquake of 1704, as they ran into shallow water. Extreme waves and freak waves are by definition very rare phenomena, however. The greatest danger faced by yachtsmen is that of steep-to breaking waves, which need not be of exceptional height to roll over a modern, light-displacement yacht. This is particularly so if associated with such meteorological conditions as produce cross-seas, such as the passing of fronts in a deep depression, or from the circulating winds of a revolving tropical storm; a group of such waves can be particularly dangerous.

Conclusion

It sounds obvious, but a yacht at sea is subject not to average conditions, but to prevalent conditions; and for any given average value, conditions worse than average are prevalent about 50% of the time. A glance at almost any weather map of the North Atlantic, for example, will show a web of isobars, fronts, highs and lows. The sea underneath the weather at any one place will be a mixture of the sea's response to the current weather, the left-overs of its response to the 'previous' weather, plus any swell propagating in from elsewhere. There is no substitute for regular, accurate weather information, either from broadcast forecasts or the surmises of an experienced observer. Safety at sea can only, in the first instance, be a matter of avoiding such areas as produce dangerous conditions, by the use of good, recent climate data. If caught out, it is then up to the yacht and its crew.

32 STABILITY OF YACHTS IN LARGE BREAKING WAVES

Andrew Claughton

Causes of capsize

What causes a yacht to capsize? Sailing dinghies and lightly ballasted day sailers such as a J24 can be laid flat and capsized solely by the pressure of wind in the sails. In larger yachts the nearest equivalent of this is the broach whilst under spinnaker. In a bad broach, the mast can be pressed down as far as the water but once the heeling influence of the spinnaker is removed, the yacht recovers to an upright position. Experience shows us that, in flat water, gusts alone cannot capsize a yacht. Even encountering high and steep waves the story remains the same. The action of wave slope in heeling a dinghy or day sailer may assist the wind in producing a capsize, but a conventional yacht's stability is such that it cannot be capsized by even the combined action of wind and waves, no matter how high or steep.

It is breaking waves that cause capsize. If the yacht is caught beam-on to breaking waves of sufficient size, then the exaggerated steepness of the breaking wave front, coupled with the impact of the jet-like torrent of the breaking crest, will knock the yacht down to a point where the mast is well immersed. At this point, the yacht's fate is decided by its stability characteristics; it will either return to an upright position or carry on to an inverted position, where the boat may remain for some time until another wave disturbs the yacht sufficiently to flip itself upright. If the wave is high enough, or the encounter with it is timed appropriately, then a full 360° roll will be executed.

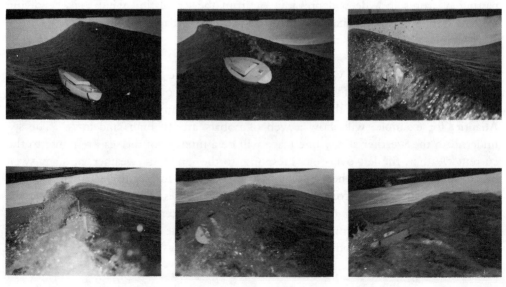

A fin keel parent model under test showing beam-on capsize.

How big do breaking waves need to be to cause this type of behaviour? Unfortunately, the answer is 'not very big'. During the model tests that were carried out to investigate the problem, when the breaking wave was 30% of the hull length high, from trough to crest, it could capsize some of the yachts, while waves to a height of 60% of the hull length would comfortably overwhelm all of the boats we tested. In real terms this means that for a 10 m boat, caught in the wrong place, when the breaking wave is 3 m high, this presents a capsize risk, and when the breaking wave is 6 m high, this appears to be a capsize certainty for any shape of boat. The word breaking is stressed because it is these that present the danger; big waves in themselves are not a problem.

As shown in the photo below left, the model tests were performed in waves that broke all along their crest at the same time, unlike the waves at sea where short lengths of crest break as the wave systems interact. Once the breaking crest at the point of impact is as wide as the boat is long, then its full effect will be felt.

How can capsize be avoided?

The simple answer to avoid capsizing is to avoid breaking waves. This does not necessarily mean staying tied to a mooring, but rather in avoiding certain sea areas in wind or tide conditions where breaking seas may be thrown up. For example, to help their small boat fishermen avoid beaking waves, the Norwegian authorities define certain no-go areas as part of their weather forecasts.

Taken a step further, even if caught out in extreme conditions of wind and wave, a technique of avoiding the breakers can be employed, but on a more local scale. During the 1979 Fastnet race, many yachts were able to keep sailing, and actively pick their way through the waves, avoiding the breaking part of the seas, much as a surfer keeps to the unbroken part of the wave by tracking across its face. Once the boat is to one side of the breaking part of the crest, the danger is over and even delaying the moment of impact until the breaking wave has dissipated some of its enegy will reduce the capsize risk. The wave is at its most destructive at the point of breaking and immediately afterwards. Active sailing also keeps the boat from being caught beam-on to the seas, which is its most vulnerable position. The risk is that a mistake in steering might cause a broach which results in the boat being beam-on to the waves. This technique does, however, need a strong and competent crew to execute for long periods of time. It is, nevertheless, a well established and successful technique for dealing with heavy weather.

As was demonstrated by crews' experiences during the 1979 Fastnet, it is not always possible to avoid capsize situations. Due to crew fatigue or plain bad luck a yacht, especially if short-handed, may encounter a capsize or knock down incident. The research carried out in the wake of the 1979 Fastnet race has been aimed at evaluating what features of hull design contribute to a safer yacht in survival conditions.

So far, I have written in general terms about stability, but we cannot go much further without explaining in more detail the physical mechanisms that keep a sailing yacht the right way up, and how things behave once the mast is below the horizontal.

Fig 32.1 shows a typical righting moment curve; this describes the variation of

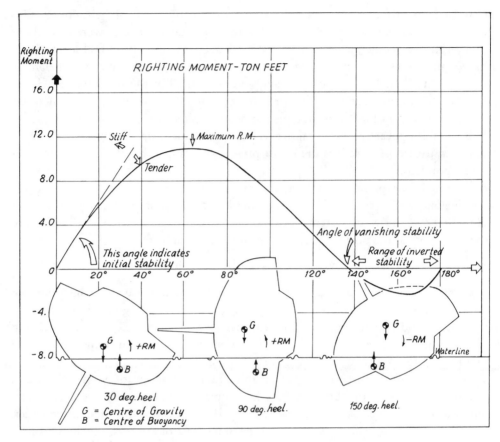

Fig 32.1. Typical righting moment curve.

righting moment as heel angle increases. All of the yacht's weight can be assumed to act vertically down through the centre of gravity, whilst the buoyancy force opposes this through the centre of buoyancy. The righting moment is the torque generated by the increasing misalignment of the yacht's centre of gravity, which remains fixed on the hull centreline – unless the crew sit out – and the centre of buoyancy which shifts to leeward as the yacht heels.

Intuitively one can see that in the normal sailing range of up to an angle of say 45°, an adequate righting moment to resist the heeling moment of the sails can be achieved either by a wide beam, so that the centre of buoyancy shifts outboard more, or by a low centre of gravity, so that it is further away from the centre of buoyancy. Whichever way the designer does it, a yacht has to have an adequate righting moment at 30–40° of heel to carry its sail properly. Some boats are stiff, i.e. the righting moment rises quickly with increasing heel angle, whilst others are more tender, i.e. with a slower rise of righting moment. In the latter case such boats very often require the crew to sit out to produce an adequate righting moment, by shifting the centre of gravity outboard.

At the point of maximum righting moment, the centre of buoyancy is as far to leeward as it is going to get. It then starts to move back to the centre as heel angle increases further. This means that at 90° the righting moment can be quite low, hence the ease with which the broached yacht can be held with its mast close to the water by the flogging spinnaker. A few more degrees past this point and the centre of gravity and centre of buoyancy are in line again, but the wrong way up. This is the angle of vanishing stability (AVS). At this point the boat is balanced in unstable equilibrium like a pencil on its end and can fall either way, ie back to upright or to end up floating upside down. At 180° heel (ie upside down) the two centres are in line, and this is another stable position unless the boat is fully self-righting. Before the hydrostatic forces can act to turn the boat the right way up, some external force must push the boat back past its angle of vanishing stability.

To calculate the righting moment curve for a yacht, the position of the centre of buoyancy is calculated from the hull drawings, using a computer. The position of the centre of gravity is determined by physically inclining the yacht a few degrees and measuring the heeling moment required.

The ability of a yacht to recover from a breaking wave encounter depends on the hull and coach roof shape. This is not only for its influence on how the breaking wave affects it, but also for its influence on the shape of the stability curve.

The conclusions about capsizing are based on the results of model tests carried out by the Wolfson Unit at Southampton University. The tests were carried out using free running models in a towing tank 60 m long x 3.7 m wide x 1.8 m deep. The breaking waves were generated by using computer-controlled wave makers, and fans provided a full scale 40 knot wind over the test area. The behaviour of different hull shapes will be described in the context of the models tested at Southampton. The results from these tests were complementary to those obtained from the parallel Sailboat Committee of the Society of Naval Architects and Marine Engineers / United States Yacht Racing Union study carried out in the USA.

The three basic hulls tested were a traditional yacht form, a typical modern 'fin and skeg' yacht, and a derivative of the 'fin and skeg' yacht modified to give it higher freeboard with no coach roof. From these three parent hulls, a further six forms were derived, each identical in all respects to its parent with the exception of beam – one narrower and one wider for each type. Models were then constructed from these line plans to represent 9.75 m yachts at 1:13 scale. The line plans of the 9 models are shown in Fig 32.2.

The different characteristics of the hulls allowed individual design features to be evaluated in relation to three aspects of behaviour:

1 Hydrostatic performance, angle of vanishing stability and stiffness.
2 Response to the impact of a breaking wave.
3 Influence on controllability, i.e. how design features can help an active sailing approach in avoiding a capsize.

Beam

The beam variation tests were done on the wide and narrow versions of the fin keel parent hull. The righting moment curves for the three hulls shown in Fig 32.3 are calculated with the centre of gravity at the same distance above the hull bottom and demonstrate the strong influence of beam on stability and stiffness. The widest yacht is the stiffest and has the highest maximum righting moment, but it also has a very large range of inverted stability. By contrast the narrow yacht has virtual self-righting ability but would be hopelessly tender because of the flatter slope of the early part of the curve.

In the capsize tests, both the parent and wide hull forms could be inverted by a breaking wave of height 40% of hull overall length (LOA) whilst the narrow hull form suffered only a 120° knock down and recovered. However, a 55% LOA wave caused all the models to execute 360° rolls when caught beam-on to the wave. One of the factors influencing the behaviour of the wider beam boats was the immersion of the lee deck edge as the boat was pushed sideways by the breaking crest. This dipping of the side deck appeared to produce a tripping action that the narrow boat was able to avoid. Running before the seas, the wide hull proved quite difficult to control and did not show any more willingness to surf ahead of the wave than its narrower sisters.

Freeboard

A comparison was made between the behaviour of the low freeboard model with a typical coach roof, and the high freeboard model with a flush deck. Both models

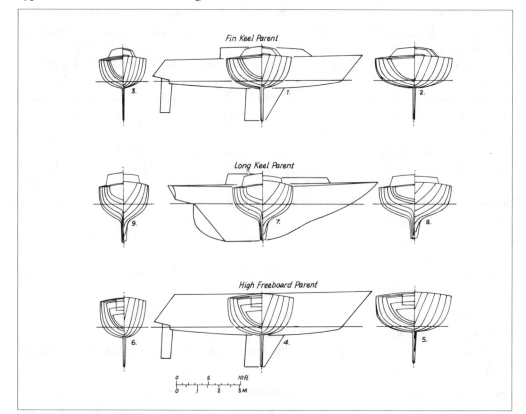

Fig 32.2. Body plans of standard series yachts. Length overall 9.75 m (32 ft).

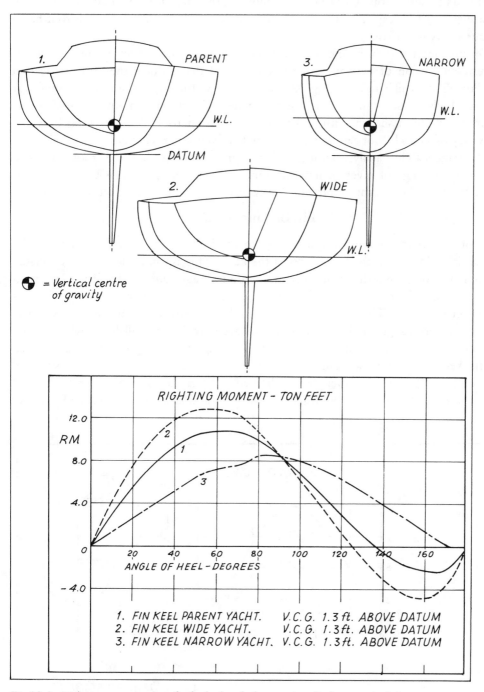

Fig 32.3. Righting moment curves for fin keel yacht beam series; displacement is 4.5 tons.

exhibited the same propensity to capsize, showing that high freeboard does not increase capsize risk. Once capsized, the yacht with the lower freeboard and coach roof had a greater ability to self-right.

On studying the righting moment curves of the two forms it became apparent that it is the contribution of buoyancy of the coach roof that reduces the area of inverted stability. This is illustrated in Fig 32.4 which shows the righting moment curves for the two hulls. Also of interest is Fig 32.5 where the righting moment curve of the traditional yacht is shown with and without its coach roof, demonstrating how the buoyancy of a coach roof can increase the angle of vanishing stability.

The static stability analysis indicated that a further increase in coach roof size could eliminate the range of inverted stability completely (a concept used with great success in the RNLI lifeboats), thus rendering even a very light and beamy craft self righting.

Fin or long keel?

One of the most obvious design developments in the last 20 years has been the substantial reduction in the lateral area of the keel. This aspect of design was evaluated not only by comparing the traditional and modern designs but also by fitting extension pieces fore and aft to the fin keel so that the keel area was approximately trebled as shown in Fig 32.6. This produced no discernible improvement in capsize resistance, and, more surprisingly, only a marginal improvement in controllability when sailing downwind. A similar result, at least as far as capsize resistance goes, was found when the fin keel and traditional design were compared at the same weight and position of the centre of gravity. Neither design showed a discernible superiority, although, of course, the traditional design with its narrow beam and larger coach roof had a higher

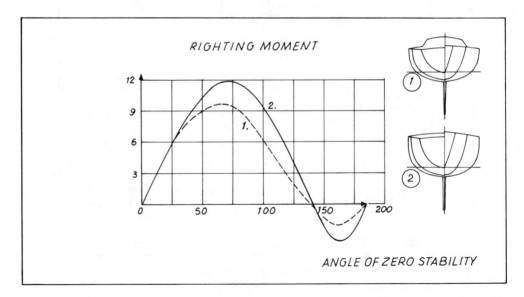

Fig 32.4. Righting moment curves for freeboard variation. 1) Fin keel parent with coachroof. 2) High freeboard without coachroof.

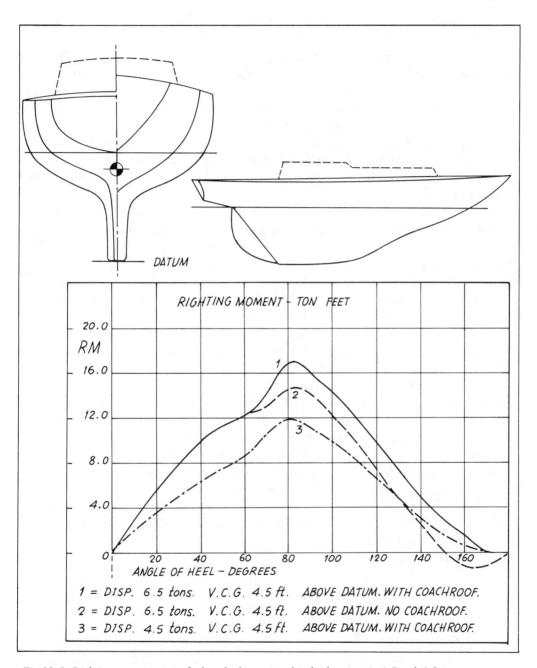

Fig 32.5. Righting moment curves for long keel parent yacht; displacement is 4.5 and 6.5 tons.

angle of vanishing stability, and therefore recovered more readily from the knockdown.

When it came to downwave controllability, however, the traditional design model was far easier to control and, despite its greater weight, it surfed very readily. The more modern design, by contrast, was hard to keep stern on to the seas, and once the hull was slightly pushed off course it broached beam-on to the breaking wave, and, in this vulnerable position, was prone to capsize. These tests indicated that lateral area, per se, does not improve capsize resistance, and downwave control is not only influenced by keel area. It was the more balanced ends of the traditional design that helped its controllability as much as the larger lateral area of the keel, because, as a wave approached from behind, the stern was lifted less, and, consequently, the bow was less immersed.

The wider-sterned modern design suffers because the passing wave lifts the stern and buries the bow deeply where it can exert a large turning moment on the hull. This will cause the yacht to broach unless the rudder can be quickly used to counter the turning moment.

The different behaviour of the three models – traditional long keel, fin keel, and increased area fin keel – gave a glimpse of the complex interaction of hull design features on controllability when running before large waves. The easy surfing and control of the traditional design model was a surprise. The results, however, should not be extended to a generalised conclusion about all traditional and modern boats. Whilst a lighter hull will be carried forward more readily by an advancing wave, to benefit from the effect of the wave, the hull shape must be such that it can be held on course easily. In strong winds, as opposed to survival conditions, the light wide-stern boats can be propelled to high speed by the sails, and any lack of buoyancy forward is compensated for by the dynamic lift generated by the forward part of the hull. This will keep the bow up and will allow the rudder to control the boat with ease. However, once the driving force of the sails is removed and the boat speed falls – the situation modelled in the tests – it was apparent that balanced ends become an important part of the control equation. As with many aspects of sailing boat performance it is the complete combination of design charateristics that is important, and herein lies the designer's skill to blend hull shape, keel area and weight successfully into a harmonious whole.

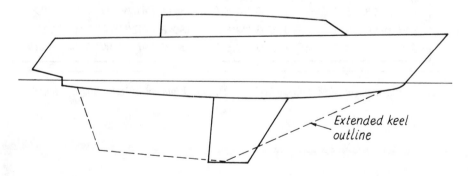

Fig 32.6. Keel extension.

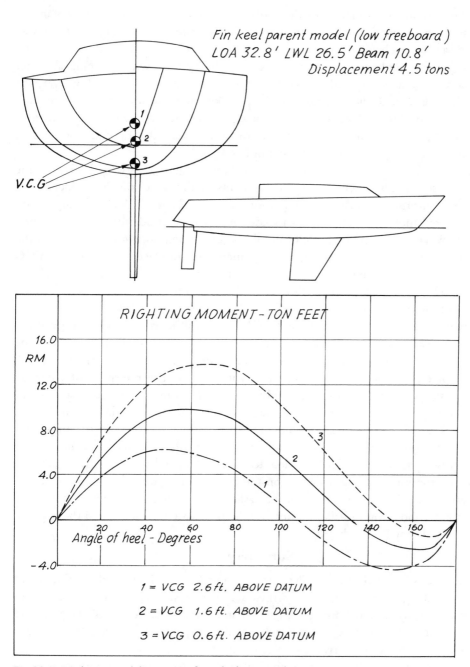

Fin keel parent model (low freeboard)
LOA 32.8' LWL 26.5' Beam 10.8'
Displacement 4.5 tons

V.C.G

RIGHTING MOMENT - TON FEET

1 = VCG 2.6 ft. ABOVE DATUM

2 = VCG 1.6 ft. ABOVE DATUM

3 = VCG 0.6 ft. ABOVE DATUM

Fig 32.7. Hydrostatic stability curves of standard series with parametric variations.

Displacement, vertical centre of gravity and roll inertia

These three parameters differ from those discussed so far because they can be altered fairly easily on existing yachts, whereas the other parameters are fixed at the design stage and cannot be changed so readily.

Increasing the displacement of the fin keel yacht by 60% while keeping all other factors constant made very little difference to its propensity to capsize; however, the increased displacement did improve the course keeping qualities of the yacht and resistance to broaching. This result is not unexpected when viewed in the context of static stability, since an increase in displacement increases the righting moment in approximately direct proportion, as shown in Fig 32.5.

Changes in the vertical location of the centre of gravity (VCG) lead to some intriguing results. The effect of large movements of the VCG on the propensity to capsize was surprisingly small; indeed in some cases the high VCG configuration actually offered more resistance to a knockdown. When considering recovery from a capsize, however, then the high VCG should be discounted as it greatly increases the range of inverted stability. Again, as with increasing displacement, a lower VCG would be beneficial to the control of the craft as it increases stiffness at normal angles of heel. Fig 32.7 demonstrates the influence of the position of the centre of gravity on the shape of the righting moment curve.

Another feature linked to weight and VCG is the roll inertia or radius of gyration. Inertia is increased by moving the component weights of the yacht further from the centre of gravity, and by so doing the yacht is made less easily disturbed by roll inducing forces, in much the same way as a skater can control the speed of a spin by bringing his arms in and out from his body. In a yacht, a high inertia can be induced by using a heavy mast and keeping the ballast as low as possible, preferably on the end of a deep keel. The capsizing tests do indicate that inertia is one of the important influences on capsize resistance. As a rough guide, increasing the inertia of the boat by 50% increases the wave height needed to cause capsize by 10 − 15%. This is discernible experimentally but its effect on the overall risk of capsize would not appear to be very great. Beware, however, of increasing inertia by use of heavy spars without a corresponding addition of ballast, otherwise the boat's VCG will rise and the angle of vanishing stability will be reduced.

Summary

The model tests and hydrostatic calculations allowed us to discern the influence of several basic design parameters on a yacht's ability to resist capsize, or recover from severe breaking wave knockdowns. Fig 32.8 summarises the results of the tests and calculations; it describes how the fundamental design parameters influence capsize − and stability − related characteristics.

Two of the strongest influences on the vulnerability to capsize are both differences in form. Firstly, a narrow craft appears to have improved resistance to capsize when beam-on to the seas, being able to slip away from the breaking wave. Also the narrow beam leads to an increase in angle of vanishing stability. Secondly, the full lateral plane and *more balanced ends* of the long keel design make it less liable to broach and capsize in

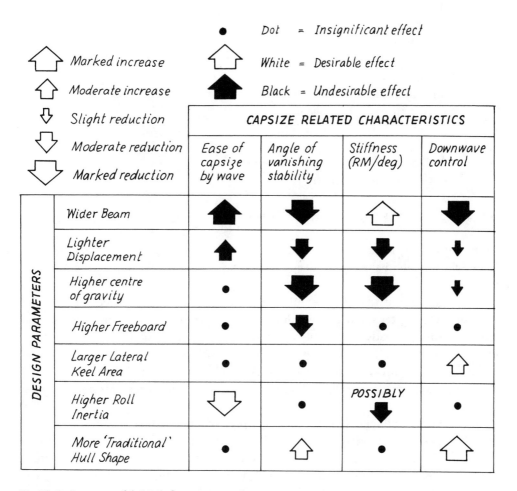

Fig 32.8. Summary of design influences on capsize.

following seas. From the tests it was apparent that those yachts having angles of vanishing stability less than 140° can be left floating upside down for a period after encountering a breaking wave. At the other end of the heel angle range, a high value of initial stability, which makes the yacht 'stiff' in sailing terms, does not provide resistance to the capsizing forces of a breaking wave, and it appears that it is the righting moment values in the range 100-130° of heel which determine the hydrostatic resistance to complete inversion and capsize.

The angle of vanishing stability represents a fundamental measure of the boat's ability to recover quickly from the impact of a breaking wave, and is the design parameter that determines whether knockdown and recovery, or full capsize results from the impact. The angle of vanishing stability figures strongly in the certification requirements for sail-training vessels and yachts and may be estimated from the formula below:

$$SV = \frac{B^2}{R \times T \times V}$$

SV = screening value
B = maximum beam (m)
R = ballast ratio: $\dfrac{\text{Keel weight}}{\text{Total weight}}$

T = hull draught @ B/8 from centreline (m)
V = displacement volume (m^3)

AVS (Angle of Vanishing Stability $\simeq$ 110 + 400 / (SV–10)

Examination of the formula shows that beam acts to reduce the range – i.e. reduces the AVS – whilst a high ballast ratio, deeper canoe body, and higher displacement increase the range.

It must be stressed that the formula is only an approximation based on the actual results of inclining tests and calculations on a number of typical sailing yachts. This means two things. Firstly it may over or under estimate the AVS by up to 10 – 15° in some cases, and secondly it will not work for unusual vessels. Nevertheless it offers a good guide. If required, the true value of AVS and a complete righting moment curve can be calculated using a well established procedure of measurements and computer calculations.

The beneficial influences of increased displacement and inertia are also apparent from the tests, but these results highlight the difficulty of designing to minimise capsizability. Many of the design features one might adopt to resist capsize actually mitigate against desirable sailing and living characteristics, which are, after all, the boat's prime function for, hopefully, all of its life. For instance, narrow beam reduces both internal volume and the power to carry sail; high inertia leads to excessive pitching in a seaway, and so on. Finally, even when all the precautions are taken in our model tests, although discernible trends in resistance to capsize have been determined, no hull form or ballasting combination consistently resisted capsize in a breaking wave with a height of 55% of the yacht's length. Moreover all the yachts could be rolled to 130° of heel by an appropriately timed encounter with the 35% LOA high breaking wave. This suggests that alterations in form which improve capsize resistance may be rendered ineffective by a relatively small increase in breaking wave height.

So what is the answer to avoiding capsize? The best and most effective way is to avoid breaking waves by avoiding sea conditions where breaking waves are prone to occur. If caught out in extreme conditions, then by keeping sailing, the boat can be kept away from the breaking crests, or positioned to surf ahead or clear of them. This avoids the dangerous beam-on condition, and delays the encounter with the breaking crest until much of its energy has been dissipated. As discussed earlier, this strategy requires active rudder control and some skill to carry out. Ultimately, the technique may well prove beyond even an experienced racing crew, let alone a short-handed cruising complement. Once the conditions are so severe that the crew are not able to remain on deck then the vessel must look after herself. Unfortunately, all yachts have a natural tendency to lie beam-on to wind and wave when left to their own devices. This is the most vulnerable position, and it is in this situation that the influence of hull design features on capsize resistance and recovery come into play, realising, of course, that no conventional yacht design can offer complete immunity from capsize under these conditions. Naturally, increasing the size of the yacht will decrease the chances of encountering a wave capable of capsizing the boat.

After our tests and calculations examining all these different design features, one must admit to a sense of disappointment that no combination of hull form or ballasting

Small craft will probably be capsized if unlucky enough to be caught beam-on to an isolated breaking wave such as this South Pacific roller. Photo: Rick Tomlinson.

Yacht capsize and drogue research: wide fin keel model.

arrangement offered a substantial improvement in capsize resistance, something, say, that doubled the breaking wave height a boat could withstand.

It is worth mentioning that in recent years the Wolfson Unit has continued its experiments, supported by the Royal Ocean Racing Club, to evaluate the influence of a drogue or sea anchor on a yacht's behaviour in breaking waves. This research again complemented similar studies in the USA by Donald Jordan. These studies showed that use of a *suitable* drogue – regarding which more details will be found in the next chapter – deployed from the stern of the yacht, will cause it to lie steadily down wind and wave, so that any breaking wave will see only the transom of the yacht and cannot exert any capsizing force.   Using an appropriately-sized drogue even the lightest, widest fin keel yacht could safely survive the 55% LOA wave with no assistance from the helmsman, whereas, lying beam-on, it would be easily rolled over by the 35% LOA breaking wave.   Indeed, it was only our inability to make higher waves in the tank that prevented us moving further up the safe wave height range.   As will be seen from the sequence of photographs on page 366, the yacht moves steadily down wave with little or no disturbance from the foaming crest.   Thus it may well be that the drogue is the simplest and most reliable way of substantially reducing the risk of capsize.

33 THE DESIGN & USE OF DROGUES IN HEAVY WEATHER

Colin McMullen

Drogues and sea anchors are closely allied; certain drogues can be used as sea anchors though the reverse is not always the case. For instance, a parachute sea anchor is unsuitable as a drogue – which is required to give a restraining influence as opposed to the near fixed anchorage provided by the parachute.

When considering the views and experiences of well known seamen, one is faced with various difficulties, as every author is naturally influenced by his own experiences. Also, the characteristics of the various boats are different – monohulls, multihulls, motor yachts and so on – as are the circumstances of the various incidents described.

Chapter 31 has described the characteristics of sea waves under various severe conditions and Chapter 32 includes descriptions of the model tests carried out by the Wolfson Unit, where the use of drogues to prevent a yacht from broaching and possible capsizing has been dramatically demonstrated.

Possible tactics in a storm

Before describing various types of drogues and sea anchors, it is of interest to consider some of the situations where either a drogue or sea anchor might add to the safety of a small craft in adverse weather conditions.

Probably the most common situation is where a small cruising yacht is running before a force 5 to 7 near gale, steering wildly, nearly broaching, and putting great

A large following sea may create conditions where a drogue may be used to advantage. This photo was taken after a force 10 storm from Chaffoteux Challenger. *Photo: Dag Pike.*

physical and mechanical strain on the crew and self-steering gear during a long sea passage. The use of a suitable drogue in such conditions could well add greatly to safety and comfort, reducing the yaw and the danger of broaching, taking the strain off the self-steering gear and adding to the comfort of the crew, all at the cost of a small reduction in speed.

In the above situation, as the wind increases to a full gale, waves start to break, and the crew may decide to heave-to, either under sail (possibly trysail and spitfire jib), or to lie a-hull. As the storm further develops, lying a-hull becomes hazardous due to the increasing danger of steep heavy breaking seas. The crew might then consider whether to run before it, or to lie bow or stern to a sea anchor or drogue.

Runnng before the gale

Both *Vito Dumas* and *Moitessier* survived extreme conditions in the Southern Oceans by surfing down breaking seas at a slight angle to prevent being pitchpoled. No drogues were used but it must be remembered that both boats were long keeled double-enders and there was no lack of searoom. Furthermore, the storms blew more or less continuously for days on end, and there was no hope of avoiding the heavy seas by slowing down and hoping the storm would overtake and gradually quieten down.

A very different situation exists if a boat is caught in a tropical storm or severe local depression. Depending on the type of boat, whether monohull or multihull, her only hope of survival in ultimate conditions may be to lie stern or bow to a drogue or sea anchor, as demonstrated so ably in the Wolfson Unit model tests.

Lack of sea room may also be one of the factors which makes the crew decide against running off down wind, and if the vessel is caught in the dangerous sector of the storm (see Fig 30.7) an unrestrained course down wind may bring the boat ever closer to the eye of the storm. Disregarding sea room, the best course of action may well be to lie to a suitable drogue or sea anchor, not only as a means of surviving the ultimate breaking seas, but also as the storm proceeds along its track, it may be hoped that it will distance itself from a vessel which is lying almost stopped in the water. At the same time the crew must try to edge towards one of the evasion tracks (page 322) away from the eye of the storm.

Although lying to a suitable drogue or sea anchor is advocated for certain boats during ultimate conditions, it must be remembered that the big 70 foot round-the-world racers seem to cope by running before heavy weather, though how many times they have suffered damage from dangerous broaches is not known. Furthermore the experience of the gallant Smeetons, who were pitchpoled twice, must not be forgotten. We must also remember that in the old square rigger days, survival depended on skilful steering, and their only course of action in extreme conditions was to run before the storm. Sadly, from time to time, one of these ships would vanish without trace.

Types of drogues and sea anchors

The RNLI drogue
The RNLI drogue shown in Fig 33.1 is an example of a heavy duty drogue which has done good service in deep displacement lifeboats where entry into harbour is often

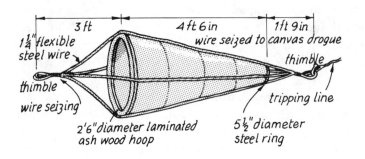

Fig 33.1. RNLI *heavy duty drogue.*

hazardous due to waves breaking on a bar or in shallow water at such places as Fraserborough, Scarborough, Whitby, Bridlington, and Salcombe. As the harbour is approached, the drogue is streamed and towed by its tripping line with the tow rope turned up, but slack. When the drogue is required, the tripping line is eased away and the drogue immediately becomes operational, although greatly reducing the lifeboat's manoeuvrability. Once the harbour entrance has been safely gained, the reverse procedure is carried out and manoeuvrability is fully restored. However, tripping lines on drogues often result in a tangle and are not normally recommended.

Icelandic drogue

As a result of the 1979 Fastnet storm, full scale trials of liferafts were conducted by the Icelandic Directorate of Shipping. The National Maritime Institute (NMI) – now called British Maritime Technology – also took an active part. As a result of these trials, it was found that not only did the stability of a liferaft depend on an efficient drogue, but that many of the drogues supplied were useless. When under strain, the forces in the drogues' warps were unacceptably high, and when the warps were slack the drogues tended to collapse and get in a muddle.

As a result of these trials, two drogues have been found to be acceptable, the Icelandic drogue and the NMI drogue. The former (Fig 33.2) is fitted with netting round the shroud lines to prevent any tangling when the strain comes off the warp.

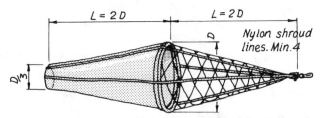

Fig 33.2. Icelandic sleeve drogue.

The NMI drogue

The NMI drogue (Fig 33.3), made of suitable porous material, has now been designated by the Department of Transport for use in ships' liferafts and ships' lifeboats, and seems very suitable for use in yachts and small craft.

The dimensions of the drogue mouth diameter recommended by Southampton University Wolfson Unit is 10% (minimum) of the vessel's waterline length, and this can result in a large unhandy drogue which is difficult to stow. Thus, the adoption of twin smaller drogues seems a better solution, either towed in tandem or individually, using separate warps brought to suitable sheet winches on each quarter. In theory, this is the preferred method as warp lengths can be adjusted easily according to the wave length, and by adopting different lengths it may be possible to maintain a continual strain in one warp, regardless of the orbital effect of the water particles in the other. Nevertheless, in certain circumstances the streaming of one drogue to reduce speed and improve steering may be sufficient. A heavy weight is required to make the drogue tow deep.

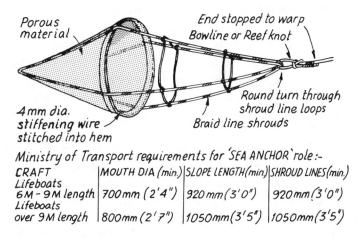

Porous material · End stopped to warp · Bowline or Reef knot · Round turn through shroud line loops · 4 mm dia. stiffening wire stitched into hem · Braid line shrouds

Ministry of Transport requirements for 'SEA ANCHOR' role:-

CRAFT	MOUTH DIA (min.)	SLOPE LENGTH(min.)	SHROUD LINES(min.)
Lifeboats 6M – 9M length	700mm (2'4")	920mm (3'0")	920mm (3'0")
Lifeboats over 9M length	800mm (2'7")	1050mm (3'5")	1050mm (3'5")

Fig 33.3. NMI perforated drogue.

The Series drogue

The Series drogue has been developed by Donald Jordan, senior lecturer at the Massachusetts Institute of Technology, in co-operation with the American Coastguard and Fig 33.4 shows a possible layout suitable for a 32 ft trimaran. It consists of a 150 ft warp into which 5 in diameter droguelets, made of light sailcloth, are spliced at 20 in intervals. A 25 lb anchor is secured to the end to keep the drogue submerged. Lines from the outer hulls of the trimaran to the drogue add another 100 ft of length.

Due to the long string of drogues, a continual restraining force is maintained in the warp, regardless of the boat's position relative to the wave crests or troughs. As a large wave approaches the stern, this restraining force will build up more quickly than with a one or two drogue rig. The prototype drogue showed good handling and durability characteristics in a full-scale trial under simulated storm conditions and appears to have significant advantages over cone or parachute type drogues.

100 ft of ¾" dia. line

75 ft. of ¾" dia. line. 45 cones.

75 ft. of ¾" dia. line
45 cones

Detail of prototype
5 in diameter cone shows
how nylon tape is stitched
to the Dacron spinnaker cloth used for the
cone & intertwined in the 2 in 1 nylon braid
tow line to hold it in position

25 lb. anchor

Fig 33.4. Series drogue.

The Attenborough sea drogue

The Attenborough sea drogue is a new device, sturdily built of stainless steel, which acts like a 'kite' and digs deep in the water. It is thus largely unaffected by the orbital water movements below the waves and exerts more of a steady restraining influence than the more conventional drogue which tows near the surface. The amount of drag can be reduced by twisting the upper chain to reduce its length. The drogue can also be used as an emergency steering system by fitting a frame and using two tow ropes. By bringing these to suitable winches on each of the vessel's quarters and veering and hauling, a measure of steering control can be achieved, though with some hull designs the drogue will probably only steer the vessel when the rudder is actually lost.

The newness of this device means that it has not yet established a 'track record'; nevertheless, due to its stability while being towed, its simple robustness and the fact that it tows deep and maintains a steady strain regardless of the wave motions, the device seems to have potential, though the report which follows suggests that adjustment of the amount of drag may be critical.

In May 1990 the 35 ft *Saecwen* , a 1961 traditional design by Alan Buchanan, was in mid-Atlantic in the latitude of the Azores on passage to England. She had been running under a storm jib for two days when, on 19 May, the westerly wind increased to storm force. Charles Watson, *Saecwen*'s skipper, said that it became increasingly difficult to keep the boat sailing downwind without broaching-to in the enormous breaking seas, topped by large overhanging crests. After being knocked down by a very large cross sea, a Size 2 Attenborough drogue was streamed on a 100 ft warp with a separate tripping line. The warp was led through a stern fairlead and secured around a genoa winch.

Charles Watson had managed to catch the warp line in his safety harness so the immediate effect of the drogue was to pull him violently into the stern where he was held by the pushpit stanchions. The moment the weight came on the winch the boat's speed abruptly slowed from 6 knots to just under 4 knots. *Saecwen* was now so efficiently tethered that the larger breaking seas effectively overwhelmed the boat. Three poopings followed within the next twenty minutes, filling the cockpit up to the coaming. The first two took place with the storm jib set and the third with it lowered.

Attenborough sea drogue with tripping line. A stainless steel drogue which acts like a kite, digging deeply into the water.

It was considered that continued poopings would endanger the boat, so the decision was made to haul in the drogue. Recovery was made with considerable difficulty due to the immense strain on the tripping line which required two people on the winch.

A decision was then made to run before the wind under bare pole, but steering was difficult in the dark for the very tired crew still shocked by the first knockdown, and twenty minutes later another enormous sea threw the boat onto her beam ends, causing minor damage on the upper deck. Following this second knock down the helm was lashed to leeward with heavy elastic cord and the boat was left to lie a-hull. While some large seas were taken aboard, the crew remained relatively comfortable for the next seven hours whereupon the conditions moderated. It may be of interest to add that when *Saecwen* put into the Azores she was amongst about thirty yachts that had been out in the same storm. Of these, six had been dismasted, one had been completely rolled over and three had lost their rudders.

After this experience, Charles Watson concluded that *Saecwen*, whether under a storm jib or with a bare pole, needed some method of preventing her broaching-to when running before the sea. While the Attenborough drogue turned potential knockdown situations into poopings, it did appear to be too efficient, causing *Saecwen* to be totally engulfed in the breaking crests of the larger waves. Perhaps, he thinks, this would have been acceptable in a yacht with a centre cockpit or higher freeboard. In the case of *Saecwen*, warps or a smaller drogue might have enabled him to continue to run off more safely. However, three days of bad weather had left the crew of three exhausted, and the option of lying a-hull with everyone securely down below seemed preferable. Charles Watson believes that low freeboard yachts, such as *Saecwen*, lie a-hull particularly well and he would in the future tend to choose this tactic.

The Galerider

A parabolic-shaped drogue made of two inch webbing on a $\frac{3}{8}$ inch stainless steel rim, known as a 'Galerider,' is available from the United States of America. It is like a large lobster pot open one end, and is supplied in four different sizes to accommodate vessels between 5 and 40 tons displacement. As with the Attenborough, it is an interesting concept but it has not yet established a track record in the United Kingdom. However, it has been heard from John Rousmaniere that, in the USA, the Galerider has an excellent reputation from those who have used it, being quite easy to store and very simple to deploy and retrieve. He says 'Frank Snyder, the former Commodore of the New York Yacht Club, deployed a Galerider from his big motorsailer in a northerly gale in the Gulf Stream, with breaking seas and all the rest. Frank Snyder said that the Galerider held the vessel's speed down and greatly eased the ride. It was necessary to adjust the line so that the drogue was completely dug in two or three wave crests back, but that was not hard with the line led to a big winch.

I have also heard of a smaller boat that was dismasted in very rough weather and deployed a Galerider with great success.'

On the other hand, John Rousmaniere thinks that the boat must have a bit of speed on for the drogue to dig in. A friend of his used one in a small gale and she said that it did not grab, though perhaps the model used was too small for the boat.

Parachute type sea anchors

Parachute-type sea anchors were originally developed in the USA by the conversion of ordinary surplus aerial parachutes. Their continued development has resulted in the design of special sea anchors to suit individual craft and the discarding of the old aerial parachute designs. There are now at least seven firms which produce various parachute sea anchors and two well known types in the USA are the 'para-anchor' and 'Shewman's sea anchor and drogue'. The para-anchor is normally supplied in diameters varying from 9 to 32 ft, and is streamed from the bow. It is important that its line is sufficiently long to ensure that a measure of elasticity is introduced into the tow and a length of at least 300 ft is advocated. Even so, very heavy strains can be exerted on the towing points and if the vessel is knocked backward by a breaking sea, the rudder can be jeopardised if not firmly secured amidships.

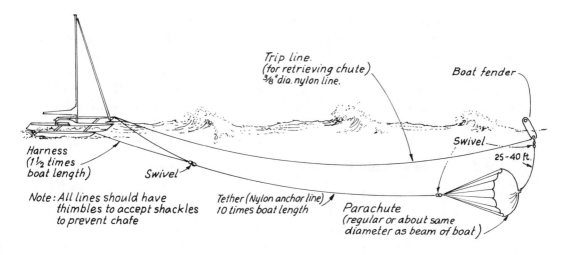

Fig 33.5. Para-anchor (as operated successfully by John Cassanova).

Shewman's sea anchor and drogue is normally supplied in diameters varying from 5 to 40 ft and is also normally streamed from the bow. When acting as a sea anchor it is operated in a similar manner to the para-anchor; when acting as a drogue, the tripping line is partially tautened. In this capacity it is known as a 'variable pull' (VP drogue sea anchor).

Parachute sea anchors have been used successfully in many multihulls, both in trimarans and catamarans. They have also been widely used in fishing boats. Due to the shallow draft of these various types of craft, they 'weather cock' roughly into the wind when tethered by the bow.

However, their use in monohulls riding by the bow has not proved so reliable, as the windage of the mast often combined with a deep keel aft and a cut away profile forward, tends to make the craft lie well off the wind.

Improvisations

Various improvised drogues have been adopted by mariners when facing heavy weather, and the most common has been the streaming of long warps either singly or towed as a loop from each quarter. The towing of motor tyres, knotted chain cable and other objects has also been adopted, with varying success. In general, although such actions have often added to safety when running before heavy weather, they may not provide sufficient restraining force to prevent broaching when overtaken by a steep breaking sea during an 'ultimate' storm.

Operational Factors

Very great strains can be experienced in the towing lines of drogues and sea anchors and especially where parachute type sea anchors are concerned. Thus anchorage points in the vessel must be sufficiently strong and every precaution taken that the crew streaming or recovering the device are not endangered.

Once lying to a drogue or sea anchor, the vessel can be battened down with the crew safely below, but here again care must be taken to avoid injuries caused by the violent motion including accelerations and decelerations when a large possibly breaking wave overtakes the vessel. Ideally, the towing lines should be made of braided rope so that 'spin' is not induced into the drogue or sea anchor and swivels are avoided. Drogues and sea anchors are subjected to great strains and stresses, often over a long period, and must be made of very robust material. Also precautions must be taken against chafe, especially where the towing line is lead over the vessel's gunwale.

Advantages and disadvantages of parachute-type sea anchors

There are many examples where multihulls owe their survival to parachute-type sea anchors streamed from the bow. The main requirements are that the diameter of the parachute should be sufficiently large and the line adequately long. Fishermen have also found paras of great value, not only in surviving storms but also allowing them to maintain their station during periods of rest, darkness or borderline bad weather conditions.

As already stated, however, para-type sea anchors are not always successful when operated by deep displacement monohulls if tethered from the bow. If it is decided to tow from the stern, then one of the types of drogues giving a restraining force, rather than the almost fixed anchorage provided by a para is probably desirable. It must be remembered also that once anchored to a para-type sea anchor, the vessel is incapable of any manoeuvre. This may be serious if endeavouring to avoid a navigational danger or attempting to avoid the eye of the storm. On the other hand, the fact that a vessel remains almost stationary in the open sea when anchored to a para-type sea anchor, may be advantageous if the vessel is threatened by a lee shore.

Advantages and disadvantages of drogues

There is good evidence that some form of restraining device over the stern can greatly add to safety when running before a heavy sea in even only moderate wind conditions. There is also no doubt that certain craft are vulnerable to broaching and possible capsize if running down wind in 'ultimate' conditions with no drag device rigged astern.

There have been numerous examples where disaster or near disaster has overtaken vessels on such occasions and these practical examples have been confirmed by the model tests carried out by the Southampton University Wolfson Unit and by similar tests undertaken by the US Coast Guard Research and Development Center. Both organisations have also demonstrated how suitable drogues streamed from the stern can prevent such disasters.

Although single or twin Icelandic or NMI drogues streamed from the stern can greatly add to safety, it is necessary for them to be towed deep by the addition of suitable weights to prevent undesirable slackness occurring in the rig as each wave crest passes over the drogue.

The series drogue, however, ensures that a steady strain exists in the towing line, regardless of the wave characteristics. Recovering the drogue can present problems due to the continual strain in the rig unless a powered capstan is available. On the other hand, the series drogue is the simplest drogue to stream. Whether the small boat owner will accept that one of his precious warps is threaded through a large number of small cones is open to doubt, though this should present no problem in bigger boats with plenty of stowage space available where a series drogue can be carried for its special duties only. Though heavy, the Attenborough drogue may also prove to be a useful device towed from the stern especially as it can be used as an emergency steering gear.

All drogues towed over the stern allow a small measure of manoeuvrability which is lacking in the para sea anchor. Their disadvantage is that the vessel's stern is presented to the waves and pooping can be expected. Thus, not only must the drogue fixture and the exposed stern parts be adequately strong, but also the cockpit must be truly watertight with locker lids, washboards and hatches securely battened down. Cockpit drains are often too small, and the crew must be prepared to bail the cockpit rapidly in an emergency.

Conclusions

It is believed that every small craft proceeding offshore should carry some form of sea anchor or drogue for use in storm conditions. The type of gear carried will depend on the class of boat. Although some boats, notably multihulls, will lie successfully to a sea anchor secured to the bow, this is not always the case, especially where monohulls are concerned. It can be guaranteed, however, that every type of boat will lie successfully to a suitable drogue or sea anchor secured to the stern. As a drogue allows a measure of manoeuvrability and does not induce such great strain in the towing line as a para-type sea anchor, a drogue over the stern will often be more suitable than a para-type sea anchor. As a cockpit aft may be vulnerable, there could be an argument for designing a deep sea cruising vessel with its cockpit amidships, also that this cockpit is provided with efficient draining arrangements.

34 HEAVY WEATHER CLOTHING

Dick Allan

Compared with the yachtsman's more obvious concerns for the seaworthiness of his boat, the problems of what to wear at sea may seem somewhat less than riveting. Yet those with real experience of prolonged exposure to heavy weather conditions will, to a man, take a different view. Indeed, so serious is the potential impact of exposure or hypothermia on the yachtsman's ability to handle his boat, and to make rational decisions under adverse circumstances, that it is sensible to regard effective foul weather clothing as an essential part of the safety equipment of the boat. An exhausted, wet and hypothermic crew member is a doubtful asset; an exhausted, wet and hypothermic skipper is a positive danger.

The fundamentals

Too often, the choice of clothing is dictated more by fashion than function; evidence that it will keep you warm and dry commands less attention than the glossy descriptions – 'attractive two-tone, made from the new, breathable wonder fabric ...'. So it would be useful to start with a straightforward description of how clothing items achieve the functional objectives of keeping you warm and dry.

A comfortable human being has a temperature deep inside of about 37°C, and mean temperature on his skin surface of around 33°C. If he gets cold, the blood vessels in his skin contract in order to reduce the blood flow and hence also the loss of heat. If he gets colder still he will begin to shiver, which is the body's way of generating heat in the muscles, to replace that being lost to the environment. If, on the other hand, he gets hot, which may happen either because he is in warm surroundings or because he is working hard physically, then the blood vessels in the skin dilate to promote heat loss. If that is insufficient he will ultimately begin to sweat, so that evaporation will further increase his heat loss. The problem facing yachtsmen is that both situations can occur in rapid succession: one minute he may be inactive and cold in the cockpit, and the next he may be hot and sweating inside his foul weather gear as he struggles to change headsails or reef the main. The perfect clothing system to cope with both these extremes has yet to be invented, but the best of modern systems are pretty good at keeping you warm and dry, if less good at permitting sweat evaporation and preventing condensation.

Clothing keeps you warm by insulation. Insulation prevents heat flow from the body in the same way as electrical insulation prevents the flow of electricity from the cable. Clothing insulation is not primarily a characteristic of the particular fibres from which the clothing is made, be they natural or artificial, but is largely determined by the thickness of the still air layer enclosed within and beneath the clothing fabric. If the air trapped within the clothing is displaced by water, either because of leaking or excessive sweating into the clothing, then the insulation is severely reduced – a few pints of sea water inside your oilies may halve the effective insulation.

The Card *power reaching under No 3 genoa and reefed mainsail in 55 knots (true) in the South Atlantic. Most waves rolled through as if the boat was not there, but it was found advisable to steer off down the bigger ones to avoid them hitting the boat's side with a heavy crash. Photo: Rick Tomlinson.*

Traditionally, insulation is measured in 'CLO' units. One CLO is approximately the insulation of normal business clothes, and 4 CLO would be the likely insulation of a full set of clothing for Polar exploration. So close is the relationship between clothing insulation and the thickness of the trapped air layer that it is possible to use a simple rule of thumb for estimating insulation – four CLO per inch of thickness.

Even a naked man standing in completely still air has an external insulation of approximately half a CLO unit, due to the layer of still air close to his skin. If the wind gets up, this layer of still air is removed and he will feel chilly, even though the temperature of the air may not have changed in any way. A similar situation arises when wind penetrates a clothing assembly. Thus you may be sitting on deck in a warm jersey, providing just the right amount of insulation for the conditions; but if the wind gets up you may immediately feel chilly. This is because the layer of still air within the jersey is driven out by the wind. Hence, of course, the value of an external wind-proof layer, however thin, and many modern insulating garments incorporate this feature.

Insulating layers

We know that staying warm is achieved by staying dry, and that is the prime function of the outer layer of the clothing assembly. On the other hand, the primary function of the clothing insulation you wear beneath your foul weather clothing is to keep you warm.

There is available a large range of excellent insulating clothing, providing a wide choice of practical garments. The essential objective is to have good flexibility in the range of insulation available from your clothing, and this is best achieved by going for one of the multi-layer systems. These generally consist of a relatively thin inner layer, with a long-sleeved vest and long-legged underpants. The use of string vests is now rather old fashioned, their capacity for trapping air and keeping it still being less than that of modern insulating clothing assemblies. Outside this inner layer comes the main insulation layer, and I strongly recommend the two-piece versions – a longjohn or similar to cover the legs and trunk but not the arms, and a separate long-sleeved jacket. This gives you the choice of wearing the longjohn on its own, thus reducing the thickness of clothing over the arms which can be helpful when there is work to be done, or adding the jacket for use in extreme conditions or when relative inactivity at night demands the full tea-cosy. These insulating garments are greatly improved by a thin, relatively wind-proof and possibly shower-proof outer layer. This serves two important functions, the first is to turn the garments into a practical set of clothing for wearing below deck, when the outer foul weather clothing is removed. Secondly, the layer can contribute much to reducing friction between your insulating clothing and the outer water-proof foul weather clothing, a feature which greatly improves the comfort of the whole assembly.

The outermost layer

Materials

A large variety of modern fabrics, with claims to varying degrees of breathability, have appeared on the market over recent years, but the more traditional neoprene coated nylon remains in my view the best choice for offshore yachtsmen. I would not like to be too dogmatic about this, since the fast-moving fabric technology may soon produce a better answer. However, currently-available breathable fabrics all tend to be adversely affected by the salt water marine environment. Some of them also suffer from the so-called wet log effect, in which areas of the clothing subjected to movement and pressure tend to leak, for example the seat of the trousers and elbows. It is surprising what high pressures can be created in water droplets trapped beneath a seated buttock! Of course it would be possible to overcome this by appropriately positioned patches of another type of fabric, but overall I remain convinced that neoprene coated nylon is the best choice for yachtsmen, for it combines qualities of waterproofness and excellent wear resistance.

Promoting ventilation

At the start of this chapter, mention was made of the need, occasionally, to promote the evaporation of sweat, and to limit condensation within the clothing system. Even when not overtly sweating, the body produces water vapour through the skin and all yachtsmen know the damp, uncomfortable feeling this can cause. In my experience the new breathable fabrics are not nearly as effective as you might be led to believe by some of the manufacturers' claims. Although they can be shown to transfer substantial amounts of water vapour, either through micropores or by a chemical process, the amounts transmitted are not usually sufficient to cope with sweating of any kind. Some of these fabrics are also fairly expensive.

For the present the best approach to the sweating/condensation problem is to promote ventilation of the inner clothing layers by appropriate outer layer design. Thus many modern sets of foul weather clothing incorporate ventilation facilities to allow air to move between the outer layer and the lining. Also, good design of head-end (by 'head-end' one means either the neck or face closure), wrist and ankle closures, besides the overlap area between the jacket and trousers, will allow these closures to be opened up with considerable ease when conditions permit. Even when they do not, and all closures are kept tightly sealed against the weather, it has been my experience that the dampness resulting, for example, from a long overnight passage, in no way compares with the disastrous consequences of a real dousing through inadequately designed foul weather clothing.

Important design features

For offshore heavy weather sailing, the standard two-piece foul weather gear, with bib trousers and a longish overlapping jacket, remains the general design of choice. One-piece suits lack the flexibility of the two-piece, and tend to be less well ventilated. Full dry suits, with neck, wrist and ankle seals are uncomfortable and totally unventilated. They probably do have a place in short-term racing conditions, but not usually for long offshore passages, even in rough weather. Nevertheless I am not inclined to be dogmatic here, because the design of modern dry suits is improving all the time. I know of yachtsmen who find them quite tolerable even over longish periods – Robin Knox-Johnston among them. I would also recommend the top half only 'storm smock', with neck and wrist seals but open at the waist. This device can be slipped over bib trousers and worn with a standard jacket on top, and thus provides excellent water exclusion for storm conditions.

When choosing an external layer of foul weather clothing, the yachtsman will be well rewarded by attention to the following details. First a most important point to check is that every seam in the garment has been stitched and taped over. This should be checked most carefully, expecially where pockets or other fittings are attached. There are some garments on the market which have doped seams in which the stitching is sealed with a form of glue. In my experience this system is wholly unsatisfactory, since the slightest pull on the stitches opens up the stitch holes and causes leaking. Unfortunately it is not always possible to inspect the inside surface of seams, because many good suits have inner linings. Ingenuity will usually overcome this, however, and you should be wary of the chandler who discourages you from looking.

No matter how good the material or seaming, a suit may still leak severely when put to the test. This is because water tightness at the wrists, ankles, zip closures, neck and hood closures is achieved by good design rather than material characteristics. It is here that some important differences between suits may be found and some of them are not at all obvious on casual inspection. Here is a checklist:

1 Check that every seam and fitting attachment has been stitched and sealed by taping. Check that knees and seat are reinforced for wear.

2 Check that there is a good overlap of jacket over trousers even when sitting, and that the trouser legs fit easily over your boots and have adjustable closures (zip or Velcro).

3 Check that the wrists have effective, comfortable and adjustable closures.

4 Check that the fly closure on the trousers is both watertight and compatible with urination. In my experience the type with a 'dam' of material, arranged as a flap behind the zip, coupled with good elastic braces meets this requirement.

5 Check that the main jacket zip closure has an adequate flap cover, also a device to prevent water which has been driven in sideways across the Velcro flap closure from entering through the zip teeth. Some designs have a double flap cover, others have a specially stitched seam forming a water barrier; both are satisfactory – but don't accept just a simple, single flap.

6 Then, and most important of all, check the neck closure and hood arrangements. Check that the neck can be closed snugly with the hood stowed – this may save you from the occasional wave when conditions generally don't necessitate the wearing of hoods. Now deploy the hood, and re-check that you can still close the neck (or face) closure snugly. Many suits have inadequate adjustment here, especially for people with rather small chins and thin faces. Make sure that adjusting the head–end closure does not cause funnelling of the material, which may be uniquely efficient at diverting water down your neck!

7 Finally check the hood. Make sure it is attached at the rear and sides by stitched and taped seams. Don't accept a detachable hood, it will inevitably leak. The hood should have an adjustable closure and a peak of some kind. Try to imagine water being poured on your head – try it if compatible with continuing friendly relations with your chandler! Will the hood divert the water outside the neck closure? Will the peak assist in keeping water off your face and eyes? Does the deployed head–end closure restrict head movement or vision?

One drawback to the excellent water-excluding properties of modern attached hoods is that they slightly reduce one's hearing. Some helmsmen consider that they also interfere with their subconscious 'feel' for the wind and weather and thus take the edge off performance. I recognise these problems but, to my mind, they detract little from the overwhelming advantages of staying dry. However, there are a few who still prefer the old fashioned Sou'wester. Though one can hear well when wearing a Sou'wester I do not believe it will keep one as dry as a well designed hood.

Taking seas 'green' over the bow is a good test of heavy weather clothing. Photo: Rick Tomlinson.

So much for keeping dry. In addition one could mention that foul weather clothing may be provided with a number of additional features and ancillary equipment. Examples are pockets, built-in devices for the attachment of lifejackets, built-in flotation and safety harnesses, etc. One cannot lay down any firm rules or even recommendations regarding these items, except that safety harnesses and lifejackets should always be available and worn when conditions demand. There is one item, however, that is sometimes overlooked and that is the use of reflective patches placed high on the foulweather jacket. These form an effective close range location aid in a night time 'man overboard' situation, and are held in high regard by lifeboatmen and other rescue agencies. It might be mentioned at this point that reflective patches can be obscured by a lifejacket, in which case the lifejacket will need reflective patches too.

Tail piece

Finally I shall indulge in a small confessional. I once chartered a boat along with seven colleagues, most of whom were highly inexperienced at offshore sailing. We left Lymington in the evening, but on our way to The Needles the wind piped up a bit and my thoughts turned to memories of The Needles channel in a strong ebb tide into wind. 'Will you take her for a while?' I asked one of the more competent members of the crew, and I disappeared down below leaving the others enjoying the evening sunshine in shorts and T-shirt. In a short while I emerged wearing my full foul weather gear and safety harness, causing a certain amount of ridicule from my colleagues. At that moment we hit it – down there by the Bridge buoy! In a trice they were soaked, and I was left alone at the wheel for a very long time. The moral of the tale is that there is nothing so useless as a good set of foul weather clothing put on too late. It is also the skipper's responsibility to see that his crew are appropriately dressed.

35 SEASICKNESS

Dick Allan

I t is to be doubted whether any topic in yachting is more surrounded by myth and legend, folklore and old wives' tales than the causes and cures of seasickness. Everyone has a view, many have real experience, but few know the scientific facts and even the scientists know that their present understanding of motion sickness is far from complete. It is with some trepidation, therefore, that I embark on an attempt to give a straightforward account of this highly complex subject.

What is known about the causes

Seasickness is but one form of a larger group of conditions known as motion sickness. This includes, for example, air sickness, car sickness, simulator sickness (well known to pilots) and space sickness. The term motion sickness is even slightly misleading, since the well known symptoms can be evoked as much by the absence of expected motion as by the presence of unfamiliar motion. For example, pictures taken from a moving helicopter and projected on to a large screen can give rise to 'cinerama sickness'.

There are not many incontrovertible facts about seasickness but here are a few:

Age Young children below the age of two years very rarely get seasick, whilst between the ages of three and twelve susceptibility increases progressively. Thereafter susceptibility for most individuals decreases progressively with age, but even the elderly are not immune. There are large differences in susceptibility between individuals.

Sex Women are approximately 70% more susceptible than men, susceptibility being highest during menstruation and increased in pregnancy.

Inner Ear Individuals without a functioning vestibular apparatus (the part of the inner ear that has to do with balance and sensing movement) do not get seasick.

Adaptation Most people can adapt to the motion causing seasickness so that they cease to be sick, and this adaptation can be maintained over a period of land lubbing. Some, however, appear unable to adapt, and a few famous names appear among them such as Charles Darwin and Lord Nelson. Having adapted to sufficiently severe abnormal motion at sea, sickness can recur when the sufferer returns to the normal motion environment ashore – a phenomenon known as *mal de debarquement*.

The most acceptable current theory on the fundamental cause of seasickness is the 'neural mismatch hypothesis'. It will be familiar to everyone that we are able to sense motion both visually and, for example when the eyes are closed, through the vestibular apparatus in the inner ear (semicircular canals and otolith organs). One facet of the neural mismatch theory is that the stimulus to nausea and vomiting arises when the information from our ears does not accord with the information from our eyes. Thus if you go below and stare at the Decca, your eyes will be telling you that there is no

The Card rolling along off the Falkland Islands in 40 knots (true) during leg 4 of the Whitbread Round the World Race 1989/90. The mizzen mast was carried away

relative motion between you and the Decca but your ears may be telling you that everything is moving abominably, and you get sick. If, however, you rush back on deck and stare at the horizon, then your eyes and ears both tell you the same story (even if it is that everything is moving horribly) and the feeling of nausea frequently disappears. So far so good, but this part of the theory does not explain why people get sick in the first place, even when looking at the horizon, nor does it explain the curative process of adaptation. After all, the mismatch of information from your eyes and ears is no less when arriving at Cherbourg in a force 6 for the tenth time, than it was on the first occasion. To cope with this difficulty, a second facet of the neural mismatch theory suggests that not only must there be a conflict of information from the eyes and ears, but that this information is not what your brain expects to receive. There is, if you like, a model of what is normal (the motion you are used to) stored in a memory device in the brain, like a program is stored in a computer. Your brain continually compares the information from your eyes and ears with the model in its memory, and if the two are sufficiently different you feel sick. The model in your brain is not a fixed unalterable one, but is capable of being updated by continuing experience of unusual motion, until what was once regarded by your brain as abnormal becomes normal, and the information from your eyes and ears is no longer provocatively different from the model in your brain.

Prevention and Treatment

Adaptation

It has to be admitted at the start that the only totally effective prevention for seasickness is to take up golf. To that may be added that the only really effective way to treat seasickness is to go offshore sailing regularly! It has to be *offshore* mind you, because that is where you are probably having trouble with seasickness, and no amount of adaptation to the neat little lop in the Solent will protect you from the seas in the Channel.

Of course, if you are highly susceptible, then the first part of the cure can be achieved in inshore waters. As a sufferer from seasickness myself, albeit rather rarely these days, I find myself slightly more susceptible at the beginning of each season than the end, and I can hasten the topping up of my immunity by deliberately causing provocative stimuli. The most obvious way of doing this is, for example, deliberately to spend time down below reading a book or navigating for example. If you can undertake some activity which moves your head around rather more, then that will probably be even more effective.

Slight truths in folklore

We have discussed the basic cause of seasickness above, but the end result of all this neural physiology is still centred on your stomach. Therefore other stimuli that tend to make you feel sick when at sea, such as diesel fumes or the smell of bacon and eggs (if that's the last thing you want) are best avoided. Nauseogenic smells (or thoughts) are rather individual matters, and each must establish for himself what it is he should avoid.

Similarly, all experienced yachtsmen will have recognised that the quirks of appetite and personal preferences for food and drink are often very different offshore from those of the same individual at home. At the planning stage, you may think you will be ravenous, sufficient to devour an Irish stew, and then find in the event that all you want is a Mars Bar and a crust of bread. I can hardly ever sell a wine gum at home, but my precious personal supply frequently disappears down the throats of the crew when offshore! Similarly, bread sticks – everybody seems to love them. These of course are all highly individual matters which each person must find out for himself by experience. But there is no doubt in my mind that there is a pattern of eating and diet for each individual, which is very positively helpful in holding seasickness at bay.

Drugs

Much research has been devoted to testing the efficacy of various drugs in the prevention and treatment of seasickness. The table below gives a list of the more common ones and probably includes your particular favourite.

Anti-Motion Sickness Drugs

(Main brand names* in the UK and USA)

Drug	Dose (mg)	Time of onset (h) (approx)	Duration action (h) (approx)
1 Hyoscine hydrobromide (UK: Kwells. US: Pamine).	0.3-0.6	0.5-1	4-6
2 Transdermal hyoscine patch (UK: Scopoderm TTS. US: Transderm Scop)	0.5	6-8	60-72
3 Cyclizine hydrochloride (UK: Valoid. US: Marezine**)	50	1-2	4-6
4 Dimenhydrinate (UK: Dramamine. US: Dramamine, Dramocen)	50-100	1-2	6-8
5 Cinnarizine (UK: Stugeron, Marzine RF**. US: not at present available)	30	1.5-2	6-8
6 Promethazine hydrochloride (UK: Phenergan, Avomine. US: Phenergan)	25	1.5-2	24-30

*Brand names in other countries may differ.
**In the UK Marezine now contains Cinnarizine instead of cyclizine. In the US Marezine still contains cyclizine.

Hyoscine is effective especially in moderate conditions and has the advantage of being quick-acting (half to one hour). Top-up doses are required at intervals of four to six hours but, in common with other drugs taken by mouth, it is of no help once vomiting is established. The new transdermal patches also contain hyoscine. You have probably seen other yachtsmen wearing them usually attached just behind an ear. To be effective they must be in place at least six hours before going to sea as it takes that long for sufficient of the drug to be absorbed through the skin. Thereafter, absorption continues at a steady rate for a period of up to seventy two hours so they remain effective without top-up doses and are likely to be helpful even once vomiting has occurred.

Anti-seasickness drugs are also available in the form of suppositories and these have found a certain amount of favour in the USA and France. The idea is to overcome the difficulty of keeping pills down when vomiting has started but many find this method distasteful – to say nothing of its plain awkwardness when heavily dressed on a small moving boat! The hassle involved is itself likely to increase nausea. If the oral preparations are taken well before going to sea, as advised, they will be absorbed and working in good time. If they fail to prevent vomiting, it is doubtful whether a suppository will reverse the situation. On the whole I believe the transdermal patches are a better way to go. The prospect of offering rectal suppositories to the average foredeck gorilla is not one to attract many skippers.

A difficulty with hyoscine, which is also found with most of the anti-histamine drugs (cinnarizine, dimenhydrinate, promethazine and cyclizine) is that troublesome side-effects, particularly sleepiness, are rather common. It is possible to demonstrate in the laboratory adverse effects on skilled performance. Hyoscine also causes blurred vision and a dry mouth. Cinnarizine is also an anti-histaminic drug but tends to have rather less side-effects than the other anti-histamines particularly with respect to sleepiness. I personally have found it to be the best although it is necessary to commence taking it well before setting sail and preferably six to eight hours before. Interestingly, the manufacturers are reluctant to stress the benefits of taking this drug long before setting sail lest they be disadvantaged by those claiming more rapid efficacy. I doubt if this would weigh heavily with the average yachtsman who will gladly do almost anything if it improves protection against seasickness. Extensive trials of hyoscine and cinnarizine have been undertaken by, or on behalf of, the Royal Navy and these are continuing right up to the present time. Comparative trials between cinnarizine and hyoscine have confirmed the less troublesome side-effects of the former. Trials of efficacy in preventing seasickness have so far not revealed marked differences between hyoscine and cinnarizine other than a lower incidence of adverse side-effects with cinnarizine.

In summary, I would suggest that cinnarizine is the drug of choice for most yachting applications with hyoscine as an alternative for crew members in whom sleepiness would not pose problems – that is they are not necessarily crucial to the running of the boat. Transdermal hyoscine patches would also be useful if sleepiness does not matter and provided they can be in place eight hours before sailing.

Sea Bands

In recent years a novel treatment has appeared on the market – the 'Sea Band'. The idea emanates from the ancient Chinese practice of acupuncture or, in this case, acupressure. A band is worn on the right wrist with a plastic button designed to put pressure on the P6 or Nei Kuan acupressure site which lies between two of the muscle tendons. In a most carefully controlled laboratory trial conducted by the Institute of Naval Medicine, the Sea Band did not provide any increase in average tolerance to motion. Although there is just room for the possibility that a different result might be found in real sea conditions, most authorities remain sceptical. My own experience with this has been disastrous – the only time my wife has been sick was when she tried one!

Managing seasick crew

Minor degrees of seasickness are common, and not necessarily incapacitating. It is quite common for an individual to vomit and then to have quite substantial periods of feeling much better. In these cases they should be encouraged to sit in the least mobile part of the boat, generally somewhere near the centre, but especially in a position where they can focus on a good distant view towards the horizon. Better still, get them to take the helm.

For more severe cases, for example when nausea is more or less continuous and vomiting frequent, then it is best to remove the sufferer from the action. He should be put below, foul weather gear and wet clothing should be removed, and he should be put in a warm sleeping bag, with his head as near to the centre of the boat as possible and given a bucket. In this situation, and with eyes closed, it is remarkable how many apparently hopeless cases are restored to reasonable comfort. To leave such people on deck in a miserable huddle in the corner of the cockpit is to invite serious problems like hypothermia, dehydration, electrolyte disturbances through excessive vomiting besides a constant risk of falling overboard. Indeed the loss of morale to near suicidal levels is well known to most yachtsmen; it is often accompanied by an unwillingness to take much action to save oneself from various threatening events. Seriously seasick crew need to be protected.

36 HEAVY WEATHER TACTICS

Peter Bruce

Military officers are fond of saying that few plans survive first contact with the enemy, and the same may be said of a small vessel encountering heavy weather. Unexpected events are likely to occur, the weather forecast is often a simplified overview, and people's behaviour may be unpredictable under duress of prolonged exposure to blinding spray, fear, cold temperature, wearying noise and violent motion. One can be fairly sure, however, that crews, boats and equipment that have done well in previous storms will do so again.

It will be clear from the many different experiences already recounted that there is no universally applicable advice to be offered. In this concluding chapter I will attempt to summarise the various tactics available to the skipper of a single-hulled sailing boat in heavy weather.

As the weather becomes more severe, options reduce, especially if sea room is limited. As yet, modern technology has not brought any dramatic overall changes to the traditional method used for coping with heavy weather, though a number of new devices exist that do show promise. Apart from use of some of the newer designs of drogues, the tactics which are described in this chapter have been employed in one or more of the accounts of heavy weather in this book and their effectiveness in actual circumstances can best be judged by referring back to them. It is not possible to specify the exact wind strength when the various options can be recommended due to the number of variables involved. Obviously, the size of the craft is of considerable relevance, and in storms, big is beautiful. But other factors count heavily too, such as the nature of the seas, the design and strength of boat, and the size and strength of the crew. In broad terms this chapter covers courses of action to consider when the wind speed exceeds force 8–9, to force 10 and above.

It is evident from Adlard Coles' own modest accounts that he was skilful at choosing the moment to shorten sail. As he advocates, this should be undertaken in advance of immediate necessity, despite a natural reluctance to do so in the hope that the weather may show an improvement. He was of the opinion that, when cruising in bad weather, it will be prudent to adjust sail area for the squalls and gusts rather than for the mean force. He was also a believer in not running with more sail than can be carried if close hauled, thereby avoiding the need for frenzied reefing action in the event of a man overboard, or a bad landfall. Obviously, the longer sail reduction is deferred the harder it is to do, and the greater the risk of damage meantime. Incidentally this applies to all heavy weather tactics, in that bold and resolute decisions are best made before a situation becomes desperate. Likewise, most problems are best dealt with as soon as they occur.

When bad weather is approaching, most people, very sensibly, make for a safe harbour. But once caught out in a storm there is much to be said for enduring it rather than continuing to run for shelter, as breaking waves in shallow water can be infinitely more of a hazard than the open sea. A poignant illustration of the dangers of shoal water can be found in the description of the loss of *Marie Galante II* given in Chapter 16.

When beset by a storm, the human objective is to remain safe and as comfortable as possible. As far as the boat is concerned, the aim is to avoid sinking, being rolled over, pitchpoled or otherwise suffering damage to hull, rig and sails. As Adlard Coles says, the wind dictates the amount of sail that can be carried and can also cause loss of sails, rigging or mast; but it is the breaking seas that do the more dangerous damage to hull and superstructure. There is great energy bound up within a wave, and large forces are at work when a boat is struck by a breaking sea. Alternatively, a boat can be picked up by a breaking sea and accelerated to an alarming speed. A moment later the boat may be flung into the still water in the trough of the wave where she will stop almost instantly. It is at this point that serious damage to crew and boat is also likely to occur. Clearly, it is desirable to avoid breaking waves but, if this is not possible, then the waves should be taken within 20° of either the bow or the stern. Before the days of self-draining cockpits, a wave breaking over the stern, ie being pooped, was a major hazard to yachtsmen. It is now less of a problem, but remains a danger to motor boats open at the stern, to yachts with unusually large cockpits, or with low bridge decks which require wash boards in place to prevent a sea sweeping down below.

As increasingly severe weather overtakes a craft, a seaman's instinct encourages him to continue to maintain his approximate course whilst reducing sail appropriately. When the destination is to windward and the crew has the determination to press on under perhaps storm jib or trysail alone, despite the uncomfortable motion, this is commendable and, incidentally, is a policy that has won many races. The helmsman's technique is to close reach across the troughs, then head up into the waves, and bear away hard on the top to avoid a slam. Likewise, when reaching, the helmsman will luff into the wave crests and bear away in the troughs.

The crew wil be heartened by continuing to go in the desired direction and the forward motion has the benefit − unlike heaving-to or lying a-hull − of providing the possibility of avoiding breaking waves. As a storm tactic, this type of action has its merits only when the crew is strong enough to persevere and while it is still possible to make progress under storm sails. As conditions deteriorate further, any size of sail set, however small, becomes overpowering. At first the helmsman will luff to reduce the angle of heel but the flogging sail will cause the whole boat to shake violently. On the other hand, when the helmsman reaches off to gather speed, the boat will be thrown on her beam ends across the seas. At first some middle ground will be found between these extremes but as wind and wave increase there comes a point when it will be prudent to consider other options, especially if the helmsman's difficulties are about to be increased by darkness.

Heaving-to

One can heave-to either by backing the headsail or by tacking and leaving the jib sheet made up, accompanied by a mainsail appropriately reefed for the conditions. Then the helm should be secured so as to hold the boat's bow towards the wind. The increase in comfort that can be derived from heaving-to in a seaway has to be experienced to be believed. Noise and motion are promptly much reduced and there is no longer a need for a helmsman. Thus heaving-to is an expedient to adopt, for example, for having a meal in comfort or when an uncertain pilotage situation demands time for thought. Unfortunately not all fin-keeled yachts will heave-to untended in a sea. Yachts that can do so will continue to fore reach at a knot or two and make nearly as much way to leeward. On no account should one try to back an overlapping headsail in order to heave-to as there is a high chance of bursting the sail on the spreaders. Besides, most yachts will not heave-to comfortably under large headsails.

Storm jibs and trysails are sometimes employed on their own, as for example, in the case of *Puffin* in the Mediterranean (Chapter 17) and variously in the 1979 Fastnet. In these instances it is probable that a helmsman will be necessary, though this was not so in the case of the long-keeled *Cohoe* off Belle Isle in 1948 (Chapter 6). Given the choice, there appears to be a balance in favour of a trysail set on its own rather than storm jib set on its own. This seems particularly likely to be so when the storm jib is set on the forestay rather than the preferable inner stay positions (see Chapter 29).

Heaving-to is a seamanlike tactic to employ up to moderately severe conditions, say force seven to eight, but, when wind and wave increase above gale force a time will come when any hoisted sail will flog so violently that something has to break, or cause a boat to be repeatedly thrown upon her beam ends. These situations are neither comfortable nor safe and again another tactic must be sought.

Lying a-hull

When heaving-to is no longer sensible it may be possible to resort to lying a-hull, ie taking off all sail, lashing the helm – usually slightly to leeward, closing all hatches securely and letting nature take her course. The ease with which types of yacht will lie a-hull is variable. Broadly speaking old-fashioned narrow beam heavy displacement yachts often lie a-hull well whilst light displacement beamy yachts do not. Nevertheless lying a-hull was the most popular survival tactic in the generally lightish displacement 1979 Fastnet fleet; Harry Whale lay a-hull off Ushant aboard *Muddle Thru* in the great English Channel storm of 1987 and one could cite many other instances. It is a tactic employed by many yachtsmen – and not always of dire necessity. In a sufficient force of wind the windage of the mast alone provides stability, like a steadying sail. As an illustration of a vessel's ability to look after herself during a storm, one can recall that most of the vessels abandoned in the Fastnet race were later found bobbing about in the swell with hatches left wide open.

Thus when survival of human life is involved, the practice of lying a-hull can be preferable to remedies involving any high risk activity. The problem, of course, is that the boat is vulnerable to breaking waves from broadside-on, and to paraphrase Andrew Claughton in Chapter 32, 'breaking waves do not have to be very big to roll any sort of small craft right over, whatever her hull features.' The consequences of a roll-over can be most dramatic down below – more so than one might imagine – and the chances of the mast being lost are quite high. This occurrence, apart from the obvious effect on mobility, can lead to all kinds of immediate problems. Not the least of these is that the lack of damping effect of the mast, without which a yacht's roll inertia is halved, results in quicker and more unpleasant motion with a much increased chance of being rolled yet again.

Whilst on the melancholy subject of capsize it may be significant that several yachts have been rolled over just as a storm seems to have abated. Examples are Bill King in *Galway Blazer* in 1968, *Sayula II* in the first Whitbread Round the World Race in 1973, Michael Richey in *Jester* in 1986 and the *Swan 46* at sea in the October 1987 storm, described in Chapter 23. A left-over sea is always confused and this may provide the explanation. Another contributory factor might be the shift of wind direction as a front goes through, when the effect of two wave trains crossing at an angle is known to produce regions of extreme wave height where the seas may break heavily.

The 36 ft schooner Halcyon *lying a-hull when caught out in an unforecast storm 300 miles east of Cape Hatteras in October 1981. Mean wind speed was 50 knots, seas were 25-30 ft, with large breakers.*

Shortly after the US Coast Guard cutter arrived on the scene the schooner was overturned by a particularly large breaking wave, and quickly sank. Her sole crew – visible by the stern of the capsized craft – was rescued by a 5.5 m Avon rigid inflatable boat.

Having pointed out the dangers of lying a-hull, to try to put matters into perspective it should be mentioned that world girdlers such as Sir Alec Rose, Dr Nicholas Davies and Alan and Kathy Webb, albeit with heavy displacement boats, have found the practice of lying a-hull entirely satisfactory for weathering the gales of a normal world circumnavigation. For such solidly-built vessels it is quite a rare event to encounter the sort of weather that does not allow lying a-hull without a high risk of capsize, but there are enough examples of yachts being rolled over whilst lying a-hull in this book to show that the tactic may not always be wise.

Use of engine

Nowadays the diesel engine is a generally reliable piece of equipment that can be used to good effect when managing under sail would be difficult, such as entering and leaving a berth. The old-fashioned view that 'to use an engine is unseamanlike' may have been based upon more than just prejudice, for early auxiliary engines were not always dependable, nor easy to start, and it was wise not to encourage over-reliance on them. Moreover, the high chance of getting a rope round the propeller in a stressful situation is well known. Nevertheless, the competent use of an engine was demonstrated in the extraordinary account of *Pendragon* in hurricane *Carol* (Chapter 20) and also Alain Catherineau's celebrated rescue feat in the 1979 Fastnet. One is not likely to be able to make directly into the sea under auxiliary engine or tack through the wind in storm conditions, but, bearing in mind that even a bare pole will provide some aerodynamic lift, it may be possible to make across the seas to windward. There may often be circumstances when it will be safer to 'heave-to under engine' rather than lie a-hull, especially when running before a storm is impractical due to a lee shore. In this case the use of an engine, either in combination with sails or without, may be highly desirable. It should not be forgotten that skippers of power craft have no option but to 'heave-to under engine' in heavy weather, and more details of the methods used, such as 'dodging', will be found in Peter Haward's chapter.

When running before the sea it may be significant that the old-fashioned British lifeboat with a speed of eight knots could not always be handled confidently without the use of a drogue in combination with the engines, whereas cox'ns of modern lifeboats, with a speed of 18 knots, find they have no such trouble as they can power their way clear of breaking seas. Bearing in mind the opening sentence of this chapter, which suggests that flexibility may be necessary when facing severe weather, yachtsmen should regard the use of an engine as another string to their bow, and be ready to use it when the situation justifies, after carefully checking for ropes over the side. On the other hand one should be very cautious about taking a tow in a seaway. Few modern vessels have adequate strong points for such a situation even when the master of the towing vessel has the patience and skill to maintain a low enough speed.

Running before the seas

There is nothing new about the tactic of letting a vessel run freely before the sea, when sea room permits. Apparent wind is reduced, motion is more comfortable and the risk of being rolled becomes less likely than when lying a'hull. When considering extreme conditions, running with full directional control is vitally important to avoid the breakers and to keep the stern at the optimum angle to the waves to avoid broaching. This is not easy, especially at night, though experienced planing dinghy helmsmen have an advantage. In a confused sea, waves can come unexpectedly from odd quarters, necessitating a quick response from the helmsman and boat. Clearly the speed of the boat has to lie within a range necessary to achieve good control. If speed is too slow, the boat does not respond quickly enough, and if too fast, especially in darkness, the helmsman may not be able to react in time to avoid pitchpoling or a broach, both of which can have very serious repercussions. Some craft undoubtably handle better than others when running before the seas, and it might be easy to assume that displacement is the controlling factor. The situation is not so clear cut, with a combination of design factors such as the degree of balance of the ends affecting the issue, as described by Olin Stephens in Chapter 25. In any case a low-geared wheel is a considerable disadvantage.

A modern lifeboat almost buried by a breaking head sea. It is easy to imagine that a small vessel might be capsized if caught beam-on to such waves. Photo: Ian Watson.

An exhilarating sleigh ride aboard The Card whilst approaching Cape Horn during the 1989/90 Whitbread Round the World Race. The 80 ft yacht is running before 35 knots (true) under heavy spinnaker, and is thus able to use speed to keep clear of large breaking waves. Nevertheless, intense concentration is necessary to avoid

The corollary is that in the case of a high-geared wheel, or a tiller, some degree of physical strength is likely to be needed on the helm. It is worth mentioning at this point that vane self-steering can work up to 50 knots of windspeed or more and can be invaluable for short-handed crews in heavy weather.

The sturdy crews of the 'Maxi' yachts in the Whitbread Round the World Race find exhilaration in the downwind sleigh-rides through the southern oceans under as much sail as they can carry. Their technique is to luff across the trough to maintain speed, then to bear away almost square to the wave, just as the crest arrives, to encourage the boat to surf down the wave. It is similar to methods used by dinghy sailors and Malibu board surfers and requires some skill, strength and a great deal of concentration, especially when carrying maximum possible sail. Such crews will have perhaps six or eight expert racing helmsmen to share the steering task, whereas the average cruising boat may not have more than one, for whom the task of steering may become extremely wearisome after some hours.

Nevertheless, though their size is advantageous, in the unlikely event of being caught out in a tropical revolving storm, perhaps even the mighty yachts designed for the Whitbread Round the World Race could be in danger of being pitchpoled in exceptionally steep and large seas. Though I know of no instance, in this case even their crews might want to consider deploying some form of drag device from the stern.

Running with warps

As already mentioned, it is important to try and find an appropriate speed for good steering when running before a storm. Towing warps as a means of keeping the stern to the seas, whilst flattening the breakers behind and helping to reduce the yacht's speed to a more desirable level may sound archaic to some, but they are still used by experienced seamen, particularly in short-handed situations. Firstly, warps, having more than one use, are usually available; they are also easy to stow and uncomplicated. Furthermore one can control the length of rope paid out to ensure a steady pull of the right amount. A steady pull is important but it is not easy to achieve as, due to the movement within the wave described in detail in Chapter 31, water particles at the crests are at this point going at much the same speed and in the same direction as the boat. A long warp, unlike a single drogue, can span a whole wave length so making the pull more constant. Not many warps intended for securing alongside will be long enough for spanning an ocean wave. Thus for ocean passages one should be looking at a length of 400 ft or more. Even experienced warp-users such as Robin Knox-Johnston relate the ease with which warps can tangle when being paid out in the sort of conditions when they can help; thus preparation and skill are necessary in their deployment. When Robin Knox-Johnson encountered heavy weather in the southern oceans he put out 600 ft of two-inch circumference warp in the form of a bight, lashed the helm tightly amidships and went to sleep: the advantage of a bight is, of course, that, when making adjustments, half the weight is taken on the fixed end. He says:

'The warp held the heavily-built *Suhaili* firmly stern to the waves. She lay very comfortably regardless of the wave height in forces of at least 12 on the Beaufort scale when the sea was white with spindrift. *Suhaili* drifted downwind at an average of about two knots in big stuff and strong blows with only a storm jib set right forward, sheeted tight amidships, so that if she tended to yaw the force on the jib increased the more she came round, tending to push her back downwind. *Suhaili* has a canoe type stern, or Norwegian type if you prefer, and this meant that there was no great resistance to the waves as they rolled past, they were just divided. The warp stretched quite a lot in surges, but that was to be expected.

Had I been able to get the warps out in 1989 (see Chapter 24) I am sure that we would have been alright, and I am intending to make a couple of rope reels, which I shall hang from the deckhead in the fo'c'sle, so in future I can lead the warps directly over the stern if I need them in a hurry.'

In Chapter 20, Adlard Coles describes Bernard Moitessier's experience with warps during a survival storm in the South Pacific in 1965/66 when the 40 ft boat *Joshua* was towing 'five long hawsers varying in length from 100 ft to 300 ft, with iron ballast attached and supplemented by a heavy net used for loading ships'. At the onset of the storm it seems possible that this arrangement was successful, but once the wave height and length reached a certain point, the yacht failed to respond to the wheel, and the warps did not prevent her from surfing down the crests. Not finding these circumstances satisfactory, eventually Moitessier cut the warps, whereupon he found himself in a much more comfortable situation.

Interestingly, Dr David Lewis, returning from the United States of America after the 1960 single-handed Transatlantic Race recounts a similar experience. He had been running under staysail alone, but after the sheet fairlead had pulled out of the deck with increasing wind strength, he lowered the sail and streamed 120 ft of warp in a bight. At once, he says, his ship became unmanageable, and even when he did succeed in steering down the seas, the breaking waves would carry the warp alongside. There are more examples of such difficulties described in the 1979 Fastnet Race account (Chapter 21).

The apparent anomaly regarding use of warps is explained to some extent by Tony Marchaj in his book entitled *Seaworthiness: The Forgotten Factor*. He believes that, in a fast-growing sea, wave height increases more quickly than wave length, producing especially steep seas. As waves develop in size there will be a time when their wavelength is twice a given boat's length. At this point the bow will arrive in the trough just as the stern is at the top of the crest. With steep enough seas as a result of the fast-growing effect, one can see that there may be a tendency for the boat in this attitude to topple, ie pitchpole or broach, especially as the water particles driven by the orbital action within a wave will be moving forward at the crest of the wave, reducing the effectiveness of the rudder. When a boat has been designed with the rudder placed well aft, the control situation may be further aggravated as a result of 'ventilation', a term used to describe the situation when the rudder may be lifted partially out of the water or its efficiency reduced by aeration.

In very short steep seas, when a boat could topple and when the rudder may be less effective than usual, many yachtsmen have felt that they would benefit from the use of

Warps being trailed from the centre of the cross beam of the 80 ft catamaran Chaffoteux Challenger in mid-Atlantic in April 1989. Structural damage had occurred, and the warps brought the boat speed under bare pole down from 10 knots to 8 knots, easing the strain on the hull. Photo: Dag Pike.

some drag device to prevent their vessel broaching or pitchpoling. Tony Marchaj suggests that when the wavelength has had time to develop further there will be a need for less drag and more speed to evade breaking crests and awkward cross-seas in the troughs. Hence a need to discard the use of warps as the wave system matures. Nevertheless, this does not explain why those such as David Lewis and Bernard Moitessier had control difficulties with warps streamed after seas have had ample time to grow up, and others, such as Robin Knox-Johnson and Geoffrey Francis (Chapter 20), have not.

The explanation seems to lie in the length and height of waves compared to the length of warp used. If Bernard Moitessier used warps, as he says, of up to only 300 ft in length in seas of wavelength 500–560 ft, one could expect, through the orbital movement within a wave, for his warps to be ineffective at times. Likewise, allowing for the bight, David Lewis's warp only extended 60 ft in waves which he estimated to be 150 ft in length.

After disposing of his warps Moitessier found that it was an advantage to luff a little at the arrival of each wave crest to take the sea 15–20° on the quarter. He was now in the same situation as Warren Brown in *Force Seven* (Chapter 19), the Webbs in *Supertramp* (Chapter 20) and using the same tactics. These three are not alone in preferring to run free with the seas slightly on the quarter. Not only does it reduce the chances of pitchpoling but also, like a surfer, by steering across the waves one may be able to avoid the worst of the breaking crests. When the face of the wave is exceptionally steep there will be a very fine balance between pitchpoling and broaching, and in this case the helmsman may have to weave his boat down the wave to optimise the situation.

The experience of Bernard Moitessier and others leads one to the view that to avoid being pitchpoled, rolled over or being pooped, boat speed must be kept at a level to give directional stability at all times. Of course, once huge seas have had the time to develop, it is usually in the troughs of the waves that problems occur. Here the boat's speed is no longer being helped by gravity and wind strength may be reduced by the shelter of the adjacent wave crest. Nonetheless, in the trough of a wave, the need for directional stability remains as vital as elsewhere. Thus warp drag has to be adjusted, or enough sail has to be set, to take the craft safely through the troughs, accepting that at other times even with warps streamed, this may mean that a boat could be surfing at what may seem to be an unnatural rate. Some boats, like *Suhaili*, may be directionally stable when towing warps with helm lashed, others may demand competent helms-manship, but probably to a lesser degree than when running completely free.

The additional thrust needed to achieve good steering control in the troughs can possibly be provided by an auxiliary engine. It might be helpful to illustrate this point with an extract from an account by Richard Clifford when cruising off Ireland in his *Warrior 35* in August 1979. As luck would have it, he found himself alone at sea in the midst of the infamous Fastnet storm.

'I dropped off my crew in Glengariff on 13 August and sailed gently down Bantry Bay, then headed south-west out into the Atlantic. At 0400 the next morning the wind was too much for the storm jib, so I handed it and lay a-hull. Very soon afterwards *Warrior Shamaal* was knocked flat by a large breaking wave. I grabbed a bucket and bailed, found the bilge pump and pumped, and fought my way to the wheel to try to get the boat to run down wind, but she would not come round. The next extra large wave again filled the cockpit and the saloon to just above the cabin sole. When we came back on even keel I noticed that the liferaft was floating in its container beside the boat, the bilge pump handle had gone and the anchor had come out of its well on the foredeck. The immediate problem was to stay afloat so I rushed below, got the heads bilge pump handle and put the washboards in positions. Still *Warrior Shamaal* would not run downwind and yet another wave bowled the boat over, filling the cockpit and pouring water below. By this time the liferaft had inflated itself and broken adrift.

I was not prepared to set my 98 sq ft storm jib as it would have been too big: my 50 sq ft spitfire jib was in my garage! So I started my 15 hp auxiliary engine and with this ticking over in gear I was able to keep the stern into the wind with an occasional burst of full throttle after my concentration had lapsed.

Suddenly we were at the top of an extra big wave. *Warrior Shamaal* hung at the top then plunged forward in a horrifying nose-down attitude. Her bows plunged into the trough and I fully expected to continue on down, or the stern to flip over, pitchpoling the yacht. With a shudder we pulled out of the dive and rushed on with the next mountainous wave.

Tactics had to be changed again, so between waves I passed a mooring line from one cockpit winch aft around the self-steering, and back into the cockpit. My 300 ft warp is coiled and seized with sailmaker's twine every 60–70 ft to avoid tangles. I attached the end of this onto the mooring line and paid out each coil until the whole bight of rope was more or less floating astern.

Still running ahead on the engine and towing the bight, I experienced no further problems.'

In this account we not only have an engine being used to good effect but we also see that lying a-hull may not be a satisfactory tactic in extreme conditions. Furthermore we see that warps were thought to be necessary to avoid broaching or pitchpoling during the period when the seas were developing rapidly in height and steepness.

On balance it appears that warps streamed astern on their own can work well in extreme storms so long as the warp arrangement is long enough to span a whole wave length of the seas. The shorter the warps are in relation to the wavelength, the lower the proportion of time when the warps are providing effective drag. Consequently a collection of short warps may slow a boat down below the speed necessary to steer, or for the boat to steer herself, yet they may provide insufficient drag to prevent wild surfing when the orbital movement within the wave is moving in the same direction as the boat.

Having established that a long length of warp would seem one form of desirable equipment for heavy weather, it is worth quoting from Michael Richey's experiences in July 1981. He was returning to the United Kingdom from Bermuda in his famous 25 ft folkboat *Jester* when overtaken by a fierce storm from the SSW. He writes:

'How to handle a boat in extreme conditions is a matter for judgement and much will depend upon one's knowledge of how the boat behaves. Perhaps there are no hard and fast rules. In *Jester* I have never felt the need to tow warps, although I have often run before gales. Now, under self-steering, we were surfing down the slopes of heavy seas and the tops were beginning to fall off. My fear was that she might bury her head and pitchpole. It seemed essential to slow the boat down. I carried on board a 5 kg Bruce anchor as a kedge and since the Bruce is reputed to be hydrodynamically stable I reckoned it should tow like a paravane. Accordingly I streamed the anchor over the starboard quarter, on some 75 ft of line, taking the end – with some difficulty as one must crawl out to it – to the cleat. This immediately slowed the boat down to about half her speed and kept the stern nicely into the seas, preventing her from slewing about. So we spent the night of 9–10 July, the storm rising in violence and with a general situation of discomfort, but *Jester* well under control.'

This interesting report encourages the view that there may be an alternative to the use of very long warps during a storm.

Use of drogues

Means of keeping bow or stern continuously into breaking waves without human assistance has long been a fascinating area of research and development. Colin McMullen, who has written on them in more detail in Chapter 33, makes a case for cruising vessels, ideally with centre cockpits, to ride out a storm stern to wind and sea, with a suitable restraining drogue streamed from astern. The tiller, or wheel, would be lashed with shock cord and most of the crew would be lying in the safety of their bunks.

Drogues have been tried over many years with varying success. The best known of them used to be the sea anchor which, when led over the bow, works well as a means of checking drift in moderate conditions, especially combined with a mizzen or other steadying sail carried aft to keep the vessel head to wind. But there does not seem to be much support for a sea anchor nowadays, perhaps because it seldom works well in severe conditions. Moreover, the strains imposed on the rudder can lead to breakage, as happened to *Nova Espero* in 1951. On the other hand, an involuntarily-occurring type of sea anchor formed from the sails and rigging after a dismasting is said to be quite effective, as the wreckage makes a wave screen of a sort. Nor should one forget the rudderless *Marionette of Wight* in the 1979 Fastnet riding comfortably to a very long warp led from her bow.

A more modern variation of a sea anchor, in the form of a sea parachute, can be much more effective in holding a craft head to sea. There are some interesting accounts of parachute drogues being used with advantage, particularly in the case of multihulls, when the parachute is usually deployed over the quarter. It seems important to use a good length of resilient (ie stretchy) line and to put it out before conditions become extreme, bearing in mind that a parachute type drogue is not an easy thing to carry about on the upper deck in a high wind.

Taken overall, reports of the usefulness of parachutes in small vessels are varied. In some cases it would seem that, in order for the parachute to hold the boat directly into the waves, the action needs to be rather severe. Others report failure to keep the bow into the waves however large the diameter of drogue used. It seems possible that vessels with high superstructures and masts are likely to shear about on a parachute and therefore are less likely to lie comfortably head to the seas. It is certain that very strong securing points will be needed; moreover heavy loads will be imposed upon the steering gear if used from the bow.

Perhaps of even more interest, bearing in mind Michael Richey's success with a small Bruce anchor, are the drogues designed to be towed astern. Once again very strong anchorage points are required and methods of deployment and stowage have to be considered carefully. In the tank tests described by Andrew Claughton at the Wolfson Unit at Southampton (Chapter 32) a double-drogue device towed astern was quite remarkably effective. The arrangement was designed to simulate the 'series drogue' idea of Donald Jordan, successfully tested on the bar across the entrance of the Columbia River in the USA. The Wolfson drogue's ability to overcome the problem of orbital wave motion ensured that a model yacht, drifting in the tank broadside to the waves, was invariably straightened out just before the onset of a large breaking wave, thus avoiding a capsize, though, one has to assume, not necessarily avoiding being pooped. Assuming the 'droguelets' are made to fold flat, so that the assembly can be easily stowed and deployed like a conventional warp, the series concept would seem likely to overcome many of the practical difficulties for small crews in managing a drogue. For example, it should be quite easy to vary the degree of drag effect to suit the prevailing situation. Of course once any drogue has been deployed, its immediate recovery is likely to be difficult. Thus, if it is not helping the situation, the only remedy may be to cut it adrift.

Opportunities to study the relative merits of the different devices that are available on the market have been few and there is insufficient evidence to judge their effectiveness for general use. It seems wise to be wary of extravagant claims by manufacturers, and work on the basis that, though there are some promising devices about, one must be prepared to experiment using one's own craft before any reliance can be put on them.

Use of oil

There has never been any doubt that oil on troubled waters provides a calming influence. Some scientists attribute the mechanism to something called 'the Marangoni Effect'. A rider of probably academic interest is that a film of vegetable oil, which has a high elasticity, is said to work better than a film of mineral oil. But there are precious few recent records of small craft using oil of any sort to their benefit during severe weather. It may be significant that whereas British lifeboats were fitted with a small tank of 'wave subduing oil' – a vegetable oil called Garnet 46 – the oil was seldom used, and modern lifeboats no longer carry it. We have the instance of HMS

Birmingham using oil in the 1987 English Channel storm to assist a vessel in distress. The captain had good reason and the opportunity to study its effectiveness; in this case the oil could be seen to be working, but the problem was to position it so as to be useful. A situation when oil certainly was useful was that of Group Captain Geoffrey Francis' typhoon described earlier (Chapter 20) where he attributes his survival to an abundant supply of rope and tractor vaporizing oil. One must give credit to him for making the best use of his resources, however, not many craft would be likely to carry around 100 gallons of oil on deck in case of a tropical revolving storm. Nevertheless in such circumstances oil can be a life saver.

★★★★★★★★★

In conclusion it can be safely said that there is as yet no simple and perfect solution for weathering storms. But we can be sure that the people who deserve to come off best are those who have chosen and prepared their vessels and crews with bad conditions in mind as well as good, and have the resolve to keep on top of the situation throughout the hardships and anxieties of heavy weather.

INDEX